AF846158

STEFAN MAIWALD

BEACH LIFE

STEFAN MAIWALD

BEACH LIFE

teNeues

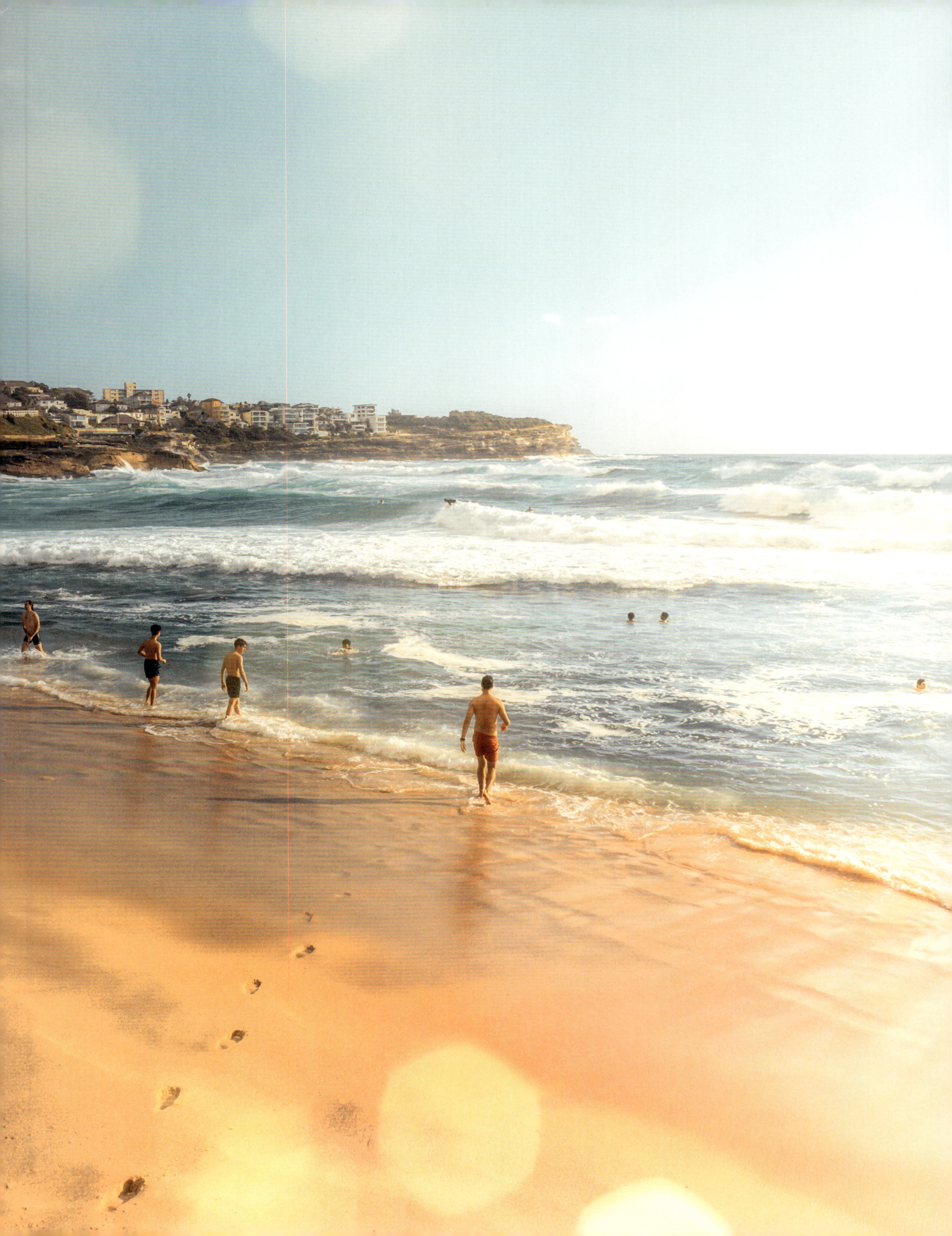

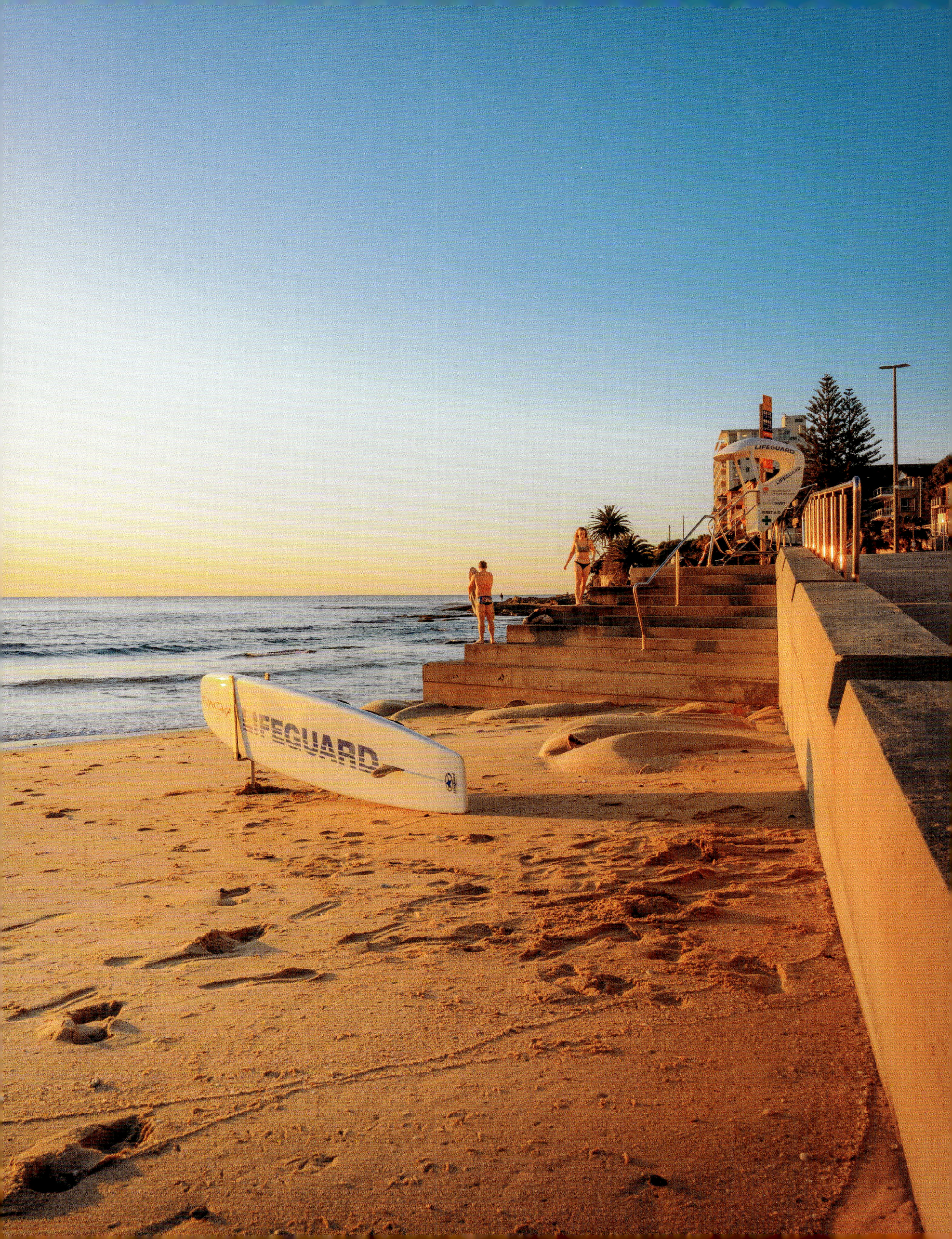
LIFEGUARD
LIFEGUARD

CONTENTS

INHALT

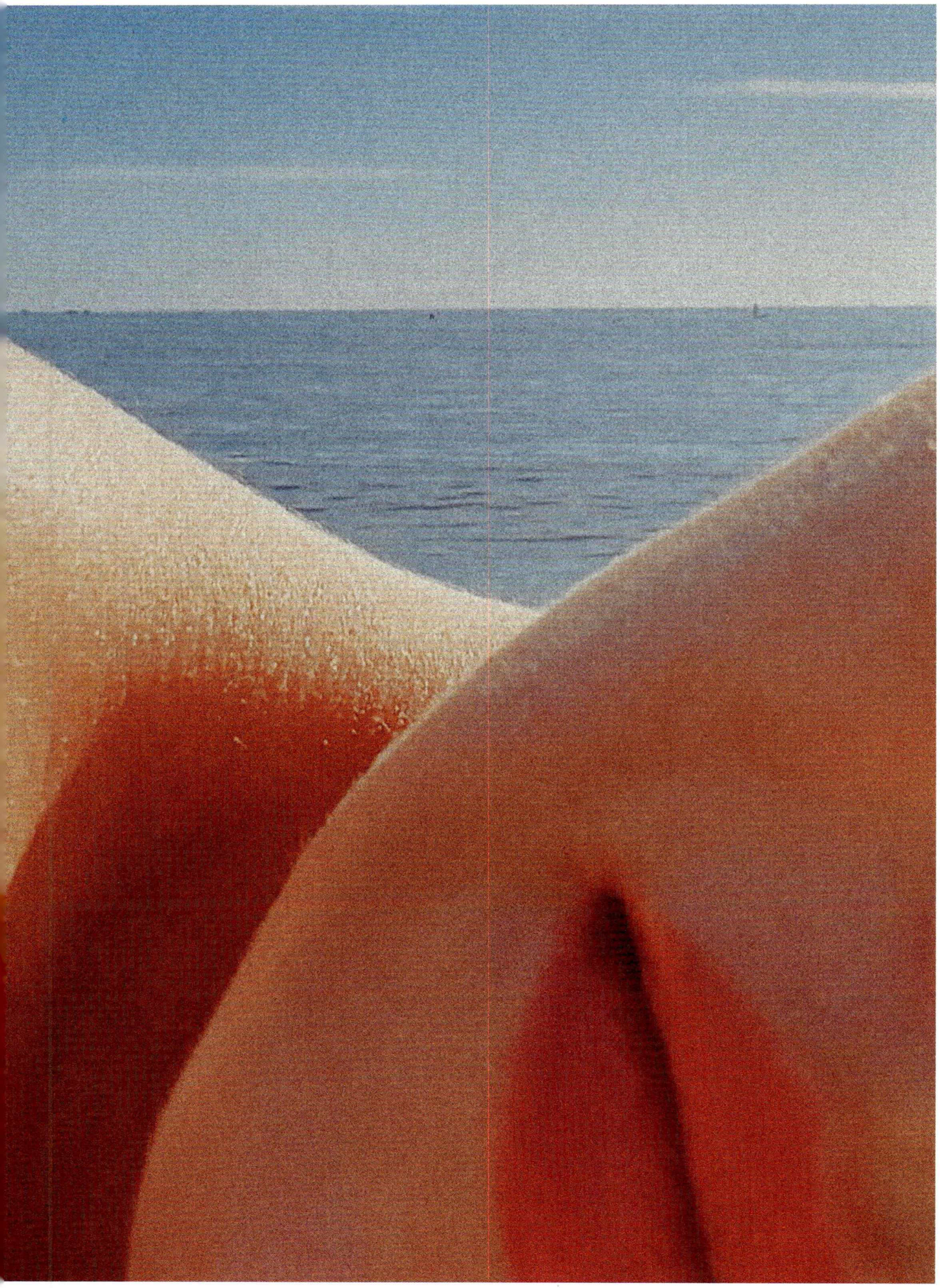
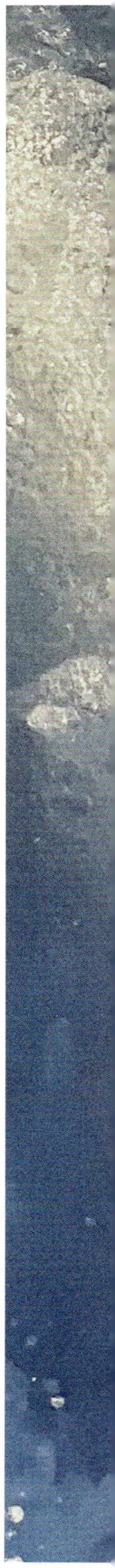

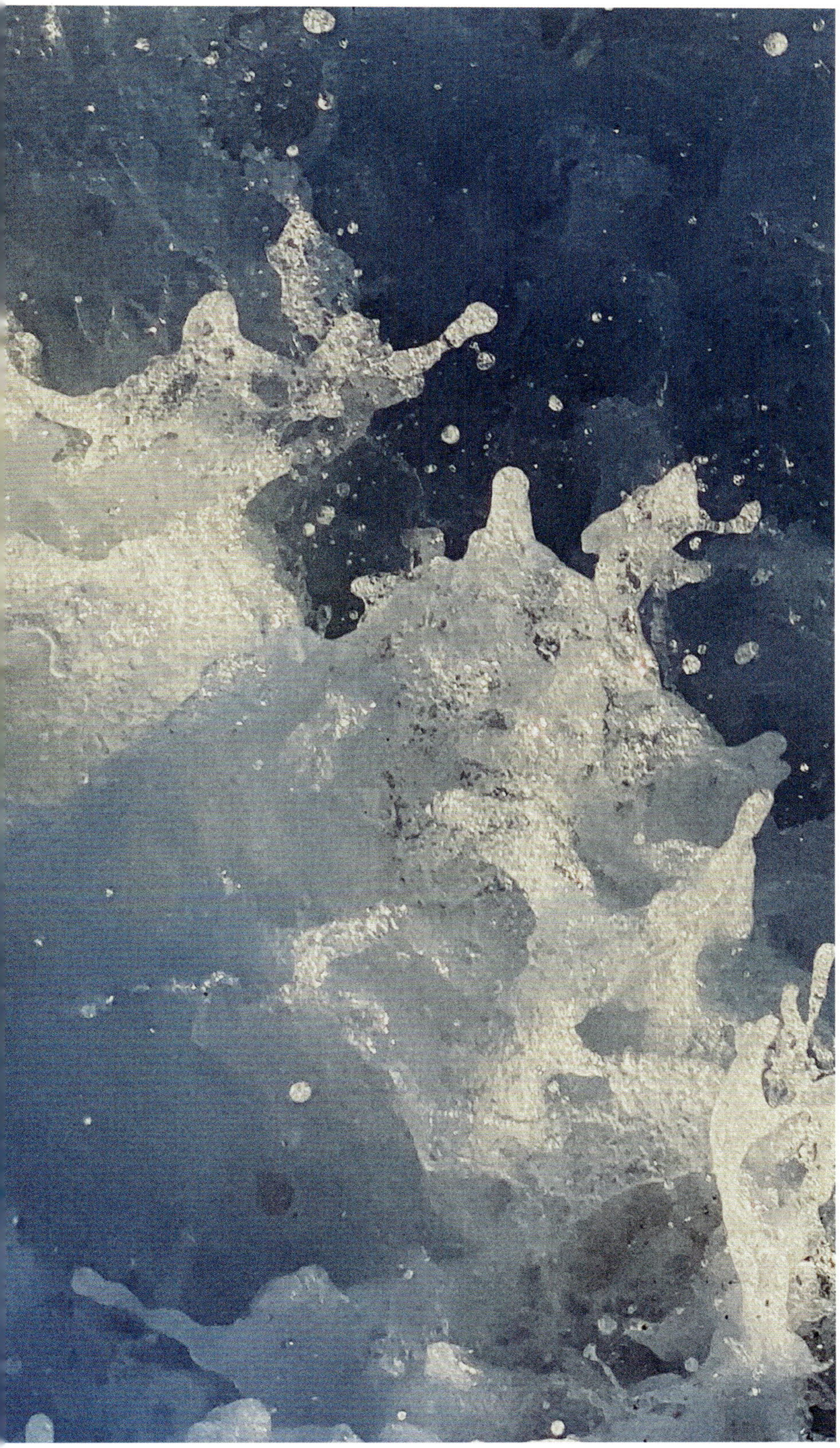

P. 4/5

Good times and tan lines.
South Cronulla Beach,
Sydney, Australia.

Sunrise at South Cronulla
Beach, Sydney, Australia.

Leichtigkeit und Sommer-
bräune. South Cronulla Beach,
Sydney, Australien.

Sonnenaufgang am South
Cronulla Beach, Sydney,
Australien.

Left: *Kiss me, Kismet.*
Marbella, Spain.

Right: *To sea, the skies.*
Marbella, Spain.

Links: *Küss mich, Kismet.*
Marbella, Spanien.

Rechts: *An die See, den
Himmel.* Marbella, Spanien.

The beach. If we could choose anywhere to be, it would always be the beach. 70 per cent of travelers worldwide spend their vacations near water–at rivers and lakes, to be sure, but first and foremost at the ocean. Note the prepositions we use here: We are *in* the mountains, but we are *at* the beach. Because, in fact, we spend almost all of our vacation time *at* the ocean, never *in* its waters for any serious length of time. Instead, we ram our beach umbrellas into the sand and set up our lounge chairs. We are relaxed observers. We are beach people. And how could it be any different for us land dwellers? When we draw closer to the coast, catching our first glimpse of the blue waters from the car, we exclaim, "The ocean!" in much the same way sailors and explorers of past centuries ecstatically shouted out, "Land ho!"

Yet the ocean itself is not our actual destination. It is the *view* of the ocean that is our target. The sound of the waves. The smell of the saltwater. Our real destination is the beach. Sunbathing, eating ice cream, playing paddle ball–in theory, we could do all these things at home in our backyard or at a local park. But would it be the same? No, because at the beach it's something totally different. The vastness of the ocean signals to us: You are here, not there. Now relax, and forget everything you left behind at home.

The beach would be unthinkable without the ocean, and vice versa. "On the border between the real and the imaginary, life on the beach unfolds as a kind of everyday utopia," writes Thomas Steinfeld in his book *Italien – Porträt eines fremden Landes* (Italy – Portrait of a Foreign Country).

Life at the beach offers something for all our senses, not just for our eyes, also for our nose, our ears: People tend to complain about crowing roosters, barking dogs, or clanging church bells. But no one has ever been annoyed by the sound of the ocean, even though it can swell to a deafening roar. It is noise that we relish, indeed even long for. The feeling of the hot sand we rest our hands on–or the wet pebbles under our feet as we carefully tip-toe into the water. And the smells of a day at the beach! Suncreen. Wet rock. Wet wood. Wet swim trunks. The fabric of the beach umbrella, warmed by the sun. The plastic of inflatable rubber boats. Seaweed. Damp towels. Cigarette smoke. Water ice. The softening ink of newspapers lying in the sun. Wet sand. Perfume. Silt. Spray-on quick tanner. Pine needles.

Innumerable oceans exist, each generation has its own. There is an ocean for children, an ocean for teenagers, and an ocean for adults. The youngest venture their first steps on the soft ground, children build sandcastles and learn to swim, young adults fall in love over a cocktail at the beach bar, mom and dad enjoy relaxed family time. At last, everyone is happy, albeit in very different ways. And if you ask a 90-year-old fellow where he would like to go one more time in his life, chances are he'll say: to the seashore.

We have always tried to tame the sea, to reclaim land from it, to break its waves with concrete. "Its inherent danger is as seductive as it is frightening, almost like the endlessness itself", writes Gabriele Romagnoli. While mountains can be conquered, the sea cannot. The oceans are an unexplored mystery. "If we knew as much about the surface of the earth as we do about the depths of the sea, we would not even have discovered elephants," said naturalist John Frederick Grassle. Mount Everest, the highest mountain on Earth, has been climbed by some 5,000 people at present, and some Sherpas have even been to its summit 20 or more times. The Mariana Trench, the deepest point of the ocean and in a way the mirror of Mount Everest, has been visited to date by five people in diving bells made of steel 70 centimeters thick.

We joyfully leap into the water, dive under its surface–a daily baptism. When we are *in* the ocean, we are happy and also a tiny bit apprehensive. We enjoy the infinity; it is both fascinating and threatening at the same time.

We probably like swimming out into the unknown, flirting with fate, with the imponderable beneath us and surrounding us. This is where we experience it physically: we are land creatures, completely helpless in the sea. The beach becomes a safe place. A base and a look-out. Land under our feet.

And at some point, after a few days, our bodies and especially our minds adapt to life at the beach. Our skin tans, we move more and more nonchalantly, less self-consciously in our beachwear, our swim trunks, our bikinis. Life slows down, perhaps naturally so, by our clumsy steps in the yielding sand. Freedom from constraints, from constrictive clothing. We can finally breathe deeply. The unaccustomed physicality is soon familiar to us, the daily routine quickly forgotten.

And those who have gone home are dreaming of being back at the beach. We want to catch a glimpse of the world beyond, where we come from and where we are going.

Der Strand: Das ist immer der Ort, an dem wir lieber wären. 70 Prozent der Reisenden weltweit verbringen ihren Urlaub am Wasser – an Flüssen und Seen, aber vor allem am Meer. Die Präposition ist hier wichtig: Wir fahren *in* die Berge, aber *ans* Meer. Denn tatsächlich verbringen wir nahezu unsere gesamte Urlaubszeit *an* der See und sind nie für längere Zeit *im* Wasser. Stattdessen rammen wir unsere Sonnenschirme in den Sand und stellen dort unsere Liegen auf. Wir sind distanzierte Beobachter. Wir sind Strandmenschen. Und wie könnte es für uns Landbewohner auch anders sein?

Wenn wir uns der Küste nähern und die blauen Fluten zum ersten Mal vom Auto aus erspähen, rufen wir: „Das Meer!" – etwa so, wie die Seefahrer und Entdecker vergangener Jahrhunderte euphorisch „Land in Sicht!" verkündet haben.

Dabei ist das Meer ist nicht das eigentliche Ziel. Das Ziel ist der *Blick* aufs Meer. Das Rauschen der Wellen. Der Duft des Salzwassers. Genauer gesagt: Unser Ziel ist der Strand. Sonnenbaden, Eis essen, Beachtennis spielen – theoretisch könnten wir all das auch im heimischen Garten oder im Stadtpark machen. Aber wäre es das Gleiche? Nein, am Meer ist es etwas ganz anderes. Die Weite des Ozeans signalisiert uns: Du bist hier und nicht dort. Du bist woanders. Jetzt entspann dich. Und vergiss alles, was du daheim zurückgelassen hast.

Der Strand wäre ohne Meer gar nicht denkbar, und umgekehrt. „An der Grenze zwischen dem Wirklichen und dem Unwirklichen angesiedelt, entfaltet sich das Leben am Strand als eine Art Utopie des Alltäglichen", schreibt Thomas Steinfeld in seinem Buch *Italien – Porträt eines fremden Landes*.

Das Leben am Strand bietet für alle Sinne etwas, nicht nur unseren Augen, sondern auch unserer Nase, unseren Ohren: Menschen beschweren sich über krähende Hähne, bellende Hunde, dröhnende Kirchenglocken. Aber über das Meeresrauschen, das bis zu einem betäubenden Tosen anschwellen kann, hat sich noch nie jemand aufgeregt. Es ist ein Lärm, den wir genießen, ja herbeisehnen. Das Gefühl des heißen Sandes, in dem wir unsere Hände abstützen – oder der feuchten Kiesel unter den Füßen, auf denen wir vorsichtig ins Wasser balancieren. Und erst der Duft eines Strandtages! Sonnencreme. Nasser Fels. Nasses Holz. Nasse Badehose. Heißer Sonnenschirmstoff. Das Plastik der aufblasbaren Gummiboote. Algen. Feuchte Handtücher. Zigarettenrauch. Wassereis. Die aufweichende Druckerschwärze der in der Sonne liegenden Zeitung. Nasser Sand. Parfüm. Schlick. Schnellbräuner zum Aufsprühen. Piniennadeln.

Es existieren unzählige Meere, jede Generation hat ihr eigenes. Es gibt ein Meer für Kinder, ein Meer für Jugendliche und ein Meer für Erwachsene. Die Kleinsten wagen ihre ersten Schritte im weichen Untergrund, Kinder bauen eine Sandburg und lernen Schwimmen, Heranwachsende verlieben sich bei einem Cocktail in der Strandbar, Mama und Papa genießen das ungezwungene Familienleben. Endlich sind einmal alle glücklich, wenn auch auf ganz unterschiedliche Art. Und wer einen 90-Jährigen fragt, wohin er noch einmal im Leben reisen möchte, dann sagt er ziemlich sicher: ans Meer.

Wir haben immer versucht, den Ozean zu zähmen, ihm Land abzutrotzen, die Wellen mit Beton zu brechen. „Die Gefahr, die in ihm schlummert, ist ebenso verführerisch wie beängstigend, beinahe wie die Unendlichkeit selbst", schreibt Gabriele Romagnoli. Die Berge sind bezwingbar, das Meer ist es nicht. Die Ozeane sind ein unerforschtes Mysterium. „Wüssten wir so viel über die Erdoberfläche wie über die Meerestiefen, hätten wir noch nicht einmal die Elefanten entdeckt", sagt der Naturforscher John Frederick Grassle. Der Mount Everest, der höchste Berg der Erde, ist bislang von etwa 5000 Menschen bestiegen worden, manche Sherpas waren mehr als 20 Mal auf dem Gipfel. Der Mariannengraben, der tiefste Punkt des Meeres und damit gewissermaßen der gespiegelte Mount Everest, wurde bislang von fünf Menschen in Tauchkapseln aus 70 Zentimeter dickem Stahl aufgesucht.

Wir springen vergnügt ins Wasser, tauchen unter – eine tägliche Taufe. Wenn wir dann mal *im* Meer sind, freuen wir uns und gruseln uns auch ein wenig. Wir genießen die Unendlichkeit, sie ist Faszinosum und Bedrohung zugleich.

Wahrscheinlich mögen wir es, dieses Hinausschwimmen ins Unbekannte, das Flirten mit dem Schicksal, unter uns und um uns herum das Unbeherrschbare. Hier merken wir es physisch: Wir sind Landwesen und im Meer völlig hilflos. Der Strand wird zum sicheren Ort. Zur Basis und zum Ausguck. Boden unter den Füßen. Und irgendwann, nach ein paar Tagen, passt sich unser Körper und vor allem unser Geist dem Leben am Strand an. Die Haut bräunt sich, immer ungezwungener bewegen wir uns in der Strandbekleidung, in der Badehose, im Bikini. Das Leben wird langsamer, vielleicht auch auf natürliche Weise ausgebremst von unseren schwerfälligen Schritten im nachgiebigen Sand. Eine Freiheit vom Zwang, von einengender Kleidung. Endlich durchatmen. Die ungewohnte Körperlichkeit ist uns bald vertraut, der Alltag bleibt schnell weit weg.

Und wer daheim ist, träumt sich wieder zurück an den Strand. Wir wollen eine Ahnung vom Jenseits erblicken, aus dem wir kommen und wohin wir gehen.

LEISURE

ERHOLUNG

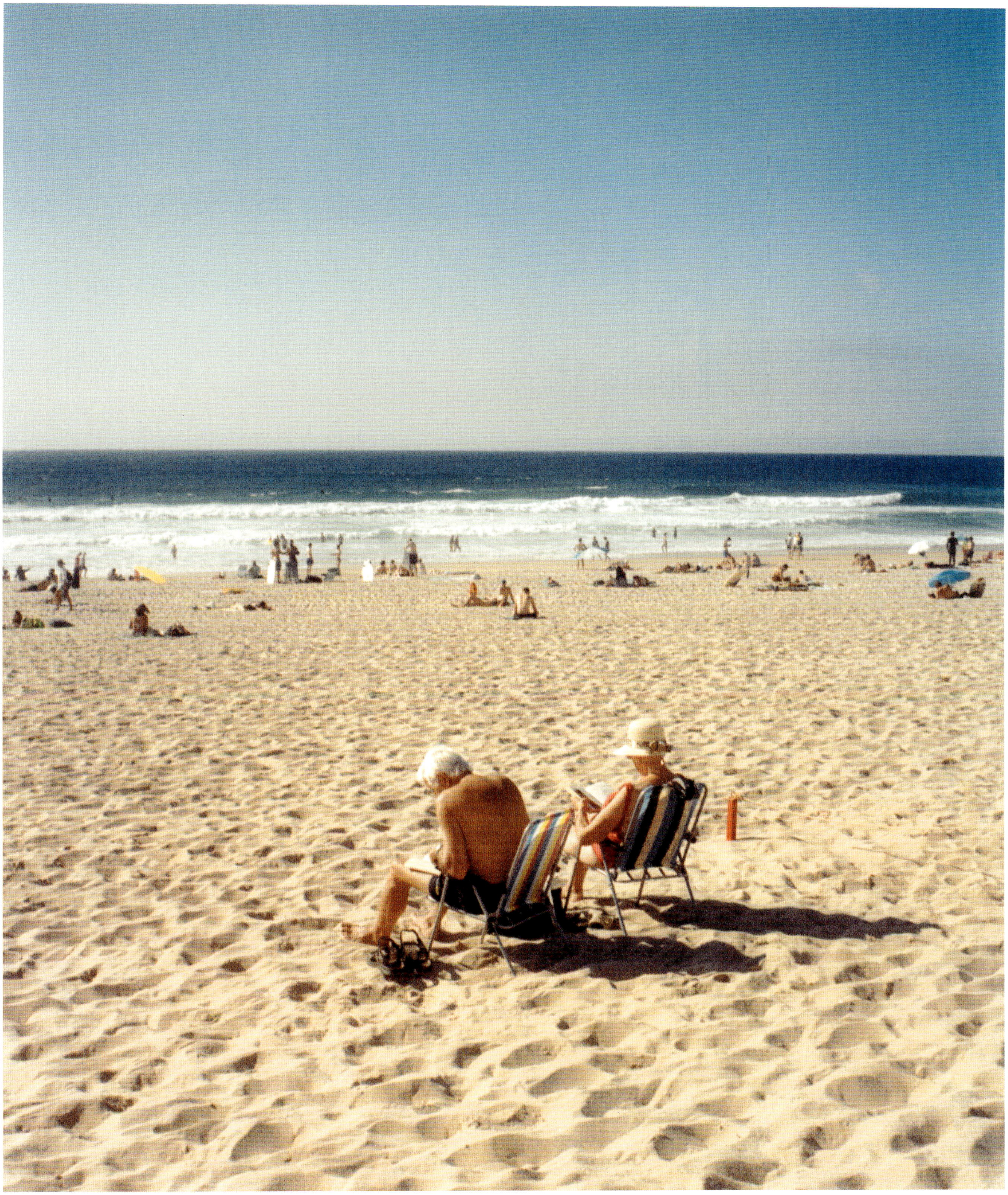

The beach is the home of lounge chairs in the sun, of beach towels in secluded coves. In any case, life at the seashore takes place largely in a horizontal position. Even the busiest, most industrious, restless people eventually give in and adopt a reclining position.
Lying down is the epitome of a positive, understated refusal to move forward. To pause instead of rushing ahead. Even the gaze itself: Instead of looking forward, the eyes are directed upward. Or at the pages of a book, on a journey into another world. A reclining person represents a harmless, elegant, horizontal middle finger raised against the hustle and bustle of the world. "Lying down was seen as the horizontal counterpart to the dreamy rambling of the melancholy *flâneur*, walking about without pursuing any goal in particular," writes Bernd Brunner in his book *The Art of Lying Down*.
Lying down is the antithesis of all of the modern demands placed on people. And it is the most comfortable position available to us. It doesn't consume any calories and doesn't require any muscular strength. That's why lying down is often confused with laziness and is frowned upon in everyday life–at least outside of the hours when we are usually sleeping. And busy people who want to take a nap coined the dreadful term *power napping* as an excuse to do so.
On the beach, there is no need for excuses. We surrender to gravity with delight. And to those who accuse us of idleness, we offer a sentence by the English writer Gilbert Keith Chesterton. The inventor of the amateur detective and priest Father Brown recognized in his essay *On Lying In Bed*, published in 1909, that the most famous work of art in history must not only have been produced while lying down, but conceived in that position as well: "I am sure that it was only because Michael Angelo was engaged in the ancient and honourable occupation of lying in bed that he ever realised how the roof of the Sistine Chapel might be made into an awful imitation of a divine drama that could only be acted in the heavens."

Der Strand ist der Ort der Sonnenliegen – oder, an einsamen Buchten, des Badetuchs. In jedem Fall findet das Leben am Meeresufer weitgehend in der Horizontalen statt. Selbst die emsigsten, unternehmungslustigsten, rastlosesten Menschen geben irgendwann nach und nehmen die Ruheposition ein.
Das Liegen ist der Inbegriff positiver, unaufdringlicher Fortschrittsverweigerung. Innehalten statt Vorwärtsstürmen. Schon der Blick: Er verweist nicht nach vorn, sondern nach oben. Oder auf die Seiten eines Buches: die Reise in eine andere Welt. Der liegende Mensch ist ein harmloser, eleganter, waagerechter Mittelfinger wider die Hektik der Welt. „Das Liegen ist das horizontale Pendant der traumartigen Spaziergänge des melancholischen Flaneurs, der, obwohl er geht, oft kein bestimmtes Ziel vor Augen hat", schreibt Bernd Brunner in seinem Buch *Die Kunst des Liegens*.
Liegen ist die Antithese zu allen modernen Anforderungen an den Menschen. Und: Es ist die bequemste Position, die uns möglich ist. Sie verbraucht keine Kalorien und verlangt keinerlei Muskelkraft. Deswegen wird das Liegen oft mit Faulheit verwechselt und ist im Alltag – jedenfalls außerhalb der Schlafenzeiten – verpönt. Und Vielbeschäftigte, die sich ein Nickerchen gönnen wollen, haben als Entschuldigung den schrecklichen Begriff des *power napping* erfunden.
Am Strand sind keinerlei Ausreden nötig. Wir ergeben uns mit Wonne der Schwerkraft. Und wer uns des Müßiggangs bezichtigt, dem halten wir einen Satz des englischen Schriftstellers Gilbert Keith Chesterton entgegen. Der Erfinder des ermittelnden Geistlichen Pater Brown erkannte in seinem Essay „On Lying In Bed", erschienen 1909, dass das berühmteste Kunstwerk der Geschichte nicht nur im Liegen erschaffen, sondern auch erdacht worden sein musste: „Nur weil sich Michelangelo der altehrwürdigen Beschäftigung des Im-Bett-Liegens hingab, kann er herausgefunden haben, wie die Decke der Sixtinischen Kapelle zu einer furchterweckenden Nachgestaltung des göttlichen Dramas umzuwandeln wäre, das nur im Himmel spielen kann."

PREVIOUS PAGE

Reading and relaxing on the Praia do Guincho in Portugal.

Entspannende Strandlektüre am Praia do Guincho in Portugal.

Lucky owner of a beach cottage in Le Havre, France.

Glücklicher Strandhäuschenbesitzer in Le Havre, Frankreich.

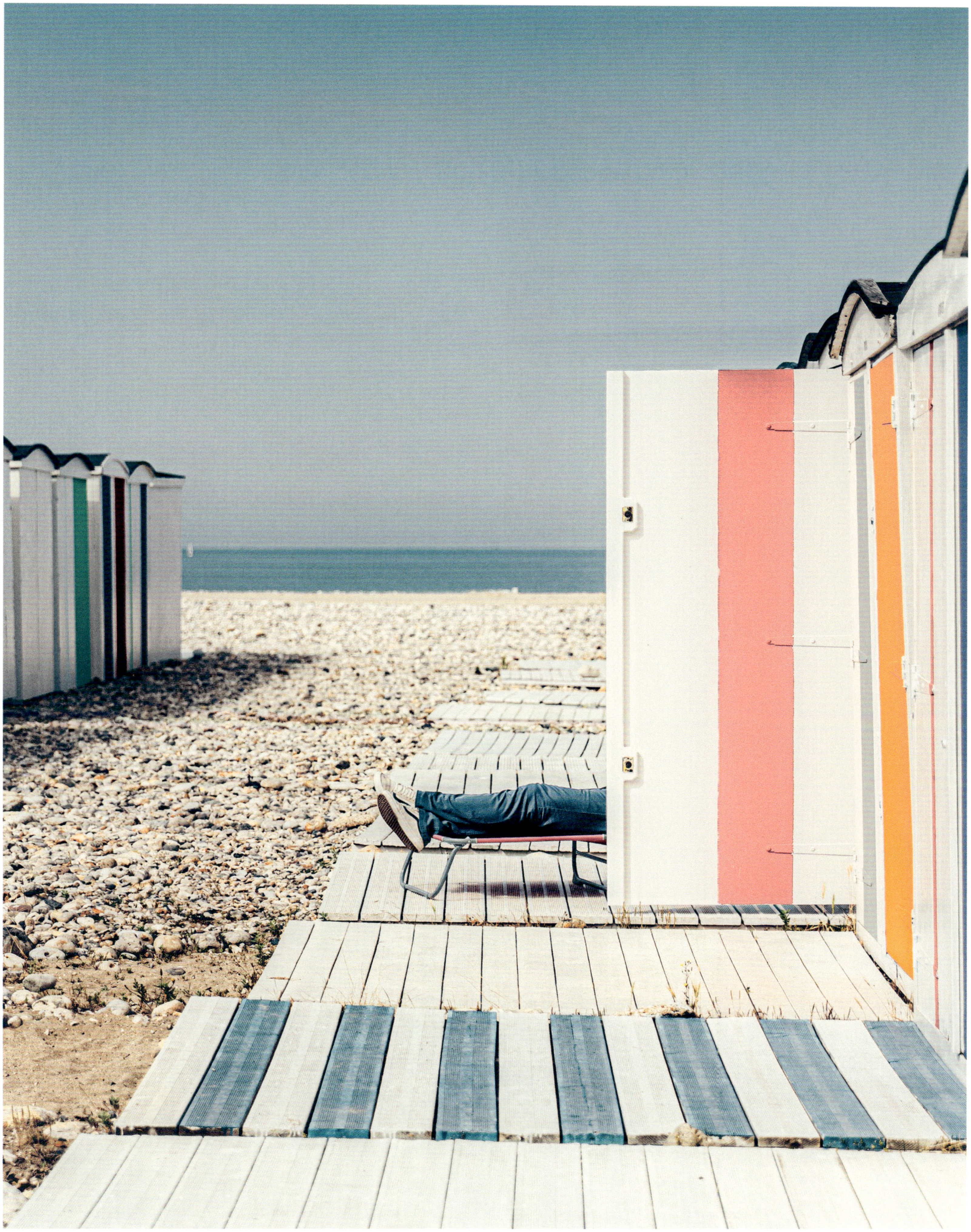

VOM VIRTUOSEN MÜSSIGGANG: DIE KUNST DES LIEGENS

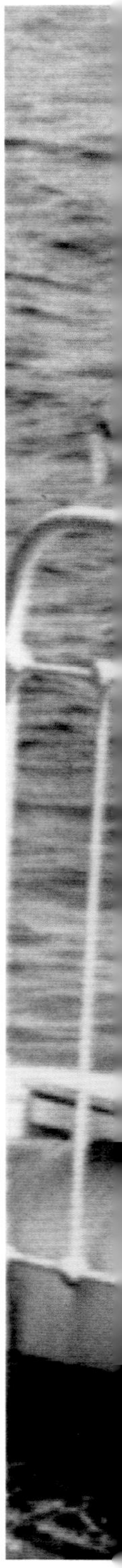

A different kind of air bed.
1967, location unknown.

Luftmatratze der anderen Art.
1967, Ort unbekannt.

MASTERFUL IDLENESS, OR THE ART OF LYING ON THE BEACH

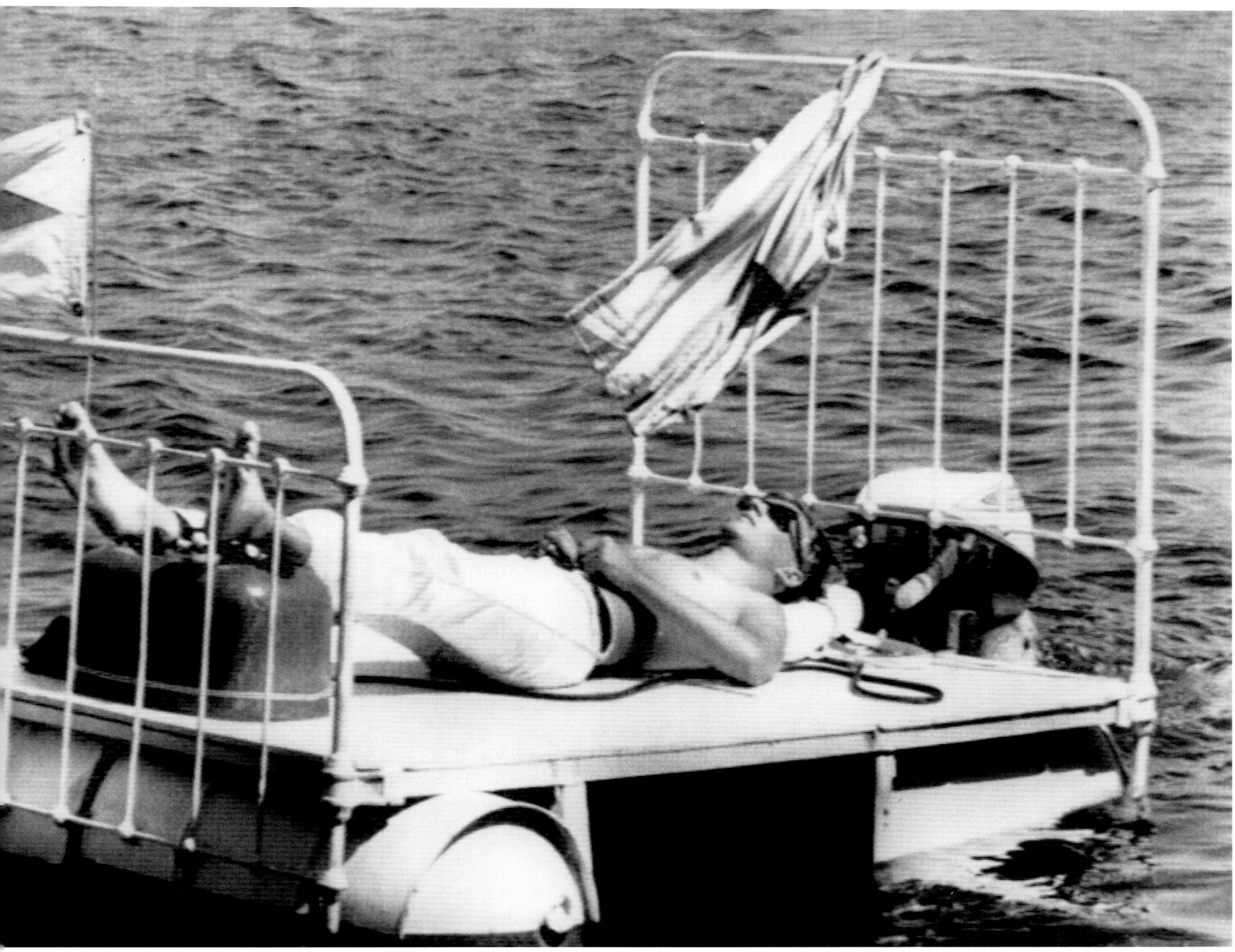

VOM VIRTUOSEN MÜSSIGGANG: DIE KUNST DES LIEGENS

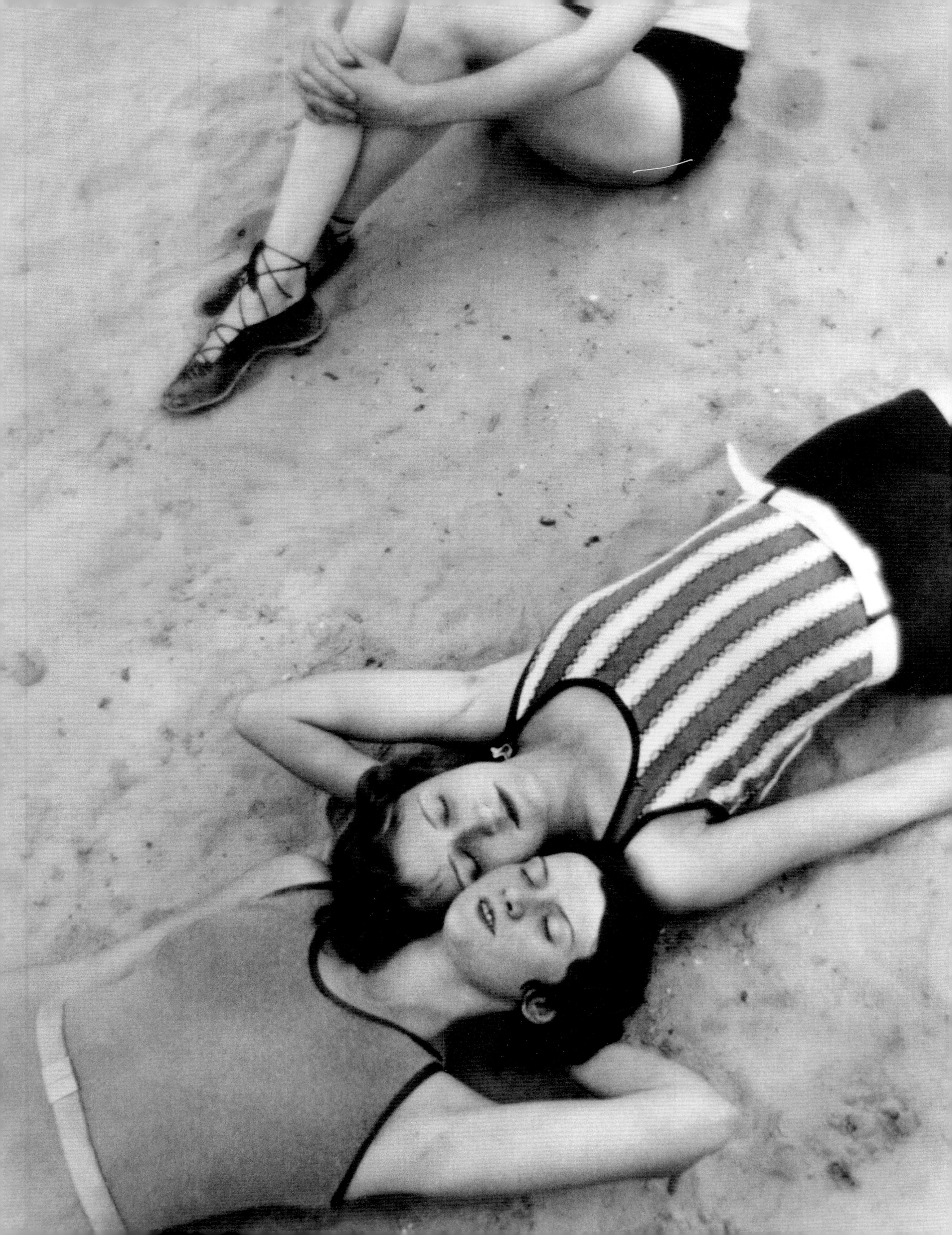

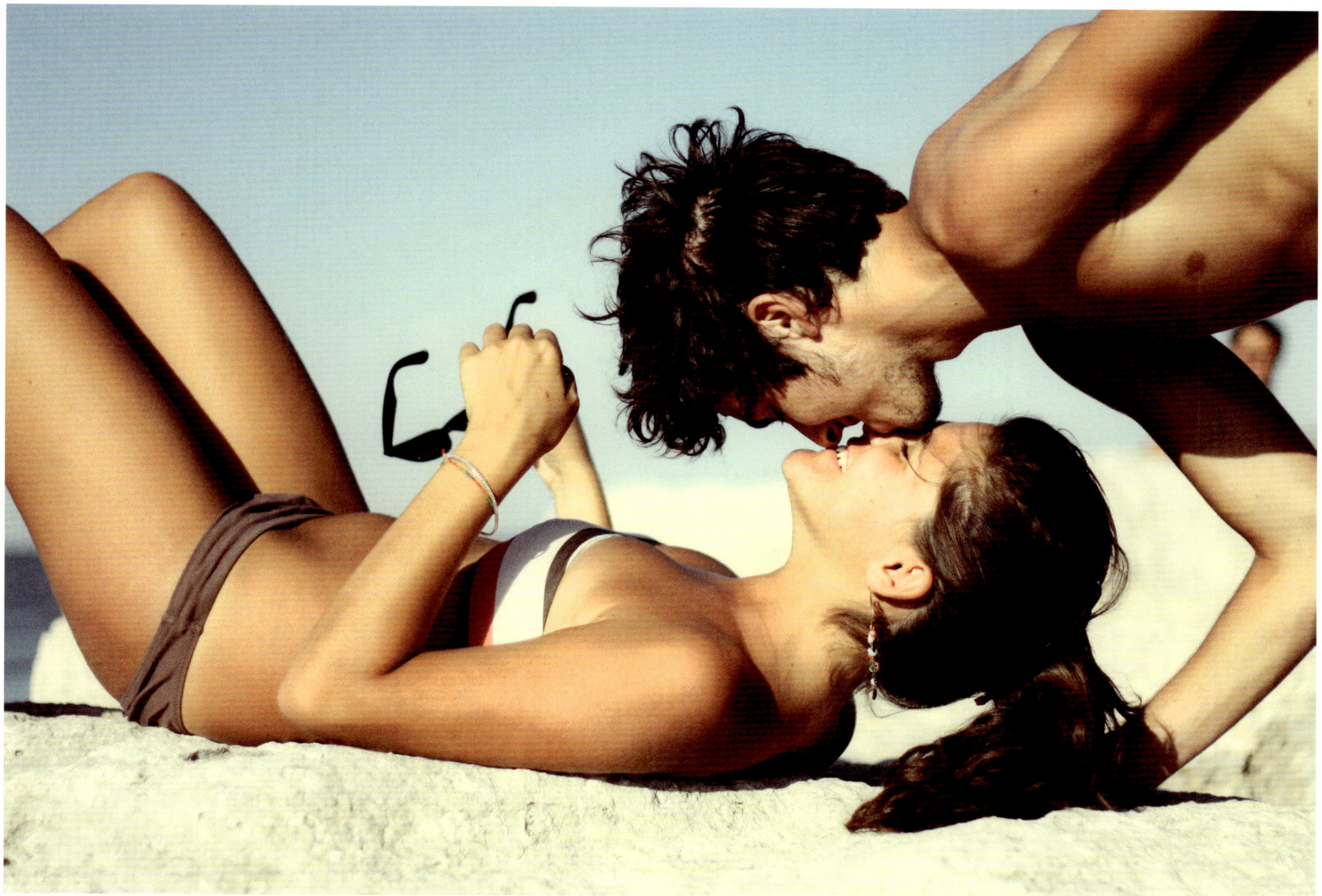

Together at the beach.
Lido degli Estensi, Ferrara, Italy.

Zusammen am Strand.
Lido degli Estensi, Ferrara, Italien.

Sunbathing together. 1929,
location unknown.

Gemeinsames Sonnenbad.
1929, Ort unbekannt.

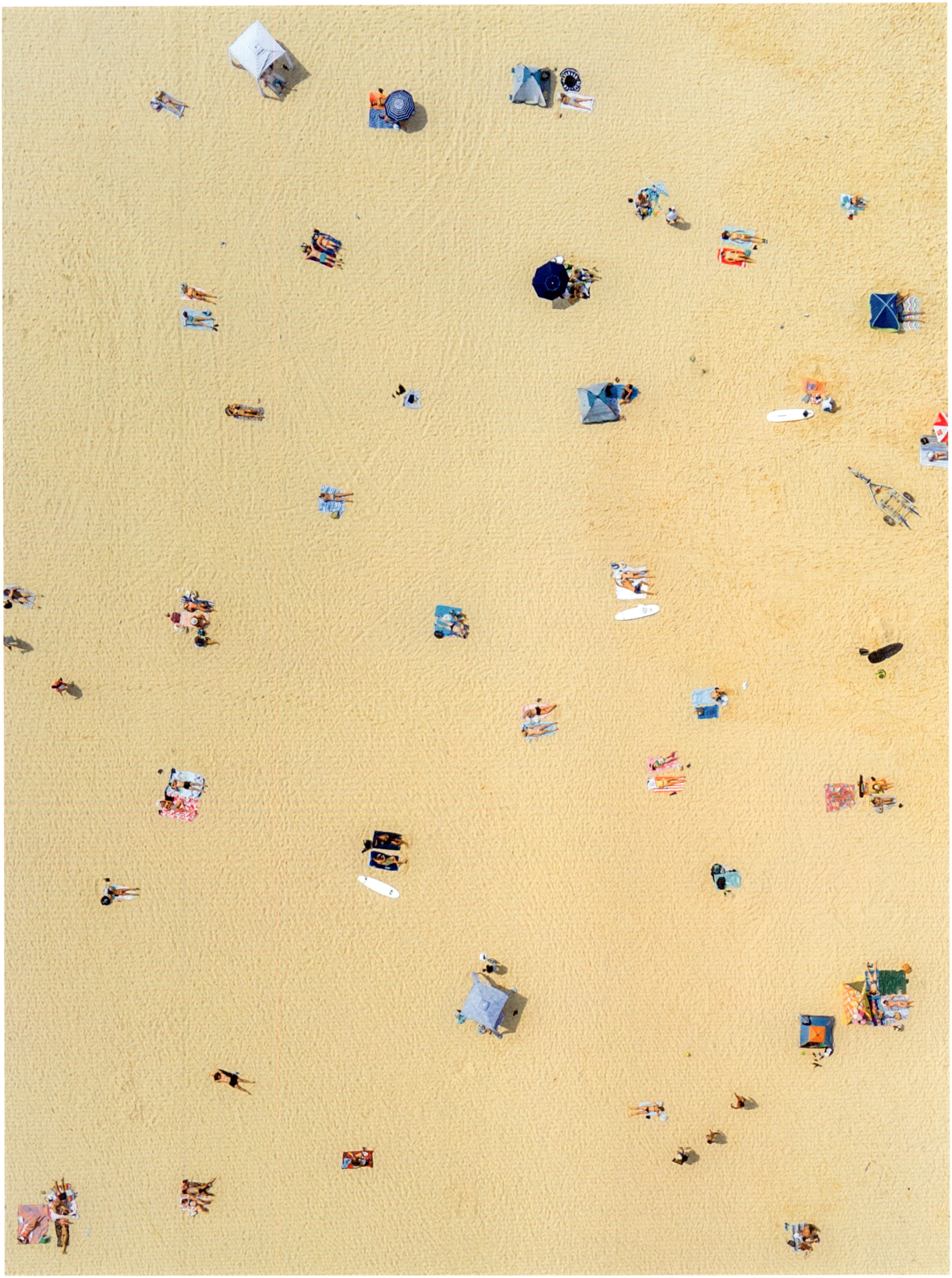

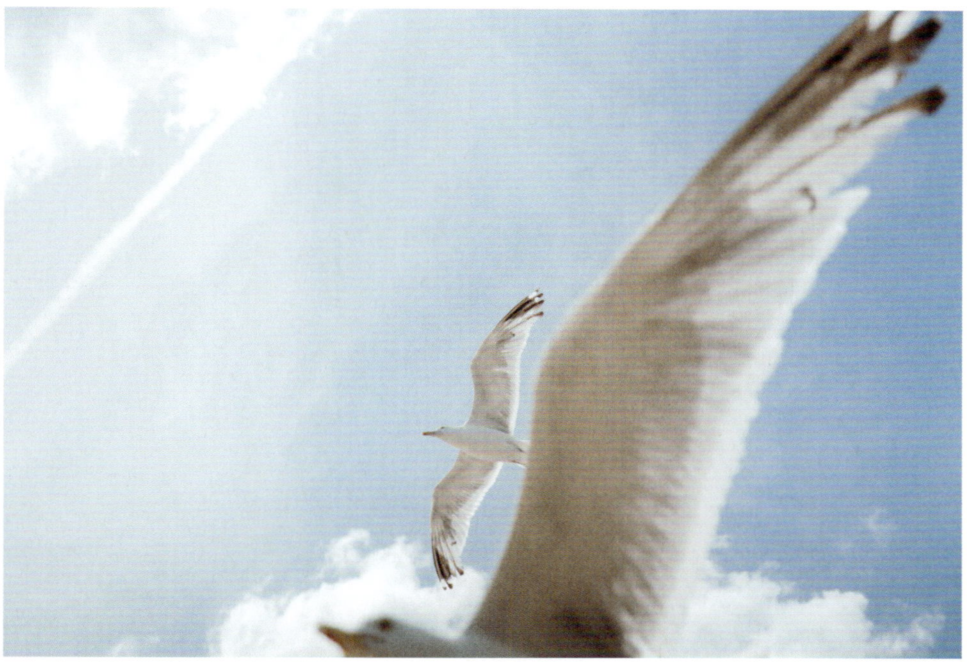

Looking up: Seagulls in the air.

Blick nach oben: Möwen am Himmel.

Looking up: Umbrella with clouds.

Blick nach oben: Sonnenschirm mit Wolken.

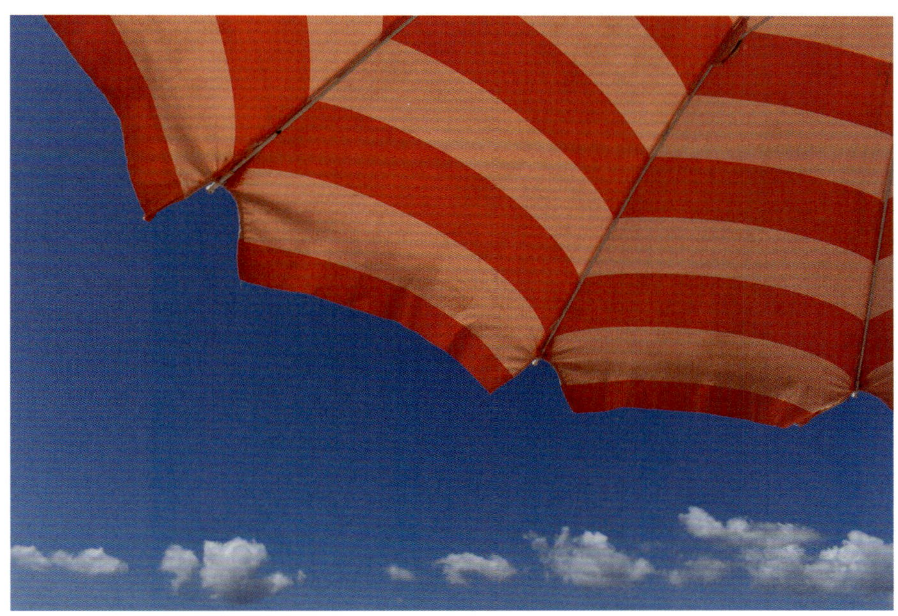

Looking down: People tanning at Maroubra Beach, Sydney, Australia.

Blick nach unten: Sonnenbaden am Maroubra Beach, Sydney, Australien.

Meet sand-body.
Sifnos, Greece.

Körper auf Sand.
Sifnos, Griechenland.

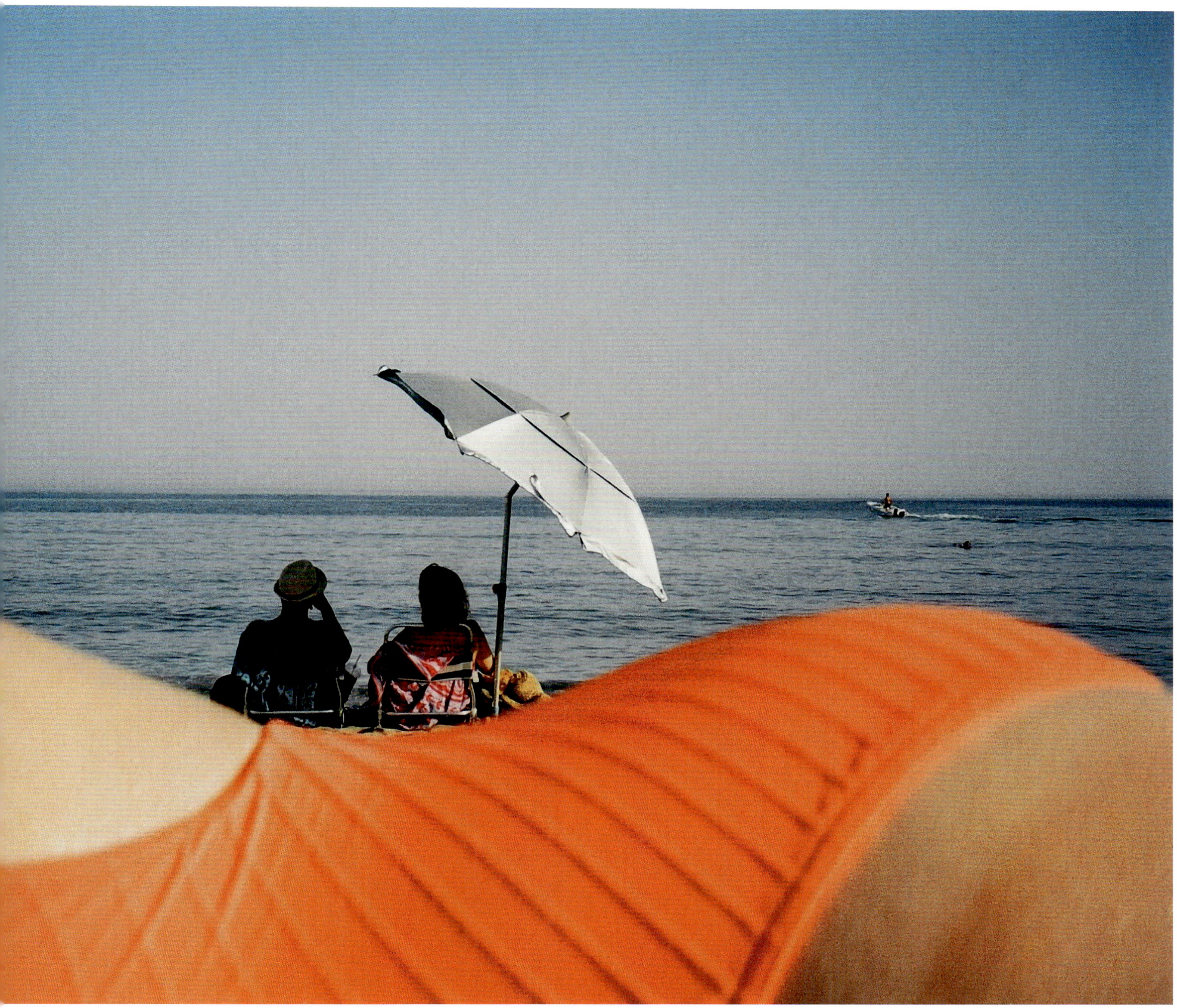

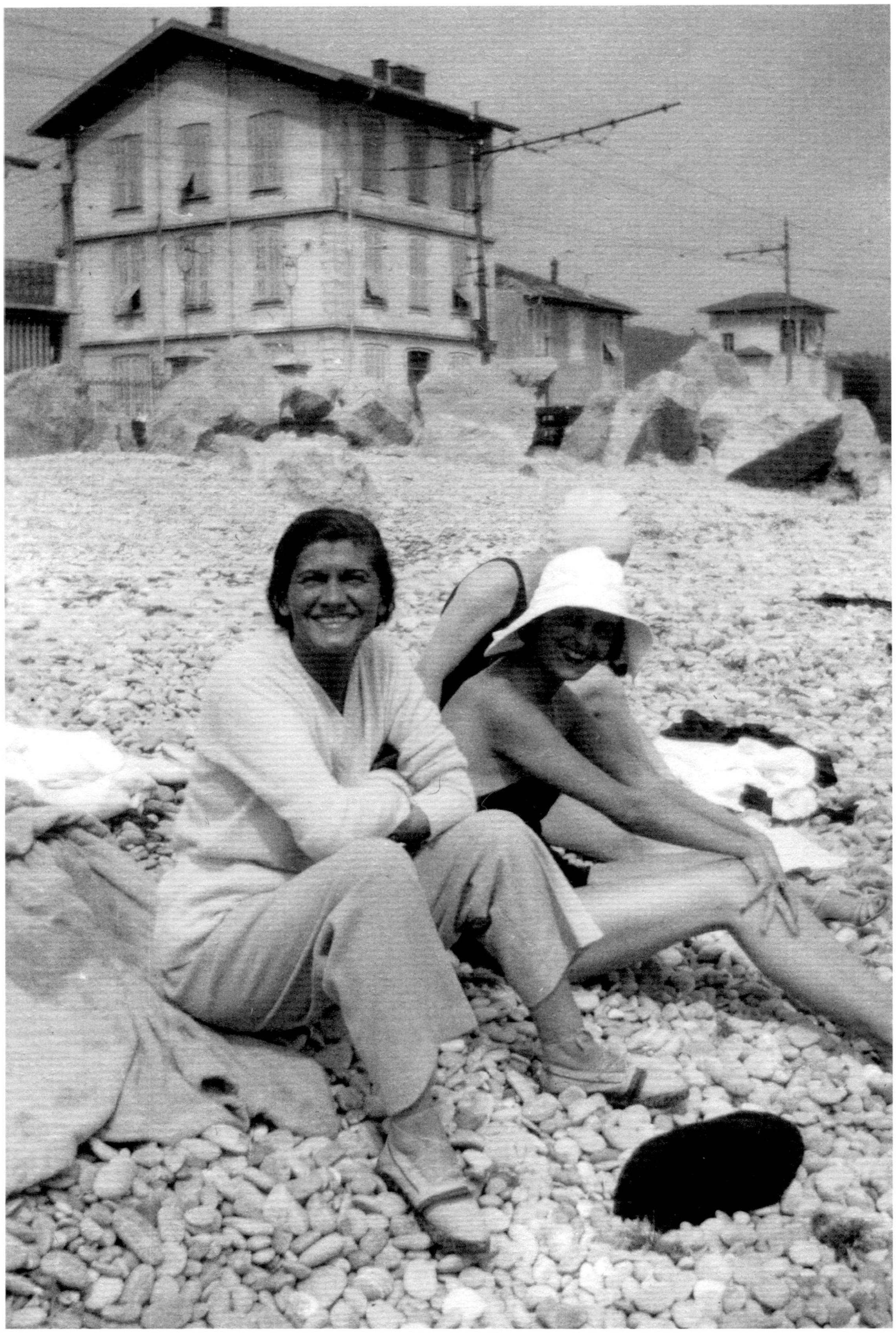

Coco Chanel and her friend Sara Murphy on the beach in the south of France, ca. 1930.

Coco Chanel und ihre Freundin Sara Murphy in Südfrankreich am Strand, ca. 1930.

A tan was considered fashionable even in ancient times, so rich Romans had patios built at their villas for sunbathing. Such patios were called *solaria* or, in the singular, *solarium*, which is why sunrooms are also called *solariums*.
Then, however, paleness became a centuries-long ideal of beauty, as it showed that a person belonged to a higher class that did not have to toil in the fields. The "blue blood" of the nobility became proverbial, because the bluish veins shone through their white skin.
Even into the early 20th century, the upper classes were careful not to get too much sun, so as not to be counted among the "ordinary" social classes.
It was Coco Chanel who changed everything. In the 1920s, she left Paris for a summer vacation at the French Riviera, sunbathed abundantly, and then upon her return proudly showed off a previously frowned-upon tan on the boulevards of Paris. Moreover, she called the color of her complexion *bronzage*, and who can resist the elegance of bronze? Soon Parisian women were imitating her, just as they wore her revolutionary bob as a hairstyle, admired her iconic short blazer, and generally emulated her seemingly carefree lifestyle. And her perfume Chanel No. 5, on the market since 1922, could be found in all the bathrooms of high society.
Thanks to Coco Chanel, it was suddenly considered fashionable to go to the seashore and sunbathe. The new trend swept away the pompous pallor of the past. No wonder, because the world of work had also changed. Farmers who toiled in the fields had become a vanishingly small minority; office jobs created a new epidemic of sallowness that was anything but genteel. A tan, now dubbed healthy, showed–and still shows to this day–that the person is not forced to devote themselves to working in stuffy indoor spaces, but apparently has a free, unconstrained, life-affirming way of spending their time.

Die Bräune galt schon in der Antike als schick; so ließen sich die reichen Römer in ihren Villen Terrassen zum Sonnenbaden anlegen. Solche Terrassen hießen *solaria* oder im Singular *solarium*, daher haben heutige Sonnenstudios ihren Namen.
Dann allerdings galt über Jahrhunderte die Blässe als Schönheitsideal, zeigte sie doch, dass man zu einer höheren Schicht gehörte, die nicht auf den Feldern schuften musste. Das „blaue Blut" der Adligen ist sprichwörtlich geworden, denn durch die weiße Haut schimmerten die bläulich wirkenden Adern durch.
Noch bis ins frühe 20. Jahrhundert achteten die Herrschaften von Welt darauf, bloß nicht zu viel Sonne abzubekommen, um nicht zu den „gewöhnlichen" Schichten gerechnet zu werden. Die Frau, die alles änderte, hieß Coco Chanel. Sie fuhr in den 1920er-Jahren zum Sommerurlaub von Paris an die Côte d'Azur, sonnte sich dort ausgiebig und zeigte anschließend die bislang verpönte Bräune auf den Pariser Boulevards stolz her. Zudem nannte sie die Farbe ihres Teints *bronzage*, und wer kann der edlen Bronze schon widerstehen? Bald machten es ihr die Pariserinnen nach, so wie sie als Frisur auch ihren revolutionären Bob trugen, den ikonisch gewordenen kurzen Blazer bewunderten und generell ihrem vermeintlich sorglosen Lebenswandel nacheiferten. Und ihr Parfüm Chanel No. 5, seit 1922 auf dem Markt, stand ohnehin in allen Badezimmern der Hautevolee.
Es galt dank Coco Chanel plötzlich als schick, ans Meer zu fahren und sich zu sonnen. Die neue Mode fegte die bislang vornehme Blässe aus der Welt. Kein Wunder, denn auch die Arbeitswelt hatte sich geändert. Bauern, die auf dem Feld schufteten, waren eine verschwindend kleine Minderheit geworden, Bürojobs sorgten für eine neue Blässe-Epidemie, die alles andere als vornehm war. Eine nunmehr als gesund betitelte Bräune zeigte – und zeigt bis heute –, dass die Person sich eben nicht der Arbeit in stickigen Innenräumen widmen muss, sondern offenbar über eine freie, ungezwungene, lebensbejahende Zeiteinteilung verfügt.

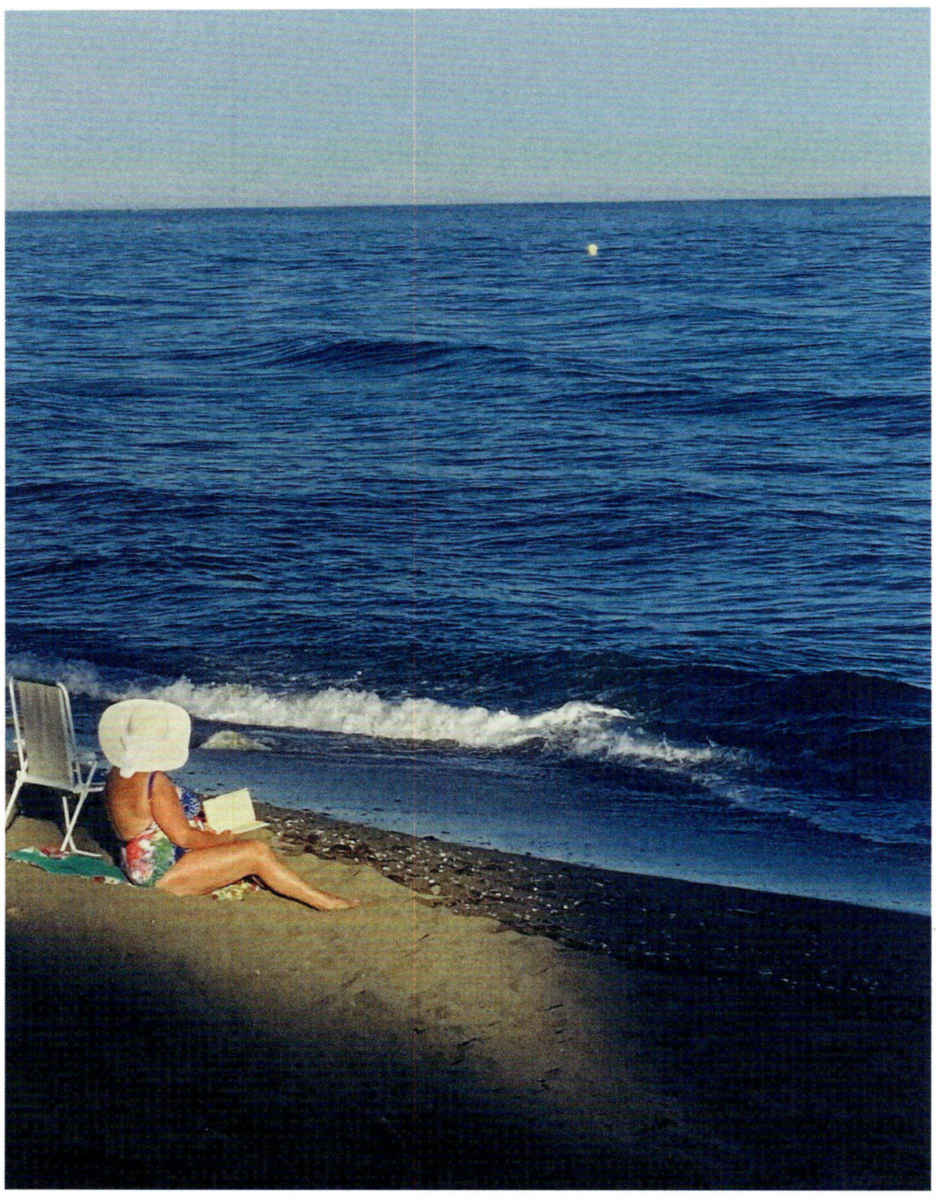

In the weighted wait, she sees the sea & reading light. From the series *Marbelland*. Marbella, Spain.

In der Schwere des Wartens sieht sie das Meer und Leselicht. Aus der Serie *Marbelland*. Marbella, Spanien.

Not to forget the sun screen.
Westerland, Sylt, Germany, 1967.

Sonnencreme nicht vergessen!
Westerland, Sylt, Deutschland, 1967.

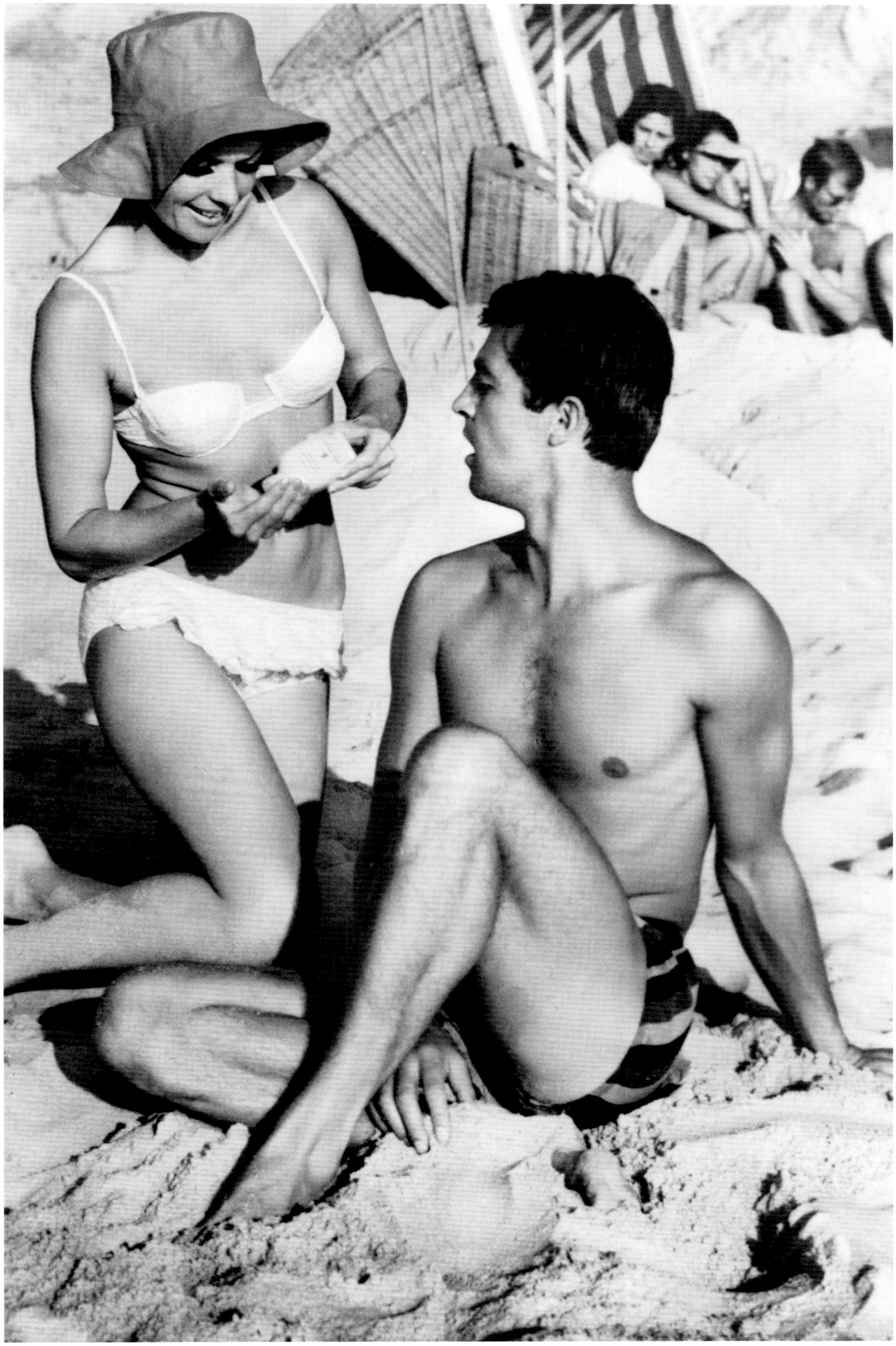

First sun rays of the year.
Lille, France.

Die ersten Sonnenstrahlen des Jahres.
Lille, Frankreich.

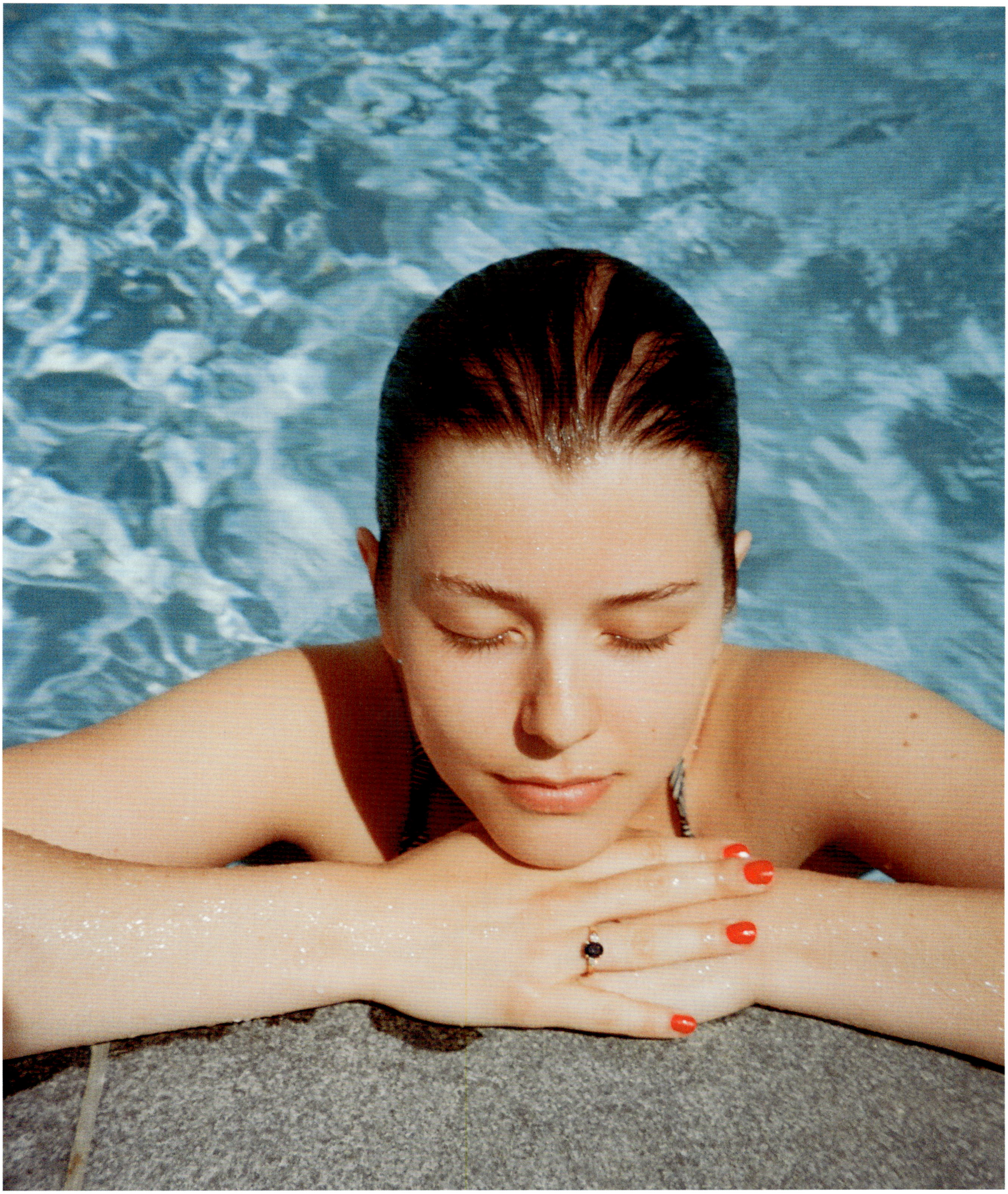

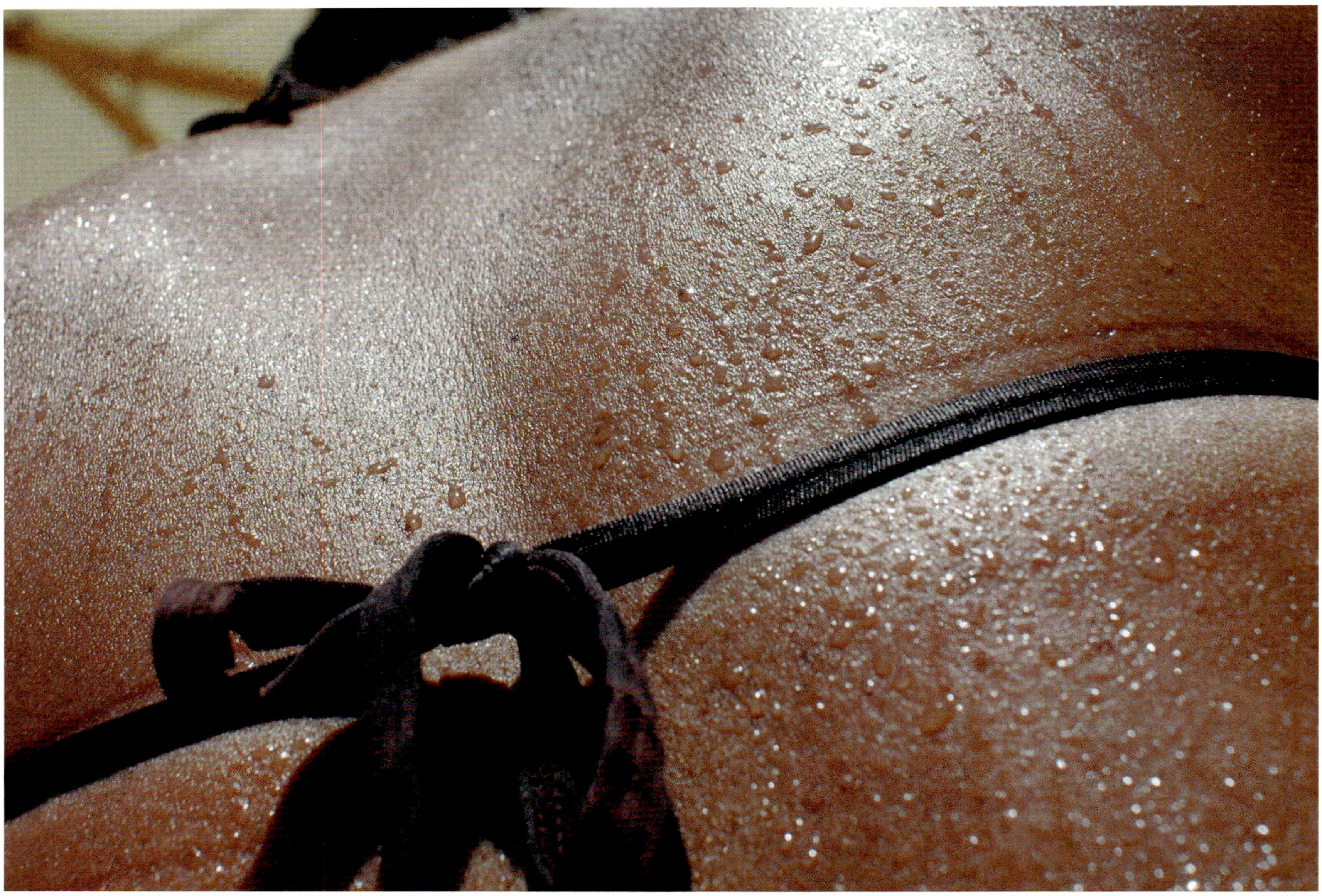

Water droplets of tanned skin.

Wassertropfen auf gebräunter Haut.

It's time for tanning.
Location unknown.

Zeit für Sommerbräune.
Ort unbekannt.

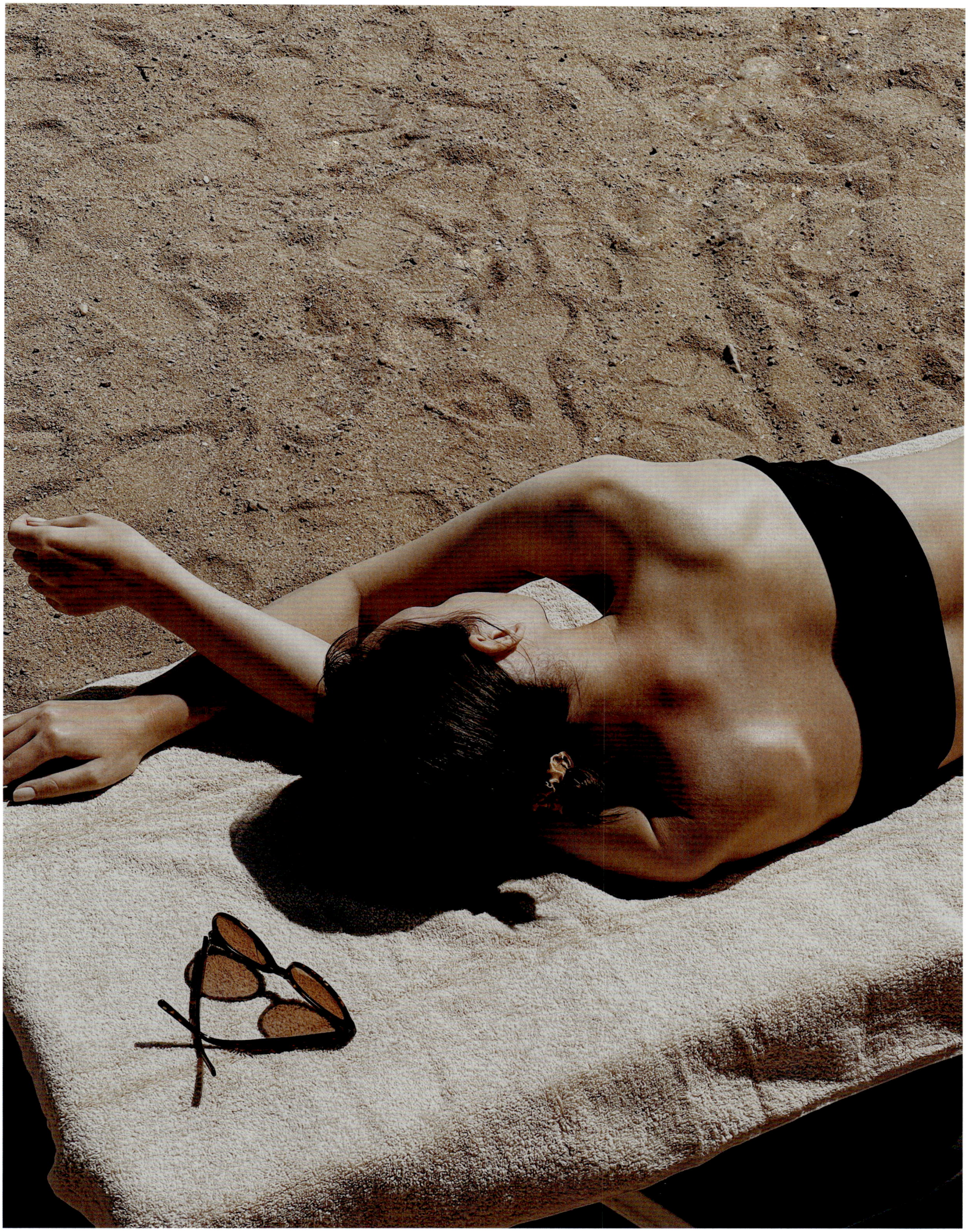

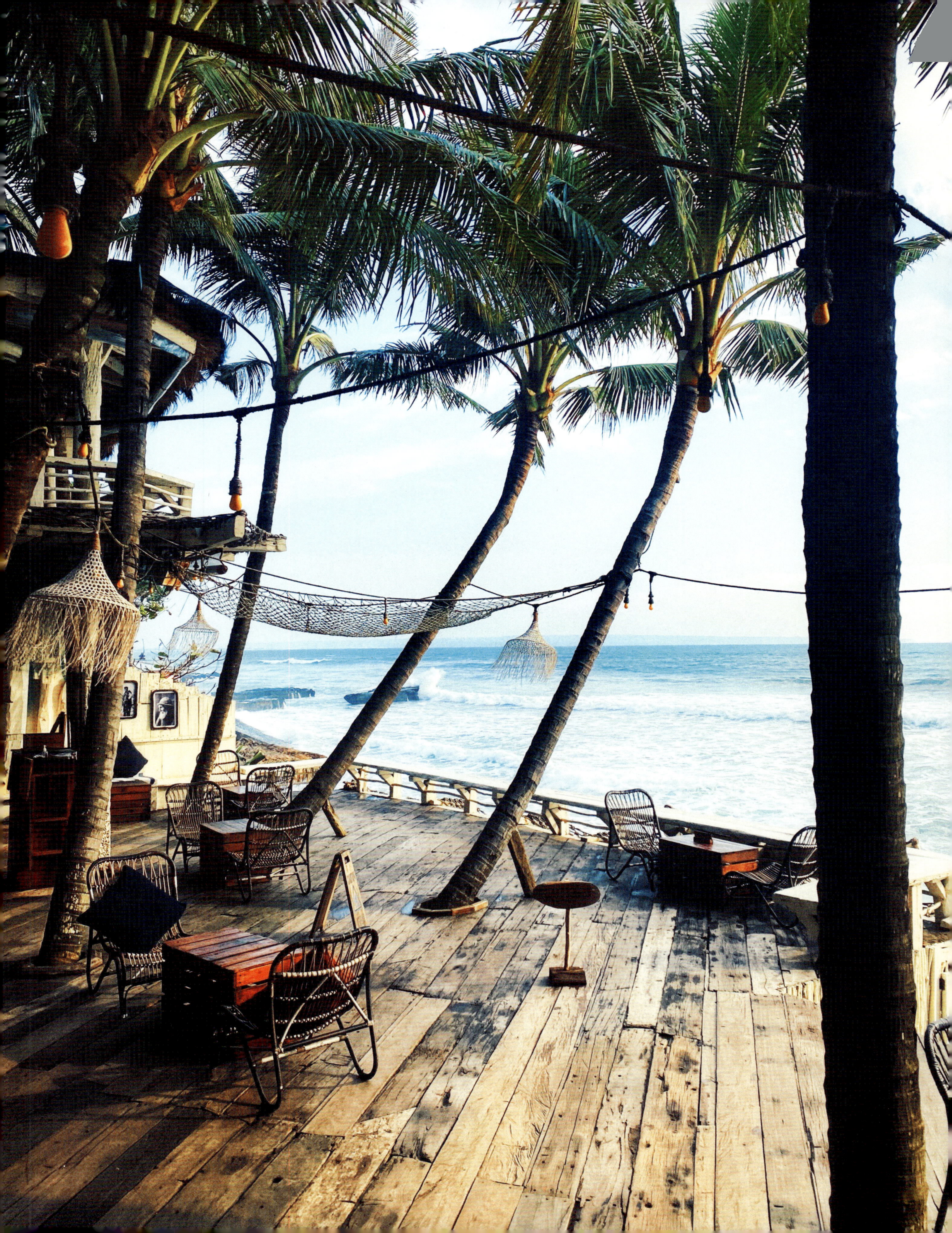

It was probably James Bond–once again–who triggered the hype. At a time when most people were just beginning to discover travel, the master spy was already lounging at the world's beach bars, which play an important role in almost every 007 episode, such as in *Never Say Never*, when the water-skiing acrobat Fatima Blush, played by Barbara Carrera, flies into the arms of Sean Connery, fresh from retirement. James catches her. Without spilling his drink, of course.
Beach bars, even more than the beach itself, are symbols of a light-hearted, happy-go-lucky life. A beach bar is all out carefree living to the ax, because in addition to having your feet in the sand, enjoying the shade of the palm fronds, and watching the sunset on the horizon, you also have alcoholic beverages and music. And where does a gin & tonic or a margarita taste better than overlooking the ocean?
Today, it's hard to imagine popular culture without beach bars. Hollywood filmed *Marley & Me*, *Burn Notice*, and *Miami Vice* at Monty's Raw Bar in Coconut Grove, Miami, and the Red Fish Grill Bar, also in Miami, aided in Cameron Diaz's breakthrough performance in *There's Something About Mary*. The Flamboyant Beach Bar in the Virgin Islands appeared in *Trading Places* with Dan Aykroyd and Eddie Murphy, and Neptune's Net in Malibu, California, has just about earned its own star on Hollywood Boulevard, having served as a set for TV series and movies for decades, most recently in *Point Break* and *Iron Man 3*. Marlene Dietrich, Orson Welles, Winston Churchill, and Charles de Gaulle have all sipped their drinks at the bar of the Hôtel du Cap-Eden-Roc on the French Riviera; Prince used the bar for his film *Under the Cherry Moon*, in which Kristin Scott Thomas made her screen debut. And Disney's TV show *Hannah Montana*, which made Miley Cyrus a global star, is essentially set at a beach bar (a fictional one, of course) run by an eleven-year-old.
The very best beach bars, on the other hand, are decidedly low-key and decline to serve as a filming location. The Dune Preserve Beach Bar on Anguilla is cobbled together from driftwood and scar from shipwrecks. Owner Bankie Banx is a reggae legend, performing there several times a week. Stars like John Mayer and Jimmy Buffett also like to drop by. It is a Caribbean dream in the colors white and blue, in an almost otherworldly idyllic setting and a deeply relaxed atmosphere. And nowhere else can you get a better rum punch than right here. What they say is true: Good beach bars make you think of retiring. Great ones make you actually do it.

Wahrscheinlich löste James Bond – wieder einmal – den Hype aus. Als die meisten Menschen erst langsam das Reisen entdeckten, lümmelte sich der Superspion bereits in den Strandbars dieser Welt, die in beinahe jeder 007-Folge eine wichtige Rolle spielen, etwa in *Sag Niemals Nie*, als dem aus dem Ruhestand zurückgekehrten Sean Connery die Wasserski-Akrobatin Fatima Blush, gespielt von Barbara Carrera, in die Arme fliegt. James fängt sie auf. Ohne dabei seinen Drink zu verschütten, *of course*.
Beachbars sind, mehr noch als der Strand selbst, Symbole fürs heitere, beschwingte Leben. Eine Strandbar ist Sorglosigkeit im Quadrat, denn zusätzlich zu den Füßen im Sand, dem Schatten der Palmwedel und dem Sonnenuntergang am Horizont gibt es auch noch alkoholische Getränke und Musik. Und wo schmecken Gin Tonic und Margarita besser als mit Meerblick?
Heute ist die Strandbar aus der Populärkultur nicht mehr wegzudenken. Hollywood filmte *Marley & ich*, *Burn Notice* und *Miami Vice* in Monty's Raw Bar in Coconut Grove, Miami, und die Red Fish Grill Bar, ebenfalls in Miami, assistierte bei Cameron Diaz' Durchbruch in *Verrückt nach Mary*. Die Flamboyant Beach Bar auf den Virgin Islands taucht in *Die Glücksritter* mit Dan Aykroyd und Eddie Murphy auf, und Neptun's Net im kalifornischen Malibu hätte beinahe schon einen eigenen Stern auf dem Hollywood Boulevard verdient, dient die Bar doch seit Jahrzehnten als Serien- und Filmkulisse, zuletzt in *Point Break* und *Iron Man 3*. In der Bar des Hôtel du Cap-Eden-Roc an der Côte d'Azur schlürften schon Marlene Dietrich, Orson Welles, Winston Churchill und Charles de Gaulle ihre Drinks; Prince nutzte die Bar für seinen Film *Under the Cherry Moon*, in dem Kristin Scott Thomas ihr Leinwanddebüt gab. Und Disneys TV-Show *Hannah Montana*, die Miley Cyrus zum Weltstar machte, spielt im Wesentlichen an einer (natürlich fiktiven) Strandbar, die von einem Elfjährigen betrieben wird.
Die allerbesten Strandbars geben sich dagegen dezidiert zurückhaltend und lehnen Dreharbeiten ab. Die Dune Preserve Beach Bar auf Anguilla ist aus Treibholz und Bootswracks zusammengezimmert. Betreiber Bankie Banx ist eine Reggae-Legende und tritt selbst mehrmals in der Woche auf. Stars wie John Mayer und Jimmy Buffett kommen ebenfalls gern vorbei. Es ist ein karibischer Traum in den Farben Weiß und Blau, in beinahe unwirklich paradiesischer Kulisse und tiefenentspannter Atmosphäre. Und nirgendwo gibt es einen besseren Rumpunsch als hier. Es stimmt, was die Leute sagen: Gute Beach Bars lassen dich übers Aussteigen nachdenken. Großartige Beach Bars lassen dich sofort aussteigen.

Beach bar on Bali, Indonesia.

Strandbar auf Bali, Indonesien.

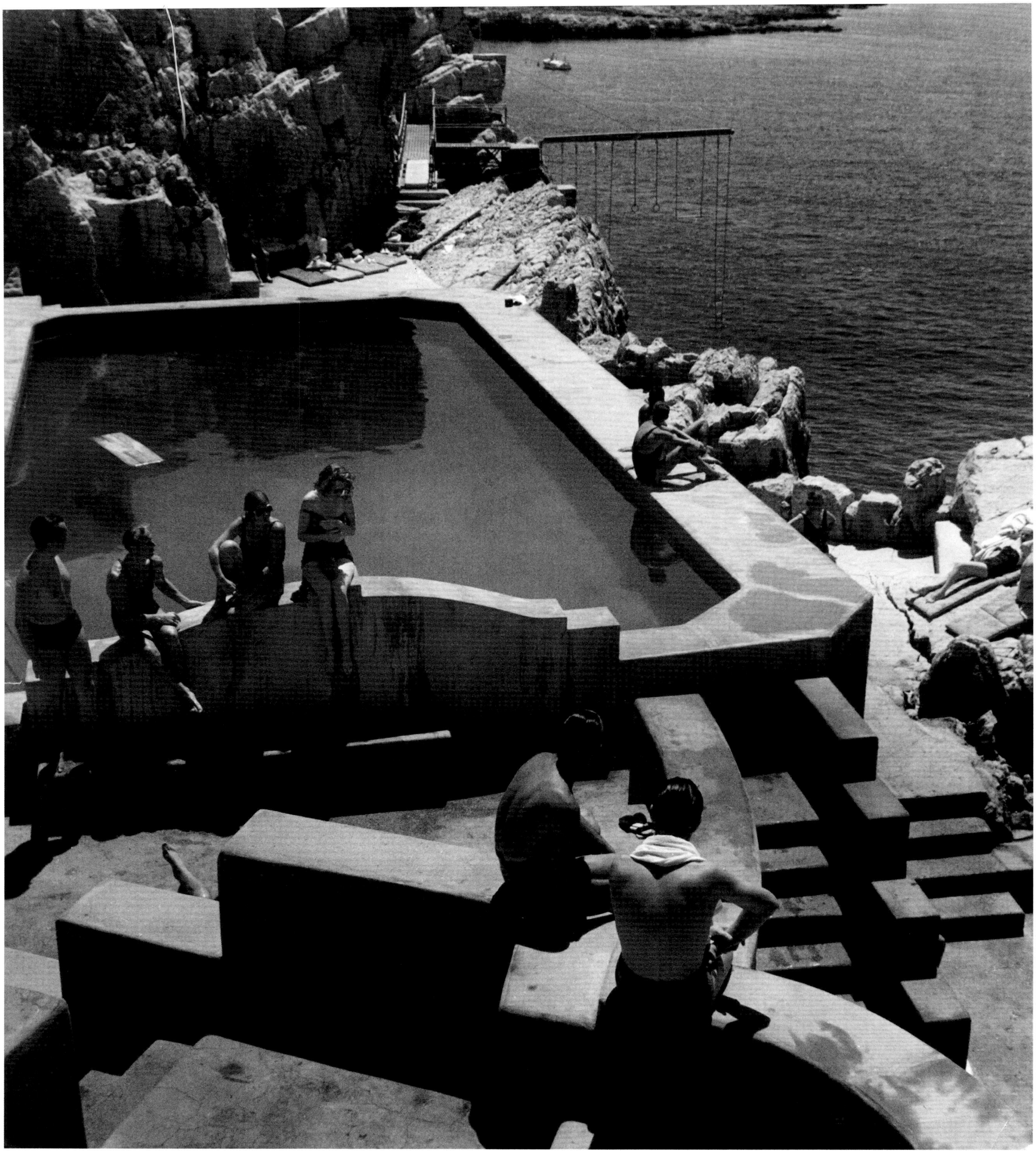

Relaxing at the pool of Hôtel du Cap-Eden-Roc. Antibes, Côte d'Azur, France, 1938.

Entspannen am Pool des Hôtel du Cap-Eden-Roc. Antibes, Côte d'Azur, Frankreich, 1938.

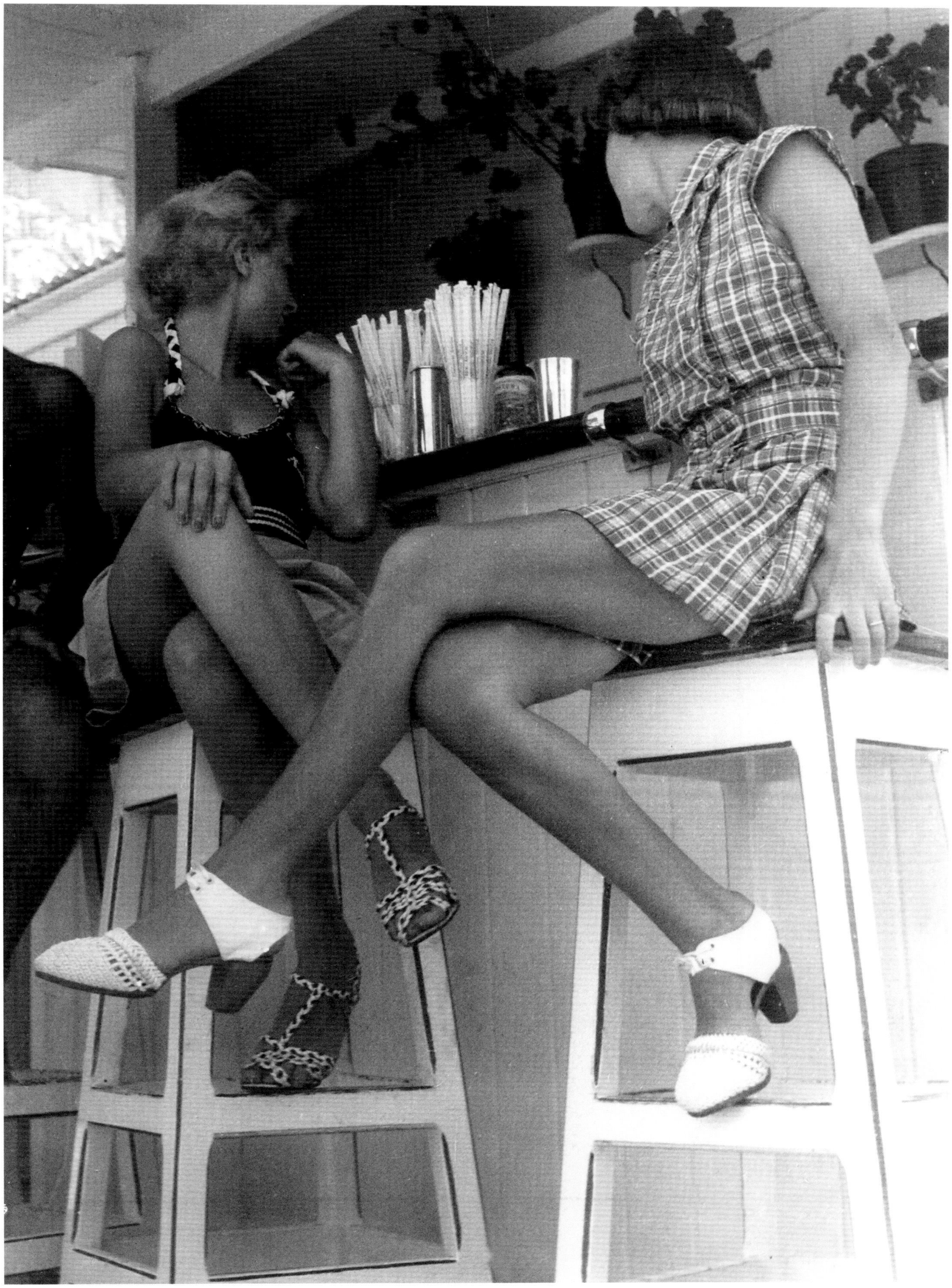

Two ladies in beach dresses at one of the bars at Hôtel du Cap-Eden-Roc. Antibes, Côte d'Azur, France, 1938.

Zwei Damen in Strandkleidern an einer der Bars des Hôtel du Cap-Eden-Roc. Antibes, Côte d'Azur, Frankreich, 1938.

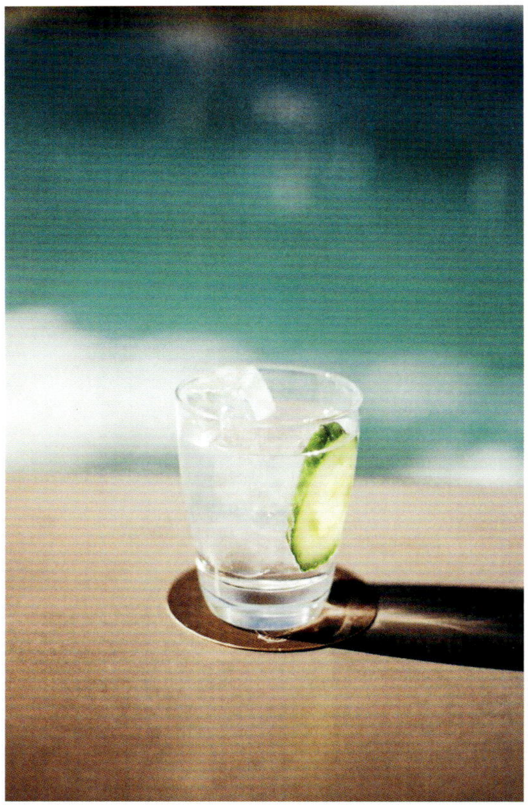

Top: Gin and Tonic at the famous Icebergs beach bar at Bondi Beach, Sydney, Australia.

Oben: Gin Tonic an der berühmten Icebergs Strandbar am Bondi Beach, Sydney, Australien.

Bottom: A glass of Margarita at a tropic beach bar.

Unten: Margarita an einer tropischen Strandbar.

Right: Mobile beach bar, Anguillara Sabazia, Italy.

Rechts: Mobile Strandbar, Anguillara Sabazia, Italien.

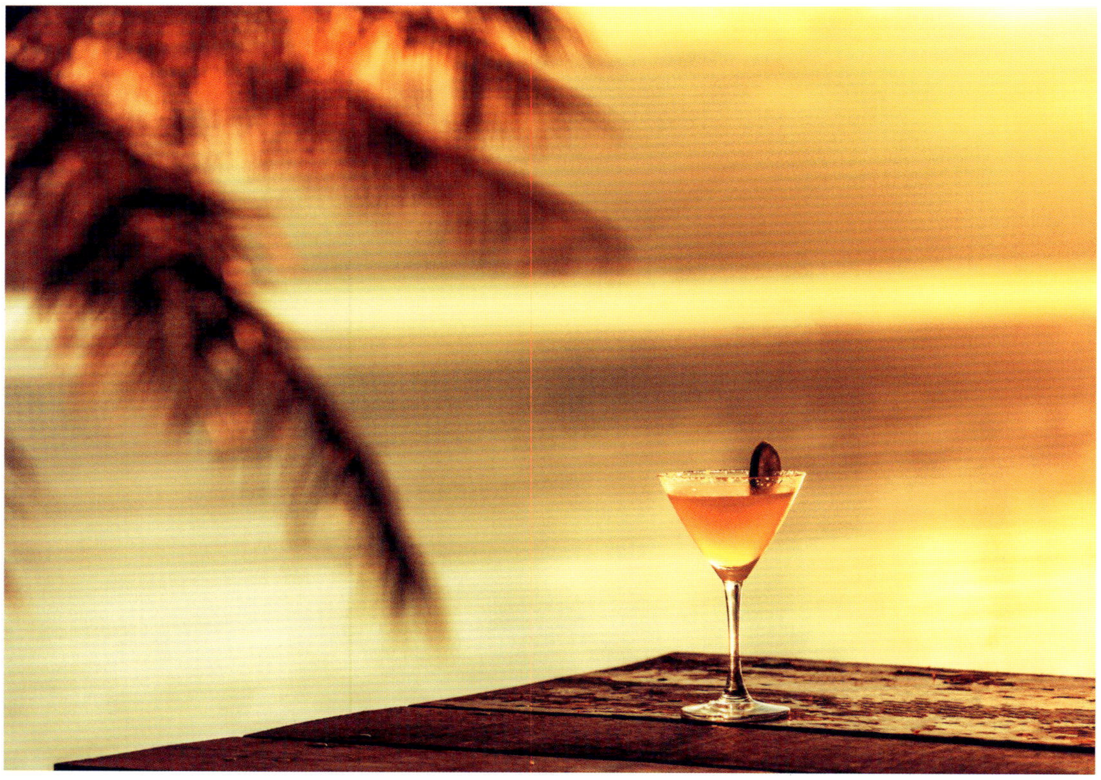

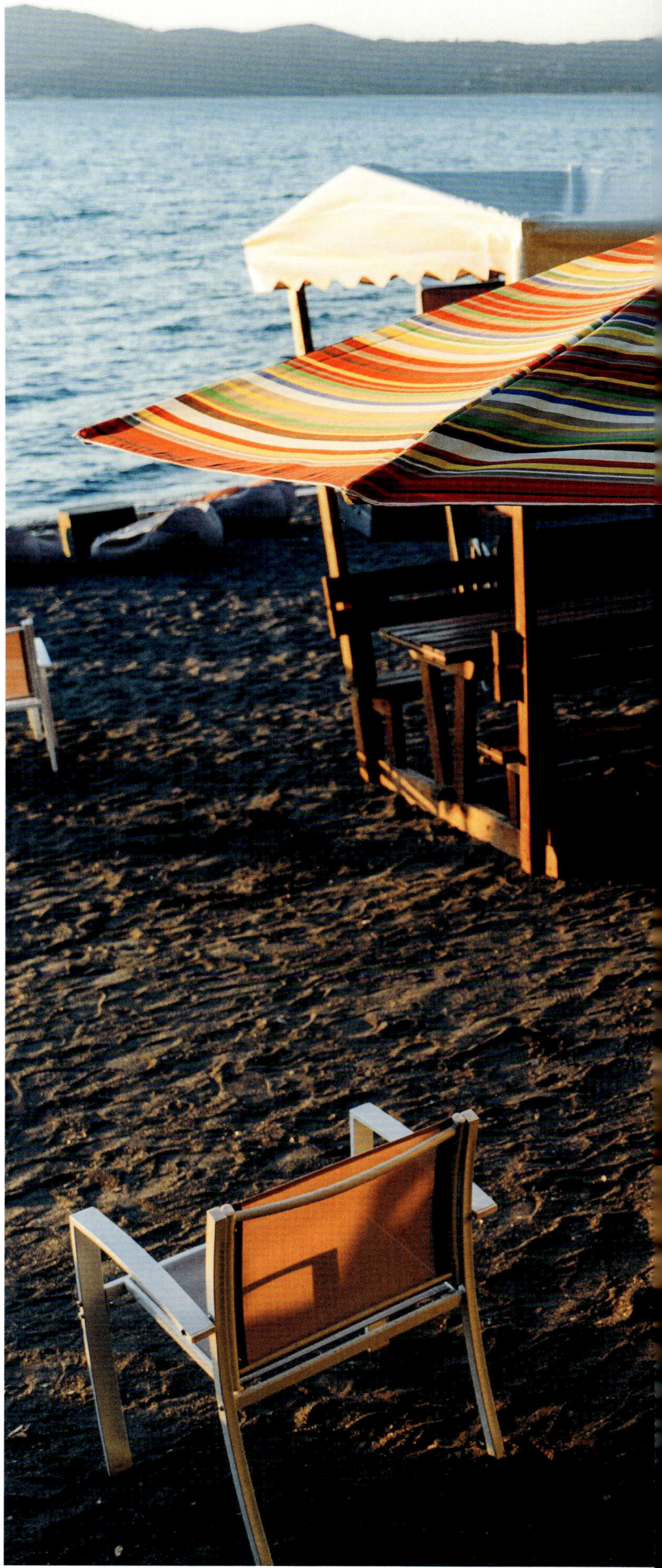

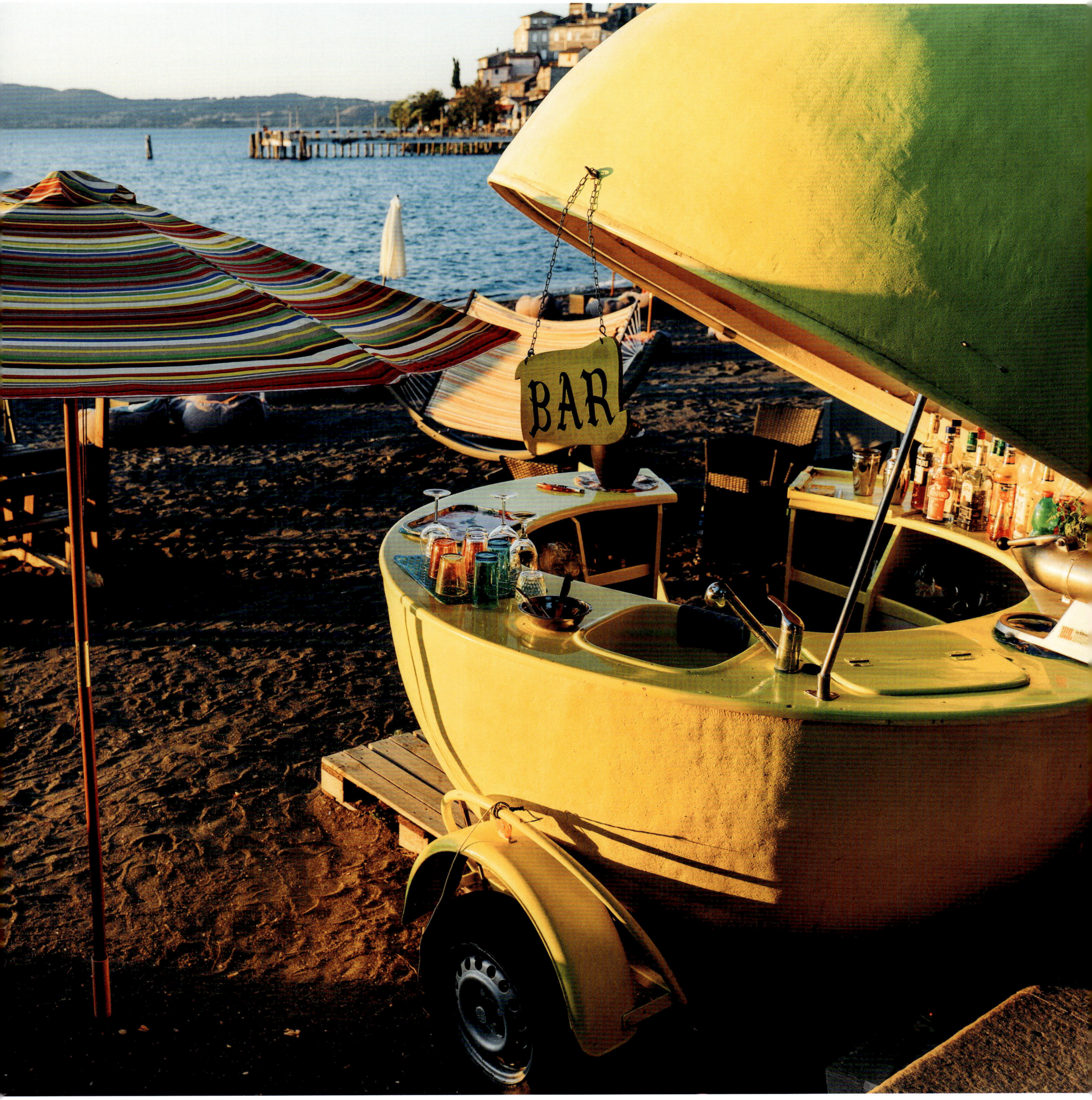
BAR

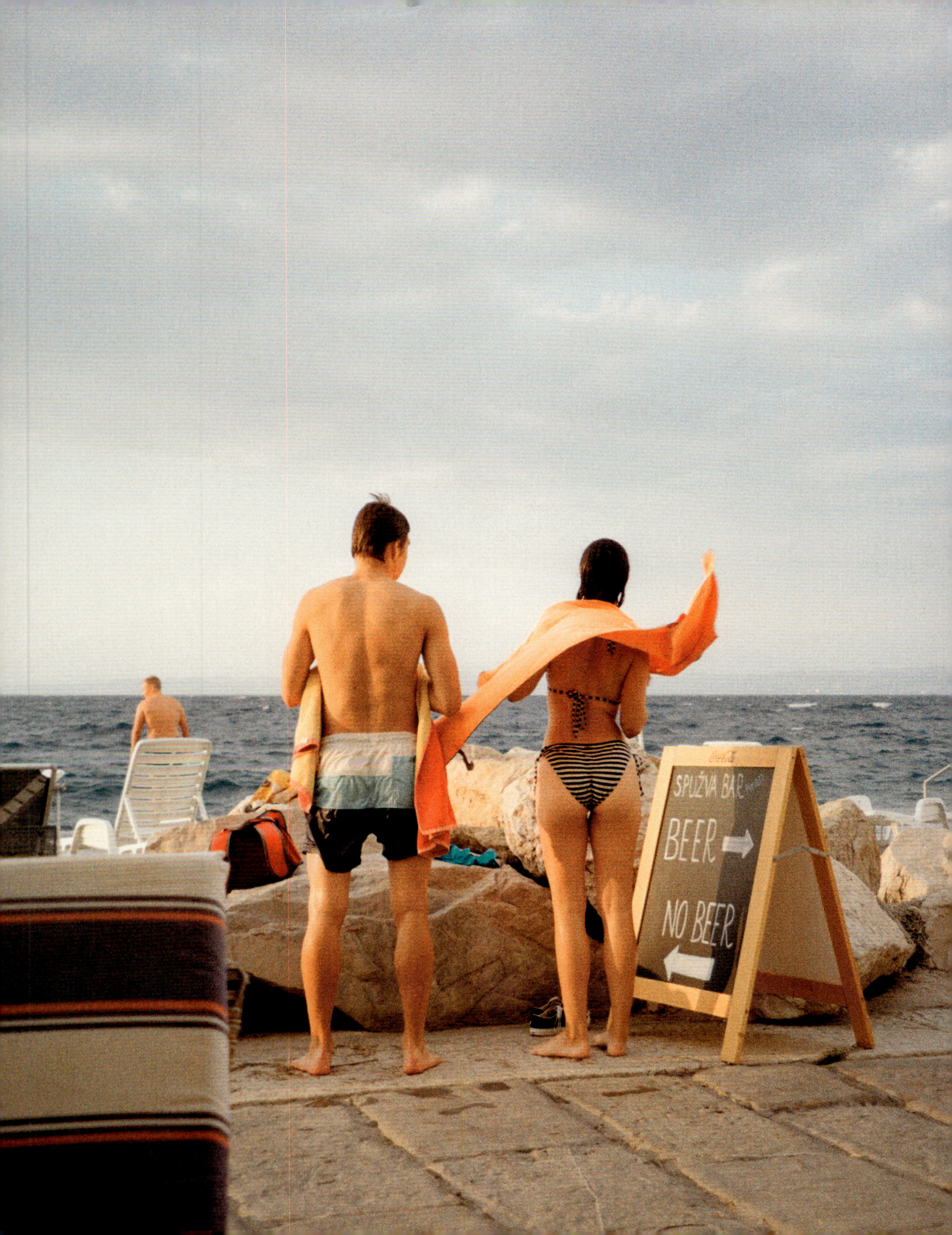
SPUŽVA BAR
BEER
NO BEER

Time to go home.
Piran, Slovenia.

Zeit, nach Hause zu gehen.
Piran, Slowenien.

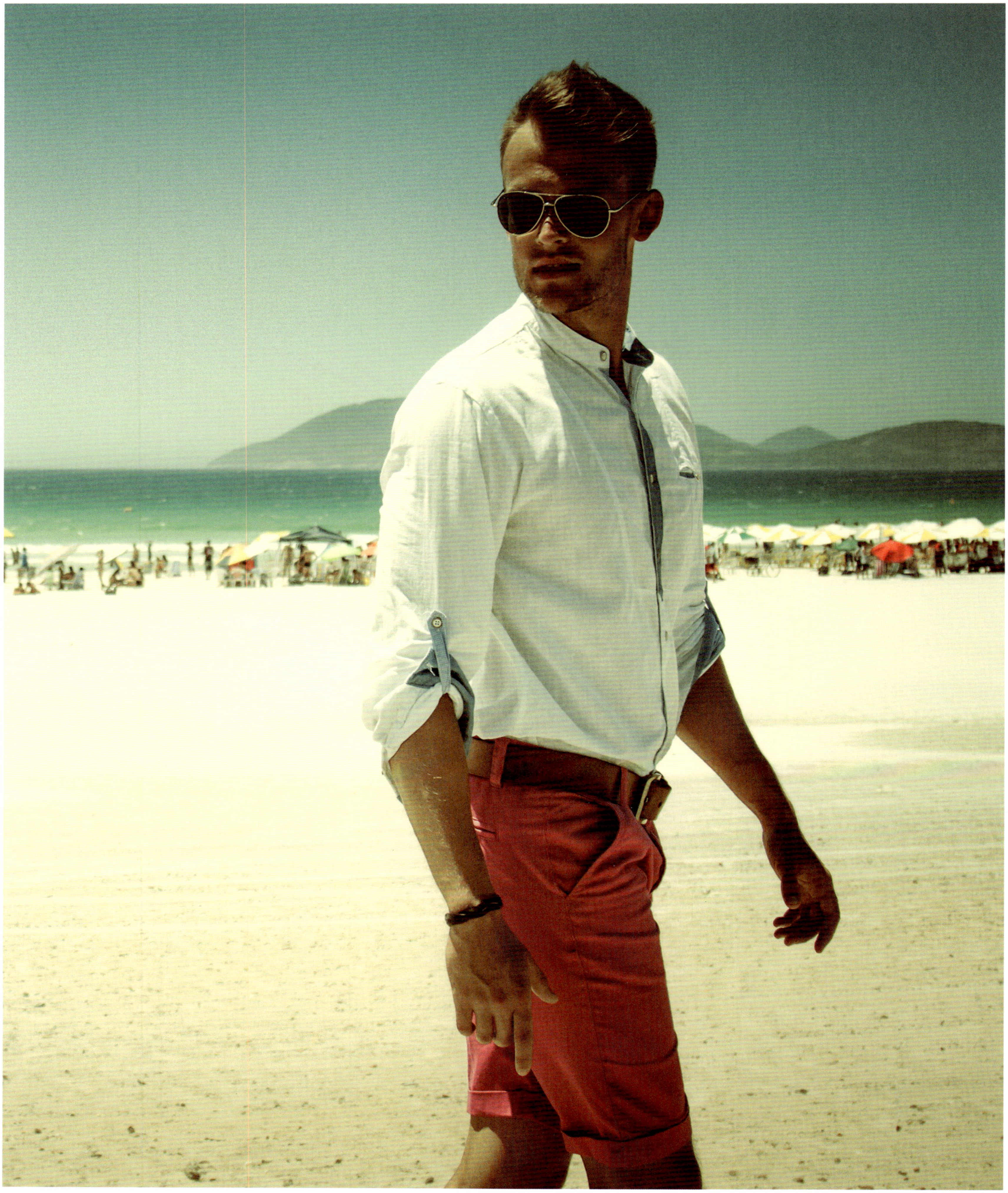

For the stereotypical Nordic traveler heading south, the vacation has to be thoroughly planned and executed, so as to get the absolute most out of the short time available. Early in the morning, people are already walking and jogging, the challenging books they finally have time for are piled up under the sun umbrella, and the sunscreens are sorted according to their sun protection factor. By 9 a.m. at the latest, vacationers from the cooler regions are arriving at the warm summery beach. It's understandable, because people from the northern regions only have two or three weeks to enjoy life at the beach. They don't want to waste a minute. Italians or Brazilians, for example, who live in prime beach countries, take it much easier. They stroll toward the ocean at some point between 11 a.m. and 2 p.m., sometimes only after enjoying a siesta. Because the beach plays such an important role in the social fabric there, they are also dressed much better than vacationers from less advantaged climates: in moccasins instead of flip-flops, and in button-down shirts instead of T-shirts, albeit unbuttoned and with their arms rolled up. They like to stand in the knee-deep water, form a small circle and chat in a lively manner, often for hours. The seashore is a refreshing extension of the *piazza*. They don't have to worry about sun protection factors.
They also play sports, but these activities are less about burning calories and more about aesthetics, such as the elegant Brazilian beach paddle game *Frescobol*. No one intends to get winded here.

And then there's the surprisingly Nordic longing for secluded coves. In magazines, beach photos from overhead ("like oily sardines") are meant to discourage mass tourism every summer. But residents of beach countries see things quite differently. They like to be sociable, they need neighbors right next door to chat with, to play cards with, to exchange daily newspapers with. For most people from southern countries, it can't be lively enough.

Für den stereotypen nordischen Reisenden, der in den Süden fährt, gilt: Der Urlaub muss durchgeplant und abgearbeitet werden, um das Maximum aus der kurzen Zeit herauszuholen. Da wird schon früh am Morgen gewalkt und gejoggt, unter dem Sonnenschirm stapeln sich die schwierigen Bücher, für die man jetzt endlich einmal Zeit hat, die Sonnencremes werden nach den Lichtschutzfaktoren geordnet. Früh um 9 Uhr spätestens finden sich die Urlauber aus den kühleren Regionen am sommerlich warmen Strand ein. Ein bisschen ist es ja auch verständlich, denn Menschen aus dem Norden haben nun einmal nur zwei, maximal drei Wochen zur Verfügung, um das Strandleben zu genießen. Da will man keine Minute verschwenden. Italiener oder Brasilianer etwa, Einwohner von Strandnationen par excellence, lassen es viel ruhiger angehen. Sie schlendern irgendwann zwischen 11 und 14 Uhr in Richtung Meer, manchmal auch erst nach der Siesta. Weil der Strand eine so wichtige Rolle im sozialen Gefüge spielt, sind sie auch deutlich besser angezogen als die aus benachteiligten Klimazonen zugereisten Urlauber: in Mokassins statt Badeschlappen, in Hemden statt T-Shirts, wenngleich aufgeknöpft und an den Armen hochgekrempelt. Sie stehen gern im knietiefen Wasser, bilden einen kleinen Kreis und plaudern angeregt, oft stundenlang. Das Meeresufer ist die erfrischende Verlängerung der *piazza*. Über Lichtschutzfaktoren müssen sie sich keine Sorgen machen. Sport wird auch betrieben, aber es geht weniger um den Kalorienverbrauch als um die Ästhetik, etwa beim eleganten brasilianischen Strandtennis Frescobol. Außer Atem will hier keiner kommen.
Und dann gibt es ja die erstaunlich nordische Sehnsucht nach einsamen Buchten. In Illustrierten sollen jeden Sommer Strandaufnahmen von oben („wie die Ölsardinen") vom Massentourismus abschrecken. Das sehen Bewohner von Strandnationen ganz anders. Sie mögen es gesellig, sie brauchen die Nachbarn direkt nebenan zum Plaudern, zum Kartenspielen, zum Tausch der Tageszeitungen. Für die meisten Menschen aus südlichen Ländern kann es gar nicht trubelig genug sein.

Elegance and nonchalance at the beach. Location unknown.

Eleganz und Nonchalance am Strand.

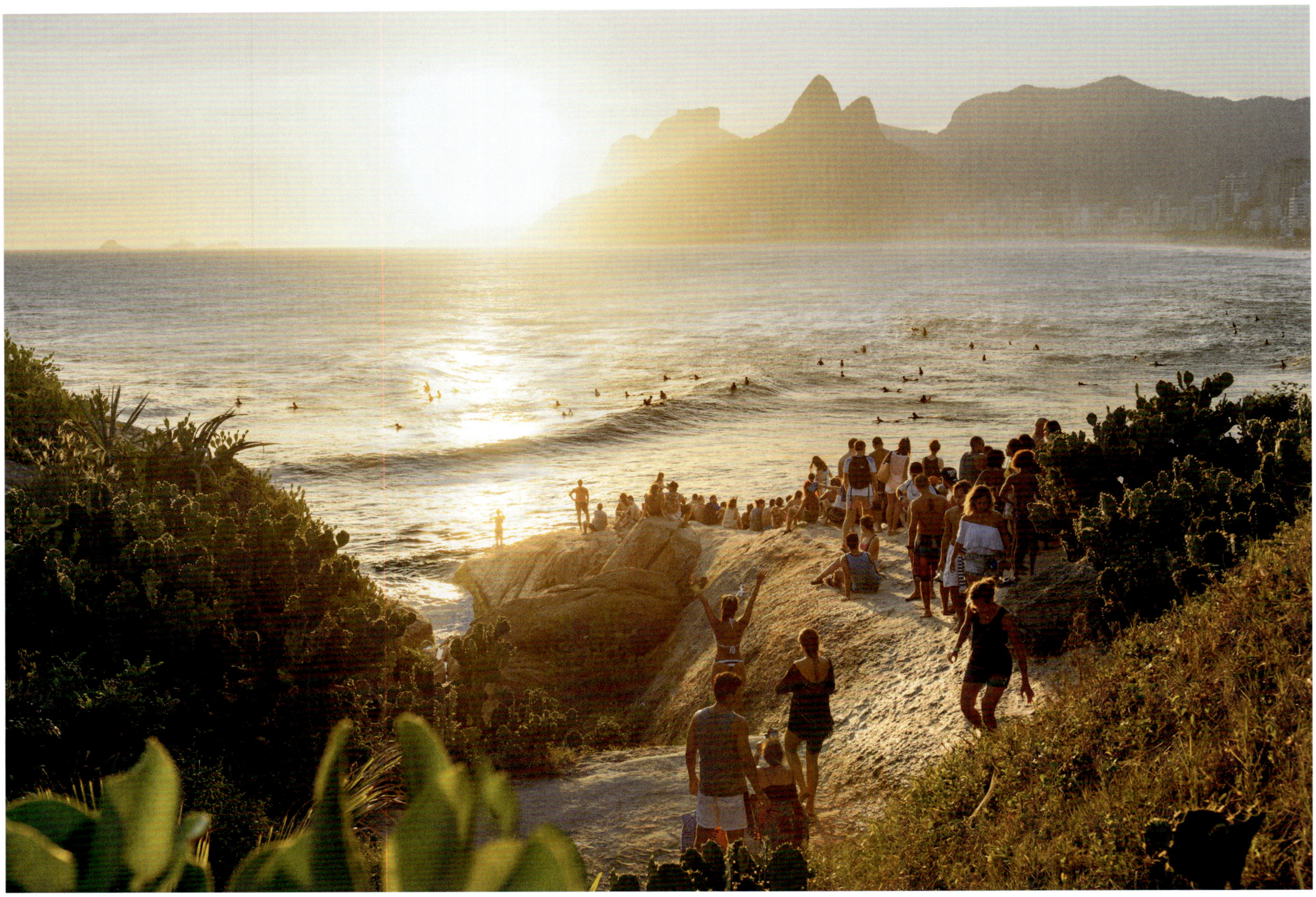

Crowd enjoying the sunset at Arpoador, Playa Ipanema, Rio de Janeiro, Brazil.

Der Sonnenuntergang am Playa Ipanema zieht zuverlässig Menschenmengen an. Arpoador, Playa Ipanema, Rio de Janeiro, Brasilien.

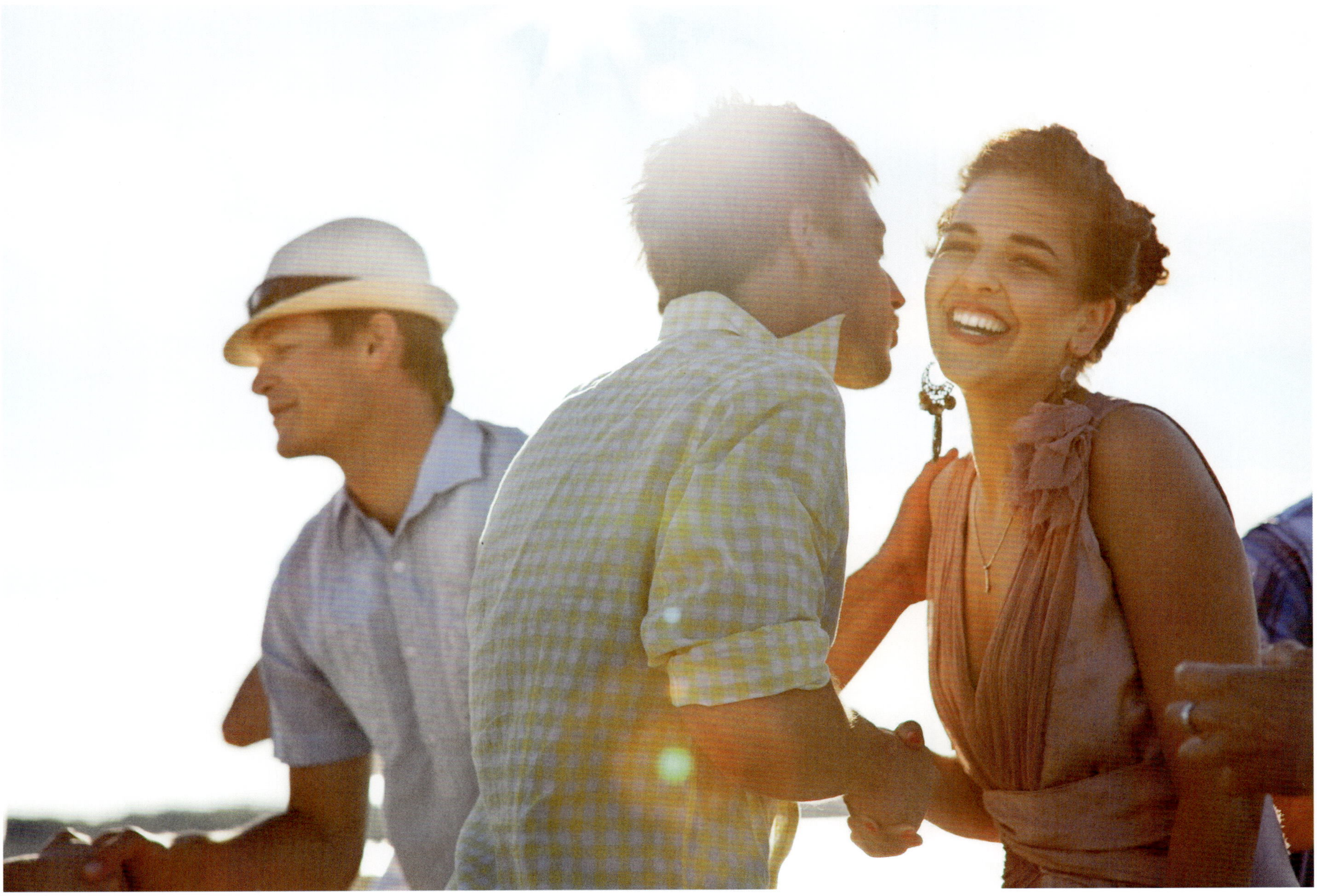

Socializing at the beach.
Location unknown.

Geselligkeit am Strand.
Ort unbekannt.

NEXT PAGE

Left: Deserted bay: Playa Joao Fernandinho in Buzios, Rio de Janeiro, Brazil.

Links: Einsame Bucht: Playa Joao Fernandinho in Buzios, Rio de Janeiro, Brasilien.

Right: Crowded beach in Polignano a Mare, Italy.

Rechts: Voller Strand in Polignano a Mare, Italien.

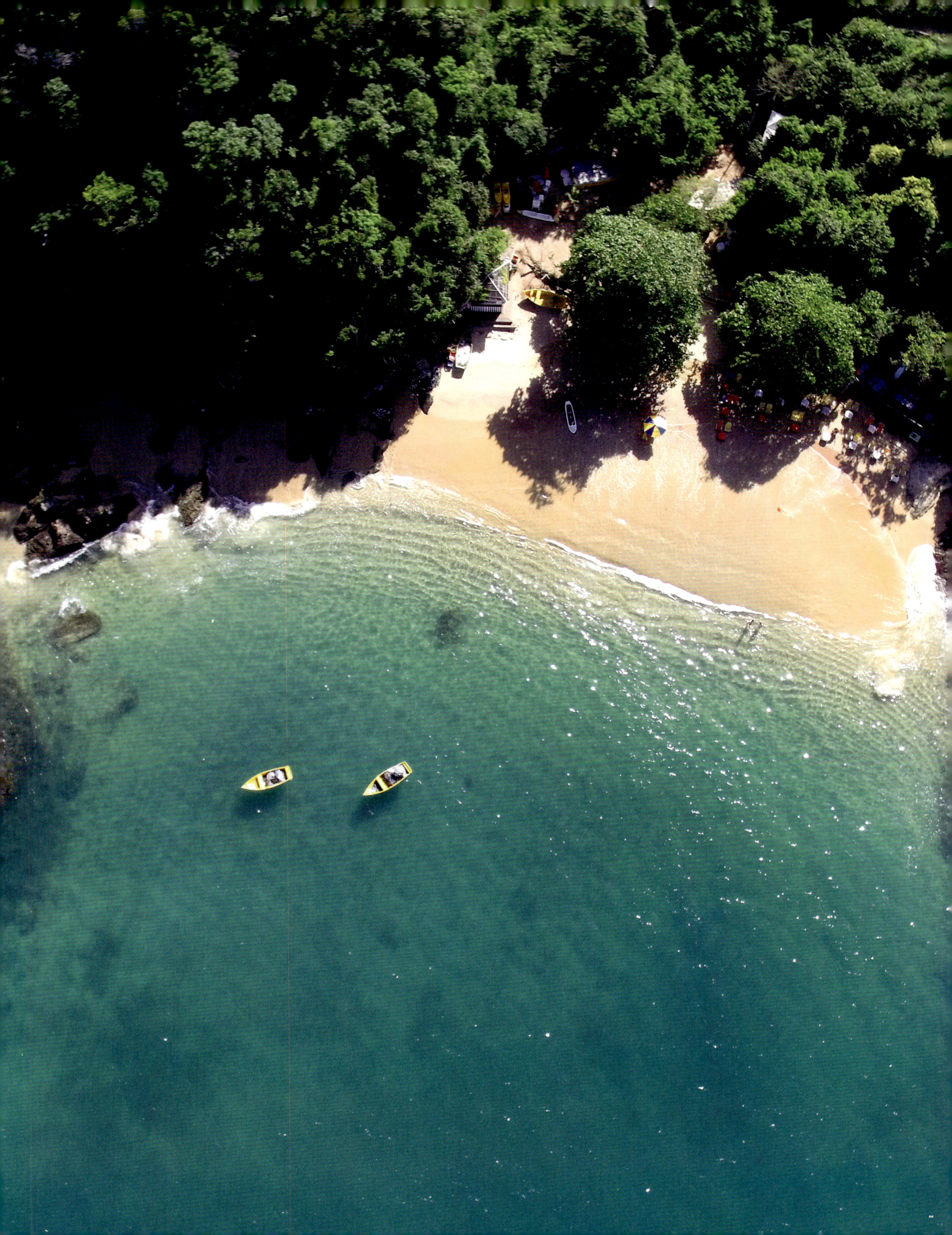

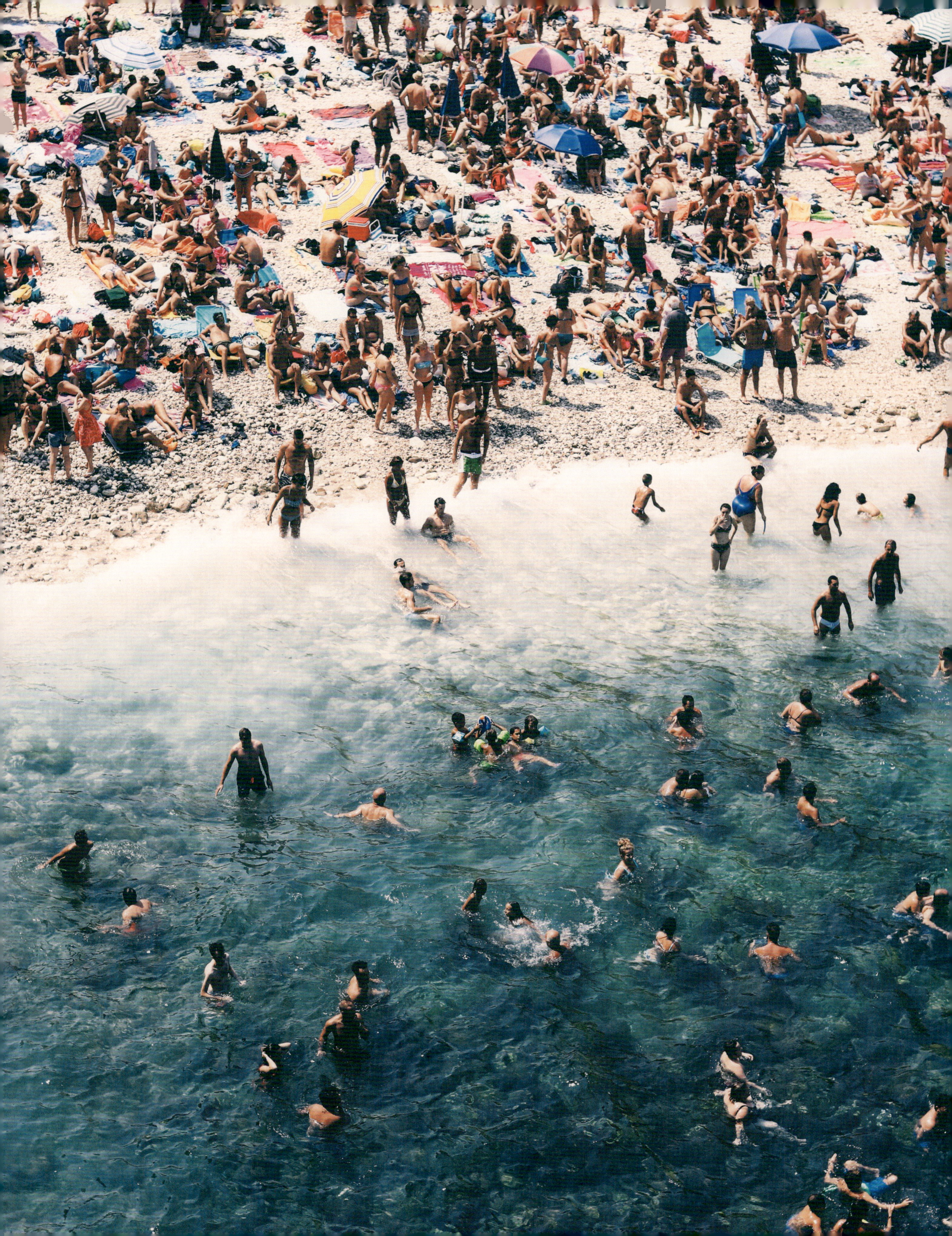

ACTIVITY

BEWEGUNG

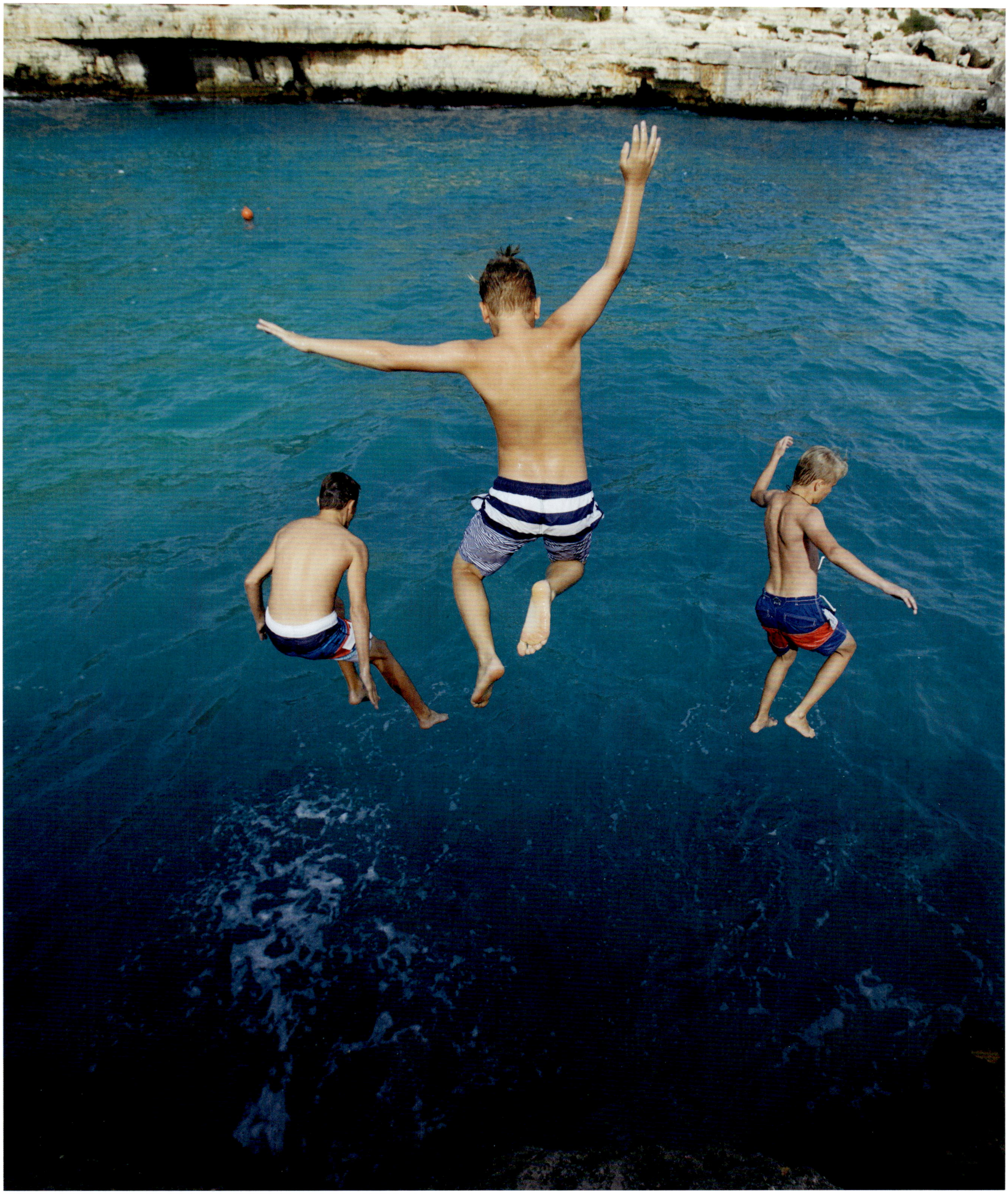

ACTIVITY

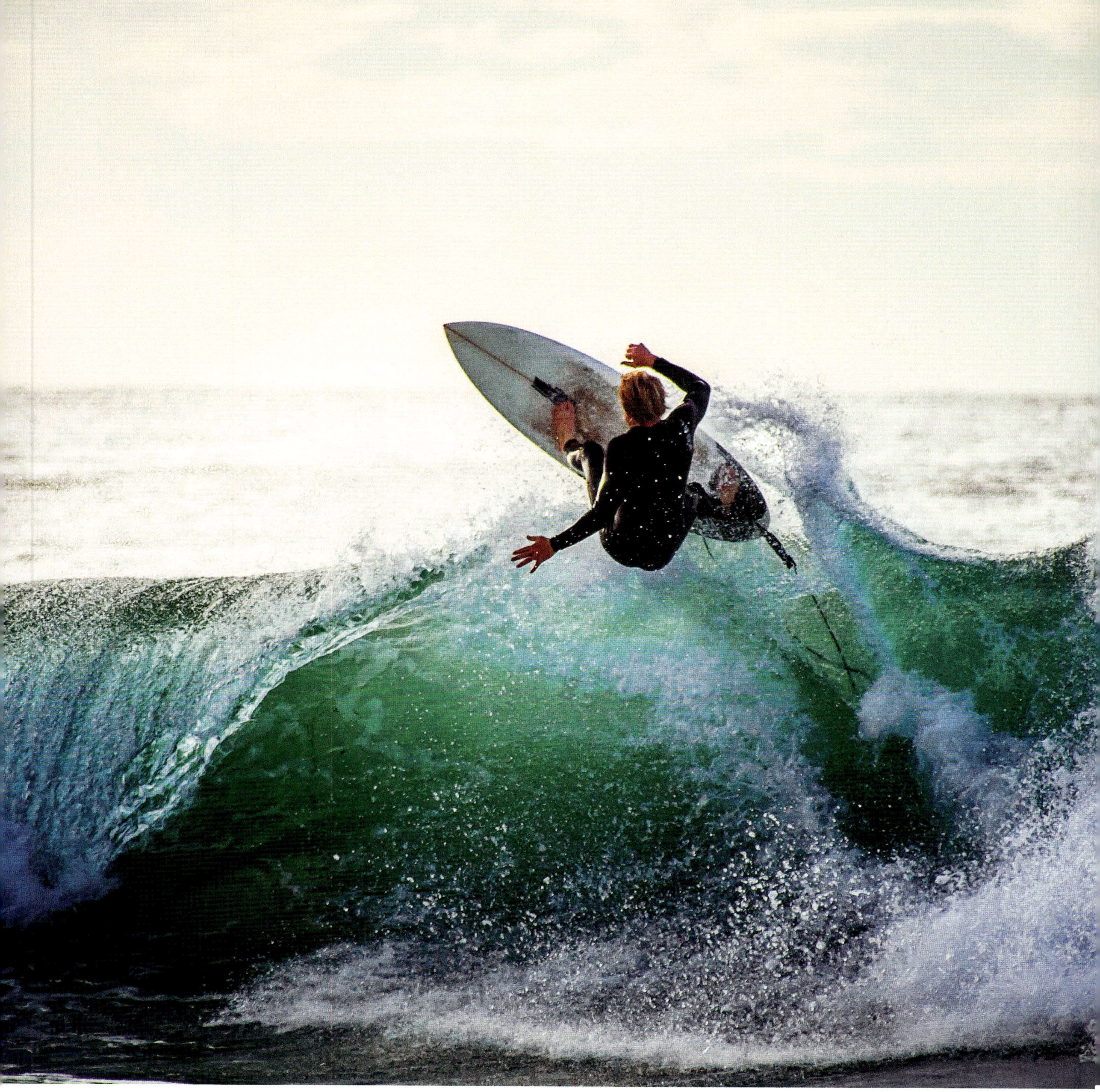

BEWEGUNG

If you pick the right wave, you can stand on your board for half an hour or more. These waves are not found in the ocean, however. They only occur in rivers whose banks slowly taper off, keeping the wave alive.
But true surfers know that the ocean is where the surfing is best. The waves break faster here, of course, but there are still places where the thrill lasts much longer than elsewhere. The surfing community's ocean hot spot is not Hawaii, however, but instead and somewhat surprisingly, it is Peru. At Chicama beach in the north of the country, just under 400 kilometers from the Peruvian capital of Lima, the waves reach up to four kilometers in length and can be surfed for three to four minutes. Peruvian pro Cristobal De Col once clocked 34 turns on a single wave there–a fabulous world record for the open sea.
On perfect days, 50 waves approach at once, neatly spaced apart. What makes them special, is that Chicama waves, despite their expanse, are relatively small (they reach about shoulder height), so they can be easily mastered by beginners as well. The best season is from May to September, and it can get very crowded, because word of this supposed insider spot spread very quickly throughout the surfing community.
Pro tip: Never surf more than one wave; even professionals have to take a break afterwards because of the long time on the board. Paddling back to the beach on the board is also discouraged–it's better to wade to the beach and walk back to the starting point from there.

Waves can also last a similar length of time at Pavones Beach in Costa Rica. In South Africa, Jeffreys Bay near Port Elizabeth is famous for its steady, wide waves, and in Australia, surfers prefer to make a pilgrimage to the Gold Coast near Brisbane to conquer the up to two-kilometer-long breakers between Snapper Rocks and Kirra.

Wer sich die richtige Welle aussucht, kann eine halbe Stunde oder länger auf dem Brett stehen. Diese Wellen kommen aber nicht im Meer vor, sondern nur in Flüssen, deren Ufer sich langsam verjüngen und somit die Welle am Leben erhalten.
Doch echte Surfer wissen: Im Meer ist es einfach am schönsten. Klar, dass hier die Wellen schneller brechen, dennoch gibt es Orte, an denen der Genuss deutlich länger anhält als anderswo. Der Ozean-Hotspot der Surfgemeinde ist nicht etwa Hawaii, sondern, etwas überraschend, Peru. Am Strand von Chicama im Norden des Landes, knapp 400 Kilometer von der Hauptstadt Lima entfernt, erreichen die Wellen eine Länge von bis zu vier Kilometern und können drei bis vier Minuten lang gesurft werden. Der peruanische Profi Cristobal De Col hat dort einmal 34 verschiedene Manöver in einer einzigen Welle geschafft – ein Fabel-Weltrekord für das offene Meer.
An perfekten Tagen rücken 50 Wellen zugleich an, in braven Abständen zueinander. Das Besondere: Die Chicama-Wellen sind trotz ihrer Ausdehnung relativ klein (sie erreichen etwa Schulterhöhe), können also auch von Anfängern gut bewältigt werden. Die beste Saison ist von Mai bis September, und es kann schon mal sehr voll werden, denn in der Surfgemeinde hat sich dieser vermeintliche Geheimtipp ganz schnell herumgesprochen.
Tipp: nie mehr als eine Welle surfen, selbst Profis müssen wegen der langen Zeit auf dem Brett danach eine Pause einlegen. Vom ermüdenden Zurückpaddeln auf dem Board wird ebenfalls abgeraten – lieber an den Strand waten und von dort zurück zum Ausgangspunkt gehen.
Auch am Strand von Pavones in Costa Rica können die Wellen ähnlich lang anhalten. In Südafrika ist Jeffreys Bay bei Port Elizabeth berühmt für seine steten, breiten Wellen, und in Australien pilgern Surfer am liebsten zur Gold Coast bei Brisbane, um die bis zu zwei Kilometer langen Brecher zwischen Snapper Rocks und Kirra zu bezwingen.

Surfing the perfect wave.
Location unknown.

Die perfekte Welle surfen.
Ort unbekannt.

PREVIOUS PAGE

Jumping into the blue. Mallorca, Spain.

Sprung ins Blaue. Mallorca, Spanien.

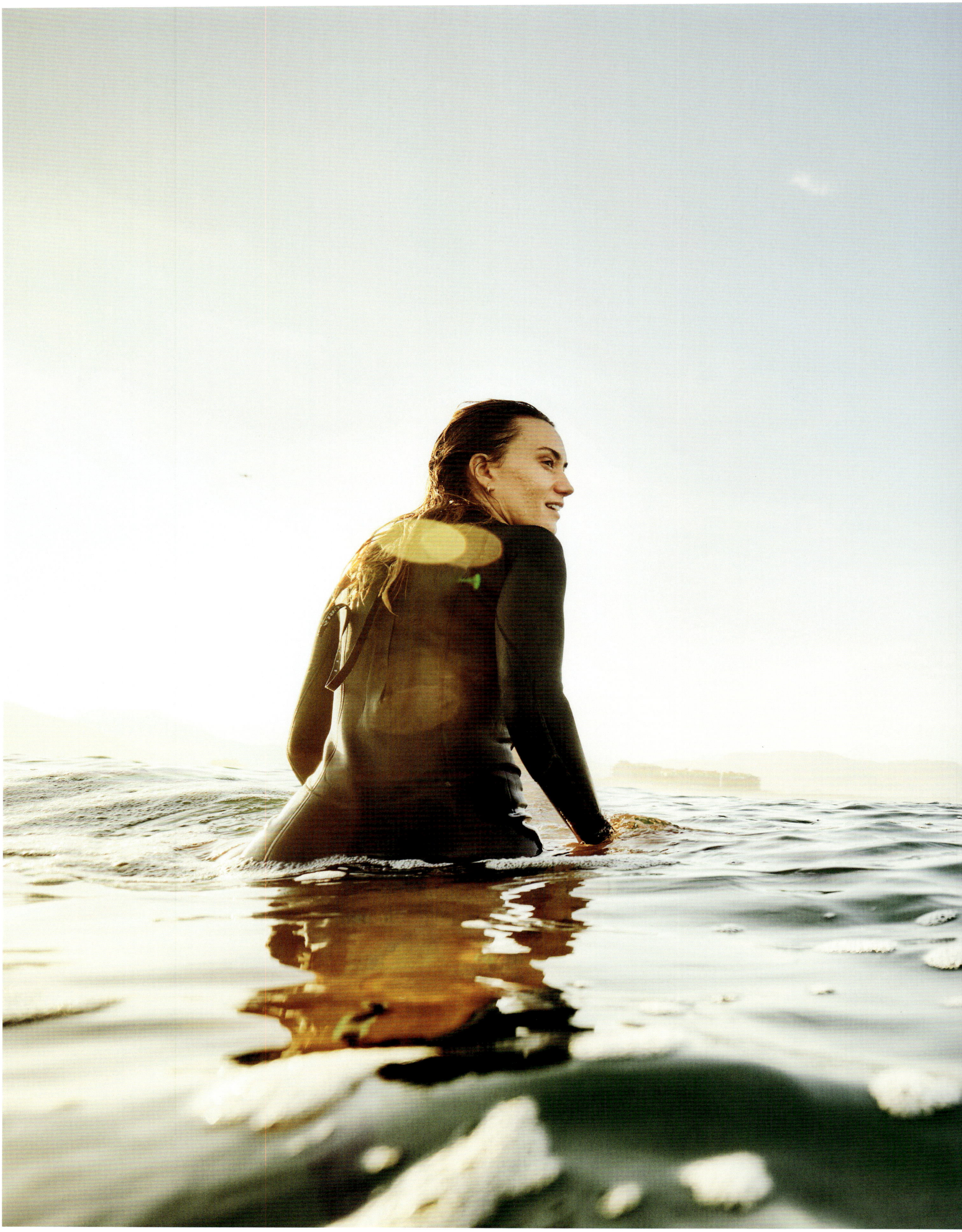

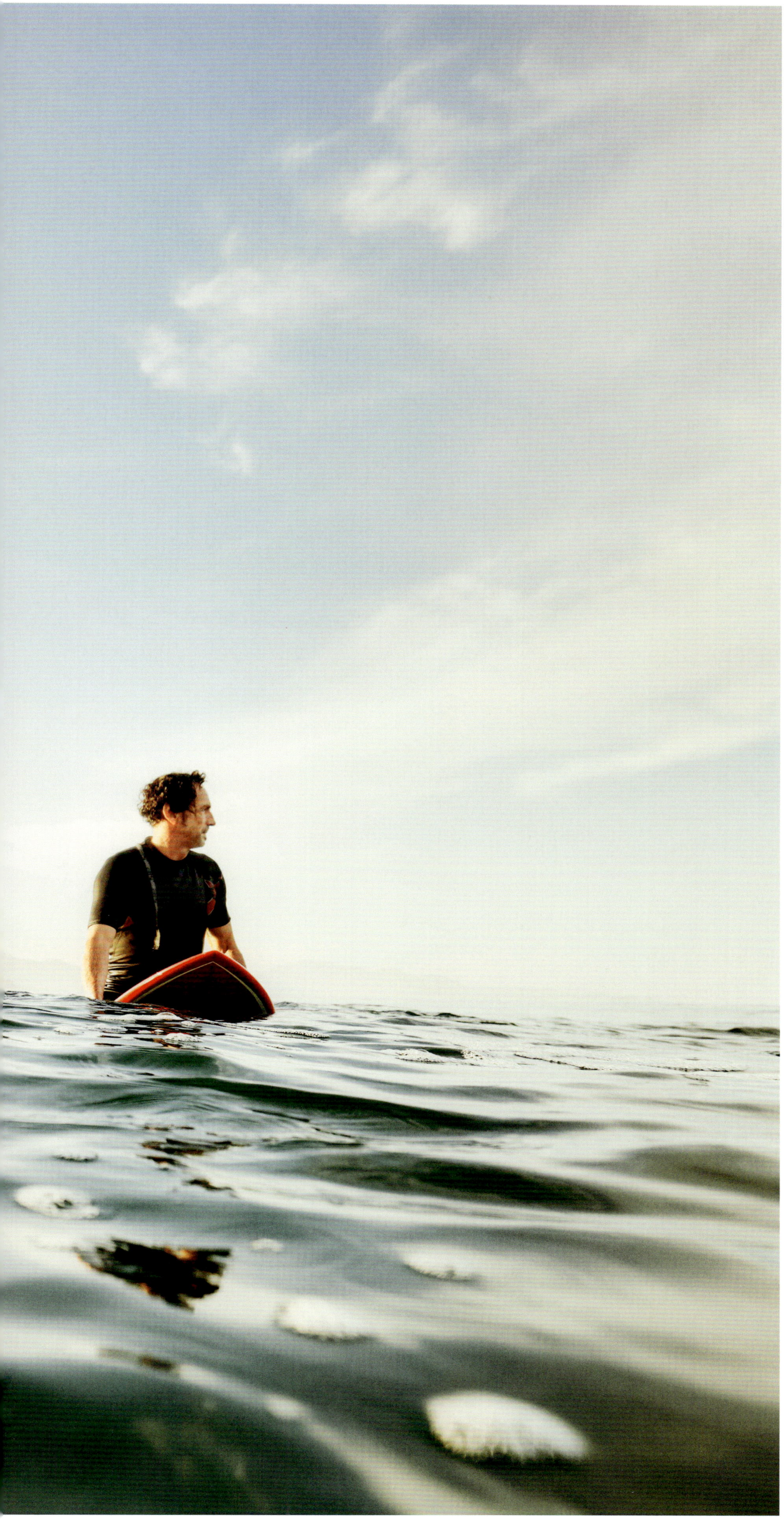

Waiting for the perfect wave.
Baja California Sur, Mexico.

Warten auf die perfekte Welle.
Baja California Sur, Mexiko.

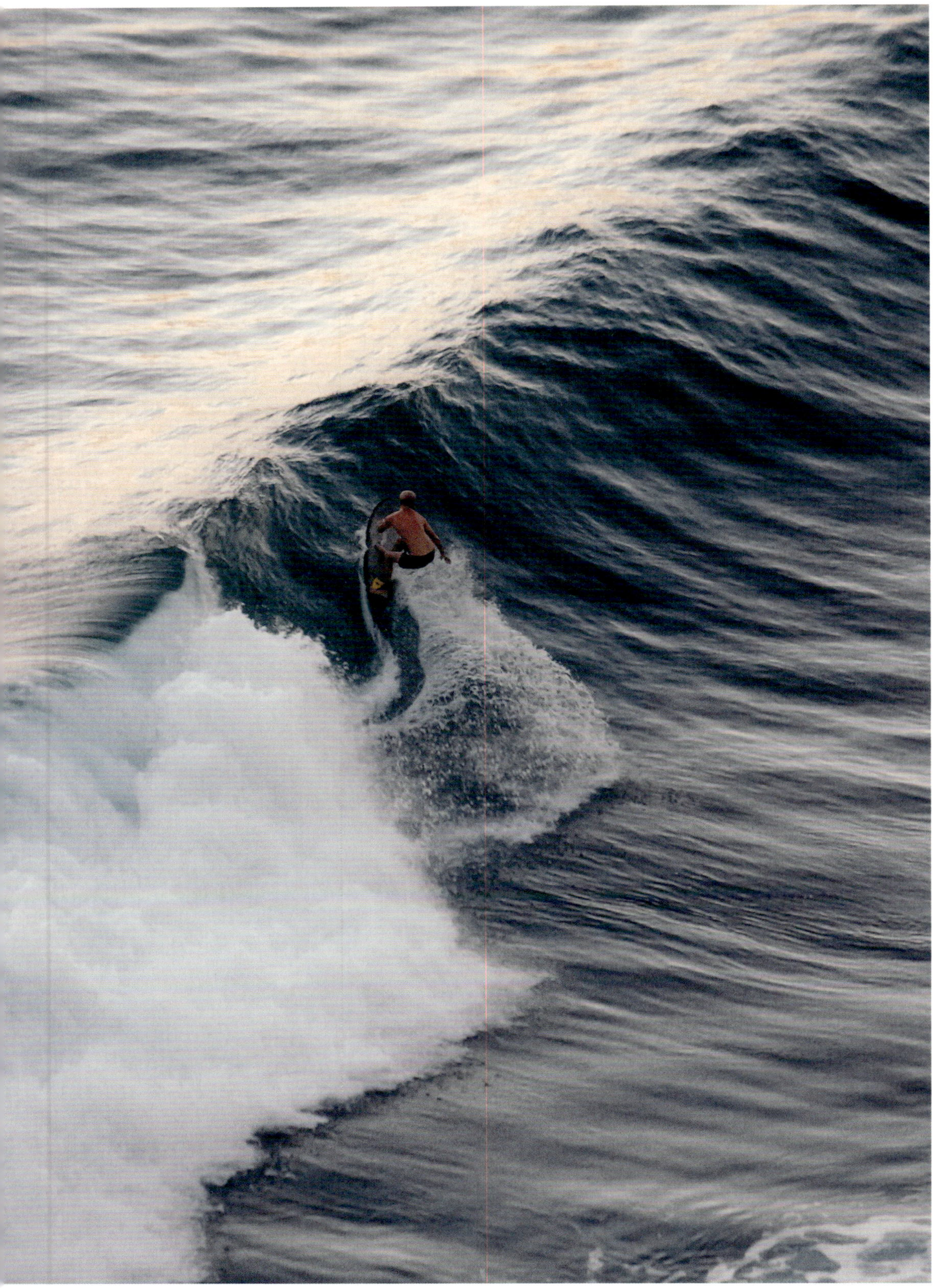

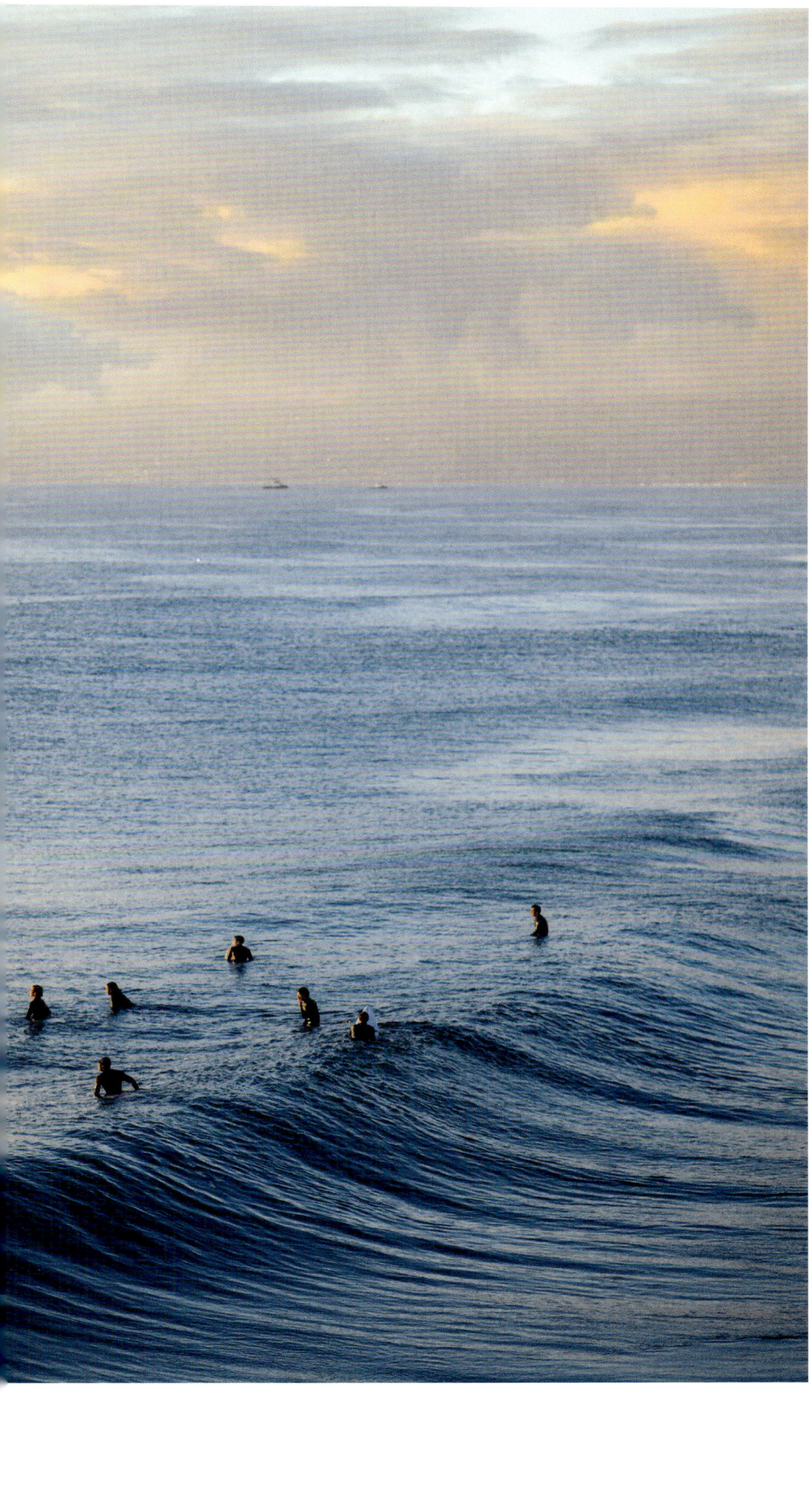

Left: The Indonesian island Bali is well-known for its beautiful surf spots.

Right: Surfers off Manhattan Beach, California, USA.

Links: Die indonesische Insel Bali ist bekannt für ihre wunderschönen Surfspots.

Rechts: Surfer vor Manhattan Beach, Kalifornien, USA.

How many of the millions of vacationers who head to the beach each year actually go swimming? Not splashing around, not bathing, not cooling off for a bit–but actually swimming? Do they swim their laps parallel to the shore, for minutes, even hours?
Yes, there are a few.
But not many.
And there are reasons for that. Swimming in the ocean as a sport is very, very strenuous. The waves and the currents make it difficult to move forward; it takes a lot more physical strength to make progress, even in seemingly calm seas. Indeed, also professional swimmers report how uncomfortable it is to swallow even the tiniest amounts of salt water. Those who are not good swimmers (and most of us are not very good swimmers) will have to swallow plenty of it while doing the crawl through the waves. Swimmers in the ocean also have to keep raising their upper body in the waves to see where they are and whether they're even swimming to where they want to go–that requires energy, too. And, in contrast to pools, there is no stopping point in the ocean. You can rest briefly at the side of a pool, but that doesn't work in the ocean.
The world records in long-distance swimming seem to have been set by people from another world: Penelope Palfrey swam 108 kilometers unassisted in the Caribbean in 2011, and André Wiersig swam almost 49 kilometers across the cold, choppy North Sea from Sankt Peter-Ording to Helgoland in 2021.

Yes, swimming in the ocean is definitely risky. But for many people, the almost meditative immersion in the weightlessness of the water, letting themselves be carried by the waves, also means the feeling of complete freedom–the ultimate swimming experience with no narrow lanes or pool sides, no screaming teenagers, no flying plastic balls, no yelling lifeguards. The famous pool at Bondi Beach in Sydney offers a wonderful option to swim *at* the ocean for those who find being in the open ocean too stormy and unpredictable.

Wie viele von den Millionen Urlaubern, die jedes Jahr an den Strand fahren, gehen wirklich schwimmen? Nicht plantschen, nicht baden, nicht sich mal kurz abkühlen – sondern richtig schwimmen? Ziehen ihre Bahnen parallel zum Ufer, Minuten, gar Stunden lang?
Ja, ein paar gibt es.
Aber es sind nicht viele.
Und dafür gibt es Gründe. Schwimmen im Meer als Sportart ist richtig, richtig anstrengend. Die Wellen und die Strömungen machen ein Fortkommen schwierig; es erfordert viel mehr Körperkraft, um voranzukommen, selbst in vermeintlich ruhiger See. Sogar Profischwimmer berichten davon, wie unangenehm es ist, auch nur kleinste Mengen Salzwasser zu schlucken. Wer nicht gut schwimmen kann (und die meisten von uns sind keine besonders guten Schwimmer), wird beim Kraulen durch die Wellen jede Menge davon schlucken müssen. Auch müssen Schwimmer im Meer im Wellengang ihren Oberkörper immer wieder aufrichten, um zu sehen, wo sie sich befinden und ob sie überhaupt dorthin schwimmen, wo sie auch hinwollen – auch das kostet Kraft. Und: Im Gegensatz zum Pool gibt es im Meer keinen Haltepunkt. Am Beckenrand kann man sich kurz ausruhen, das funktioniert im Meer nicht. Die Weltrekorde im Langstreckenschwimmen scheinen von Außerirdischen aufgestellt zu sein: Penelope Palfrey schaffte 2011 ohne fremde Hilfe in der Karibik eine Strecke von 108 Kilometern, und 2021 schwamm André Wiersig knapp 49 Kilometer quer durch die kalte, kabbelige Nordsee – von Sankt Peter-Ording nach Helgoland.
Ja, das Schwimmen im Meer ist durchaus riskant. Aber das fast schon meditative Eintauchen in die Schwerelosigkeit des Wassers, sich von den Wellen tragen zu lassen, bedeutet für viele Menschen auch das Gefühl vollkommener Freiheit – die ultimative Schwimmerfahrung ohne enge Bahnen und Beckenränder, ohne kreischende Teenager, umherfliegende Plastikbälle und bellende Bademeister. Das berühmte Schwimmbecken am Bondi Beach in Sydney bietet für alle, denen das offene Meer zu stürmisch und unwägbar ist, eine wunderbare Möglichkeit zum Schwimmen *am* Ozean.

The famous Bondi Icebergs Pool was inaugurated in 1929.

Der berühmte Bondi Icebergs Pool wurde 1929 eröffnet.

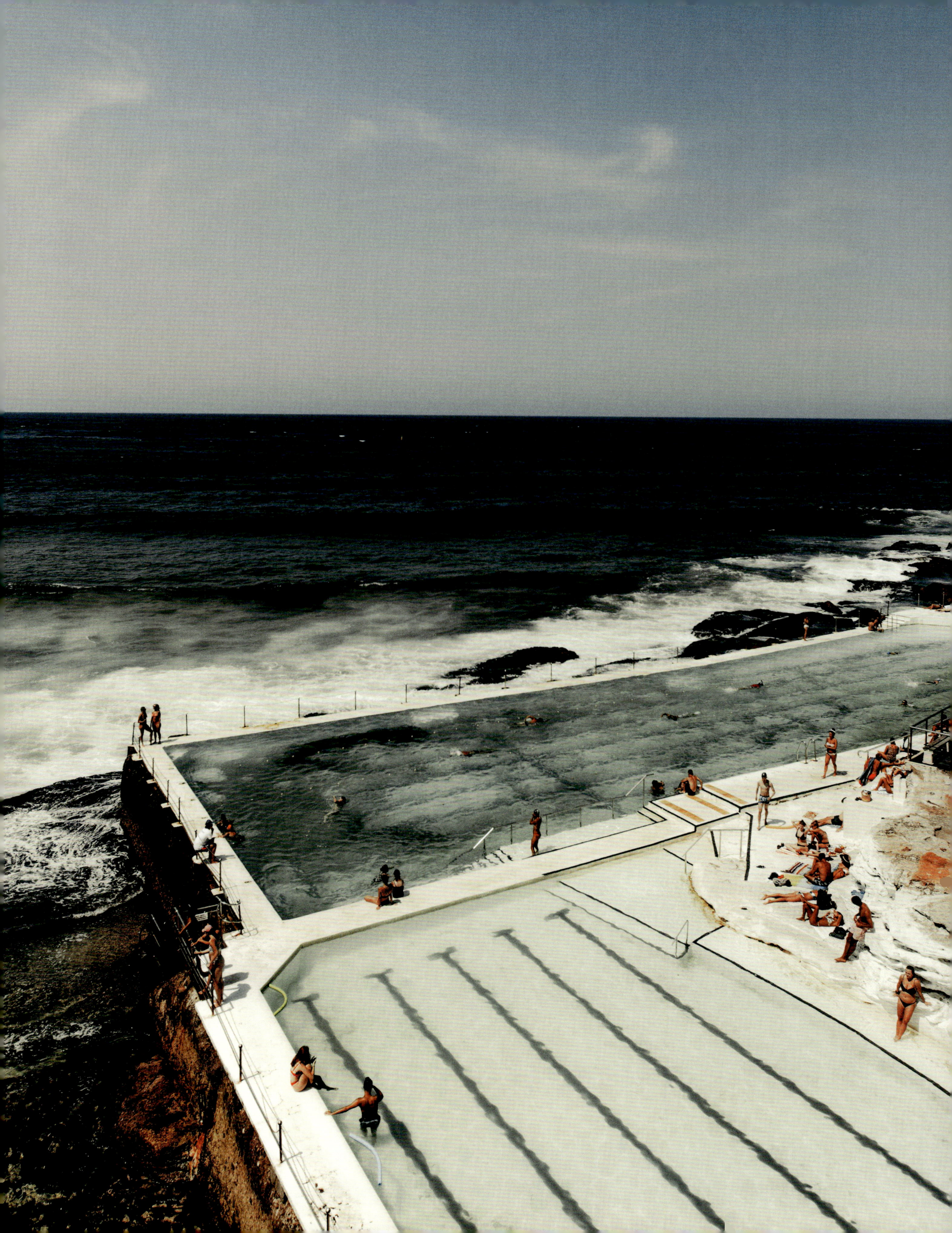

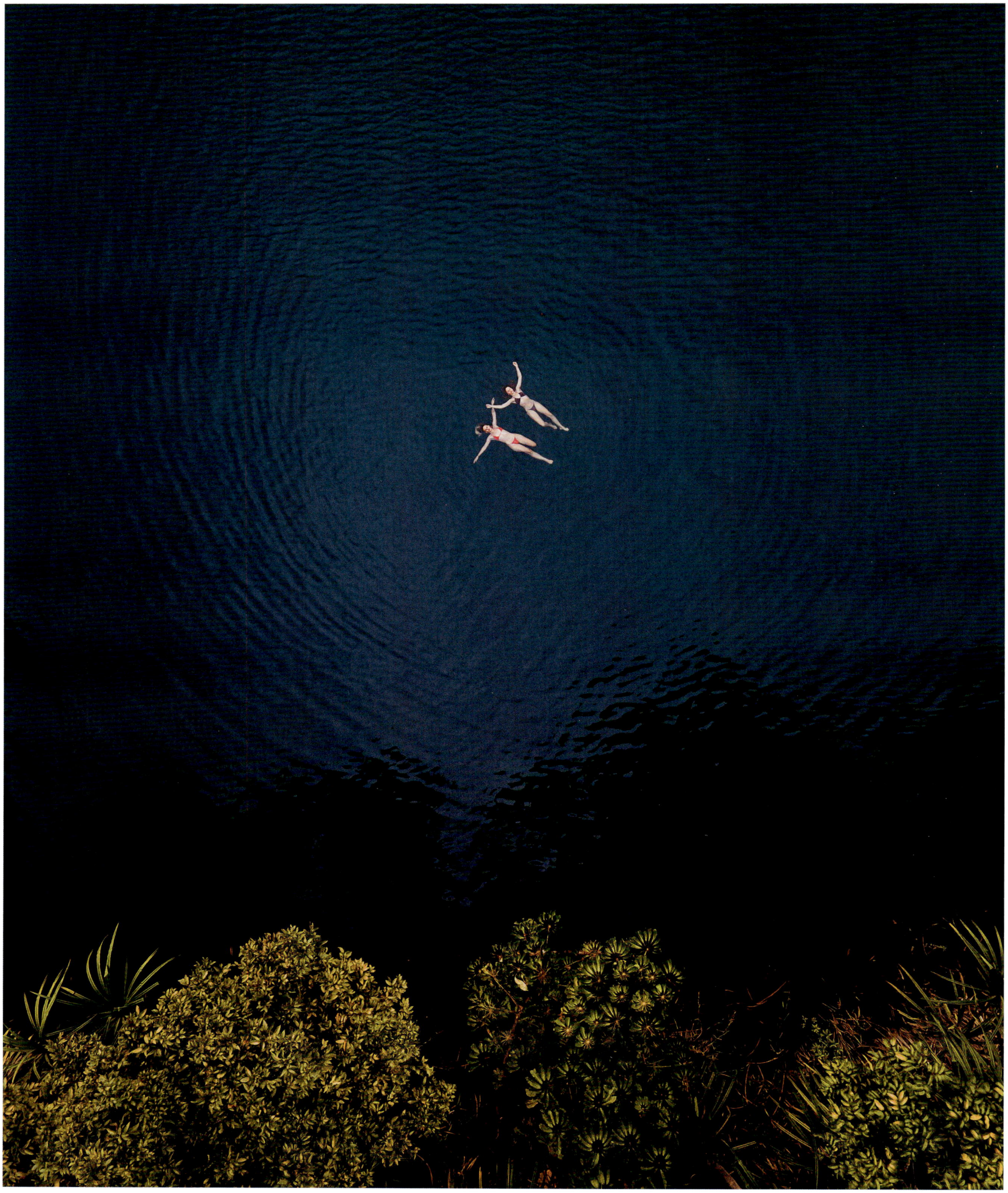

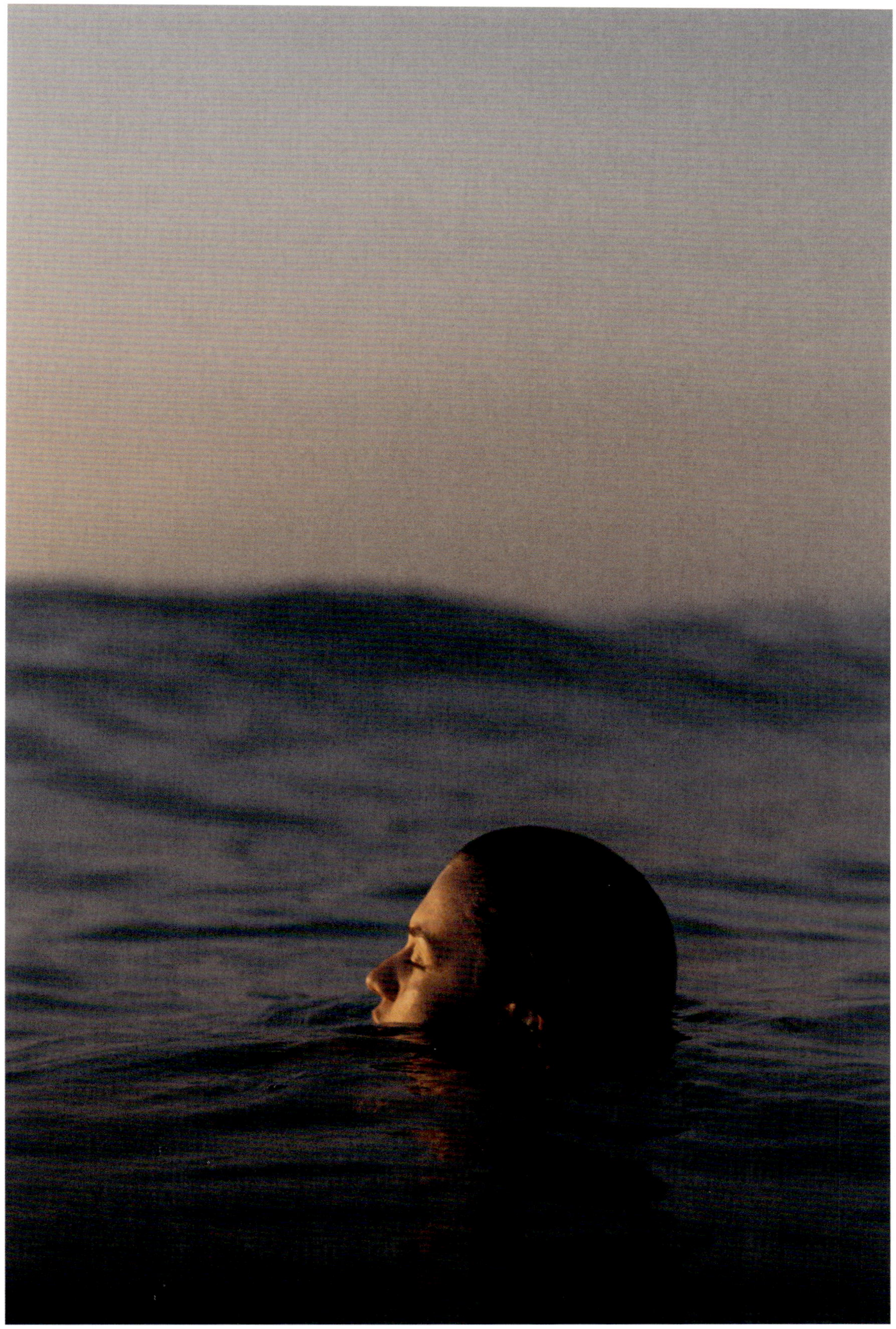

LEFT PAGE

Float Tank. This peaceful body of water makes an ideal place to swim and relax whilst surrounded by ancient rainforest. Tropical Far North Queensland, Australia.

Treiben lassen. Dieses friedliche Gewässer ist ein idealer Ort zum Schwimmen und Entspannen inmitten eines uralten Regenwaldes. Tropical Far North Queensland, Australien.

Immersing into the waves, becoming one with the ocean. Location unknown.

Meditatives Eintauchen in den Ozean, eins werden mit den Wellen. Ort unbekannt.

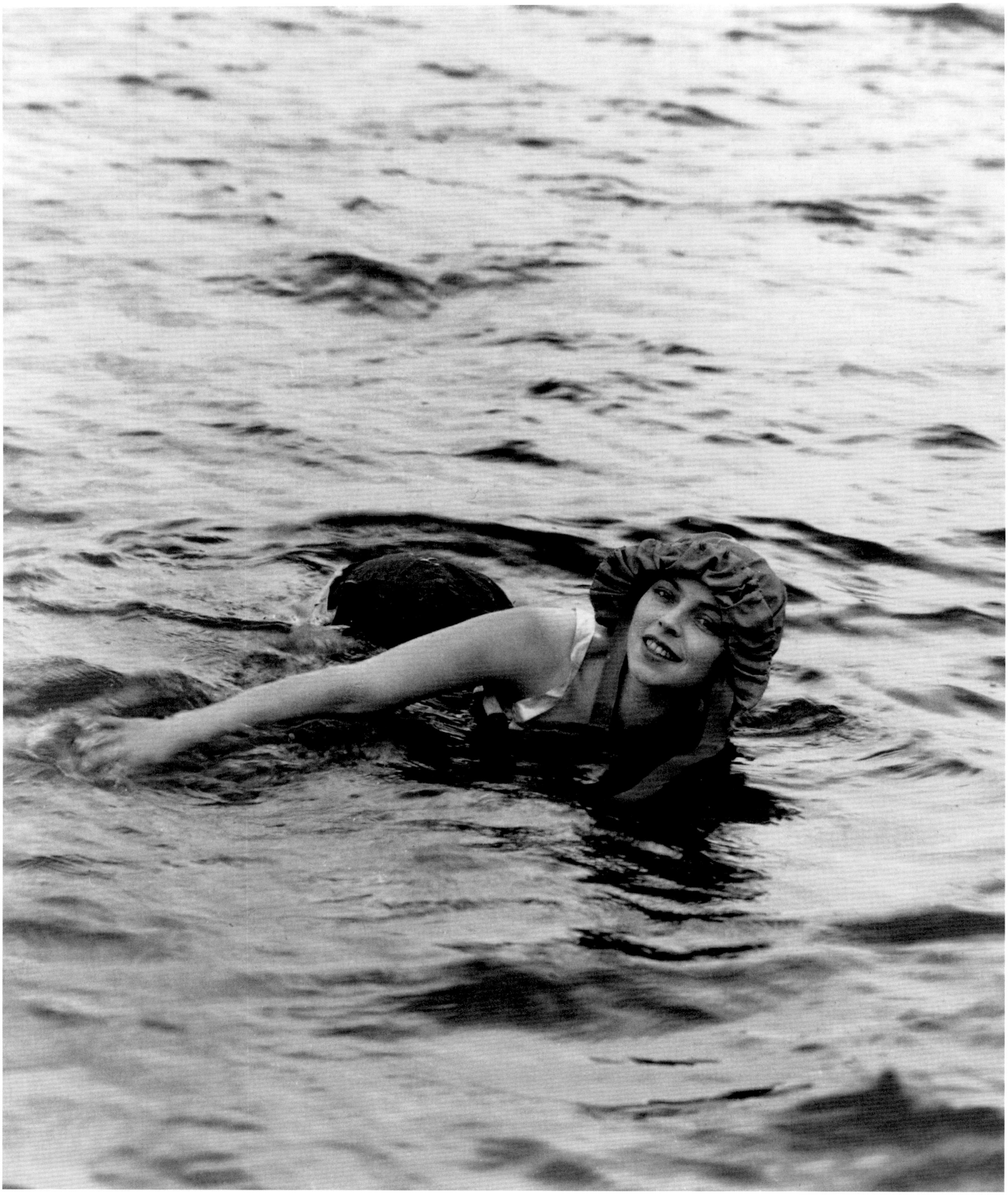

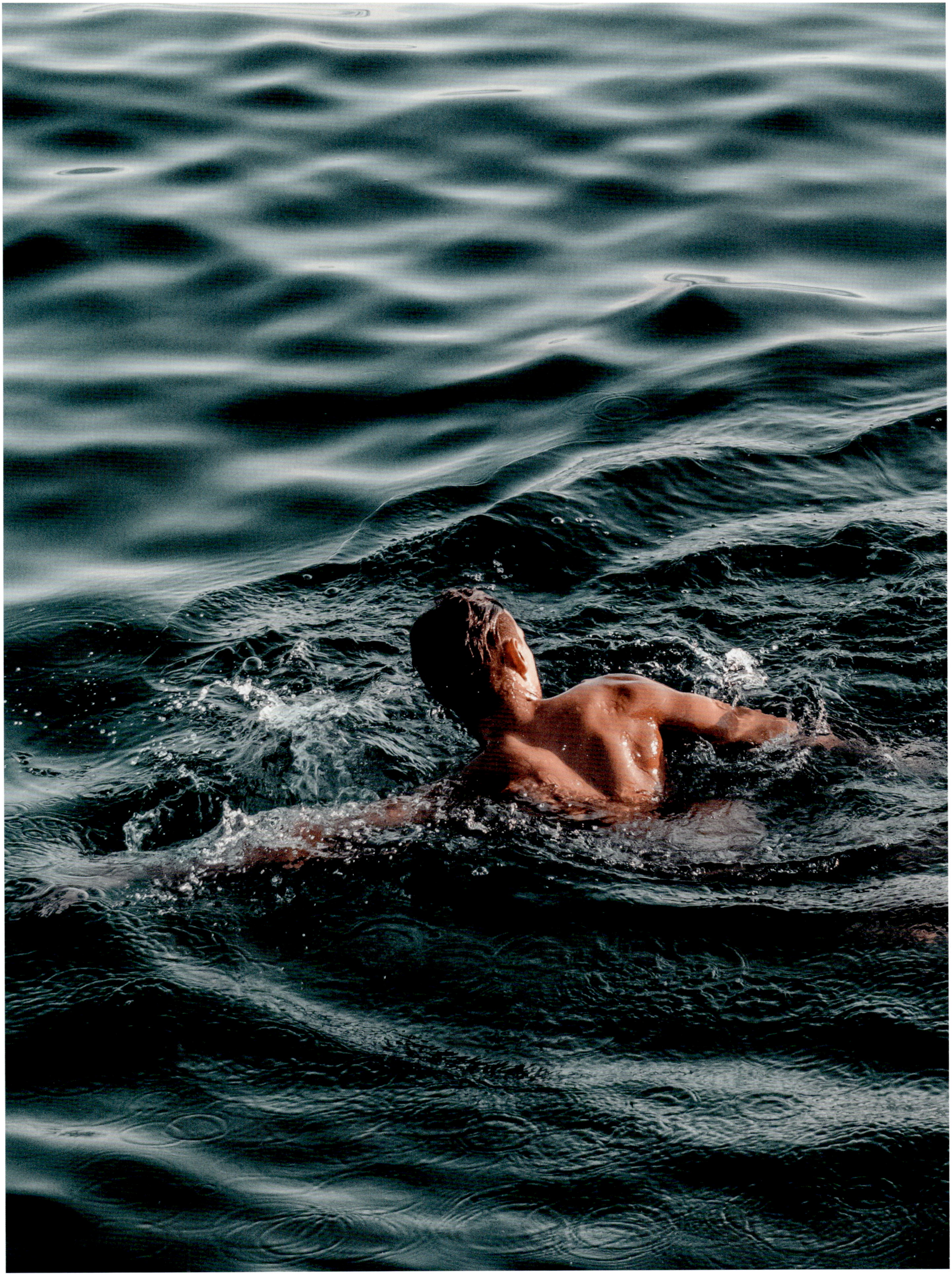

PREVIOUS PAGE

Left: Woman in the sea wearing a bathing cap, 1927. Location unknown.

Right: Swimming in the Adriatic Sea off the Croatian coast of Istria, Croatia.

Links: Frau mit Badehaube im Meer, 1927. Ort unbekannt.

Rechts: Schwimmen in der Adria vor der Küste Istriens, Kroatien.

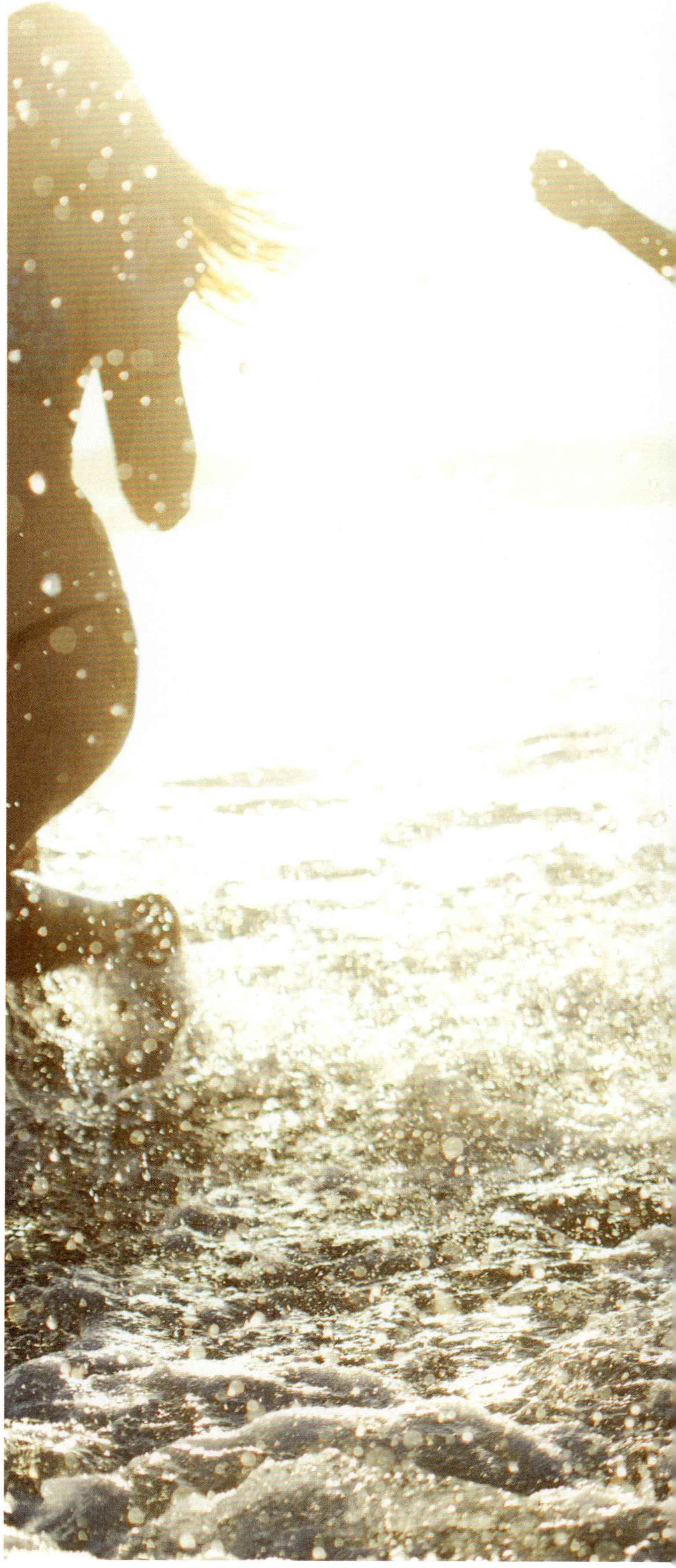

Pure joy of life. Location unknown.

Sprühende Lebensfreude. Ort unbekannt.

AT ONE WITH THE OCEAN WHILE SWIMMING IN THE SEAS

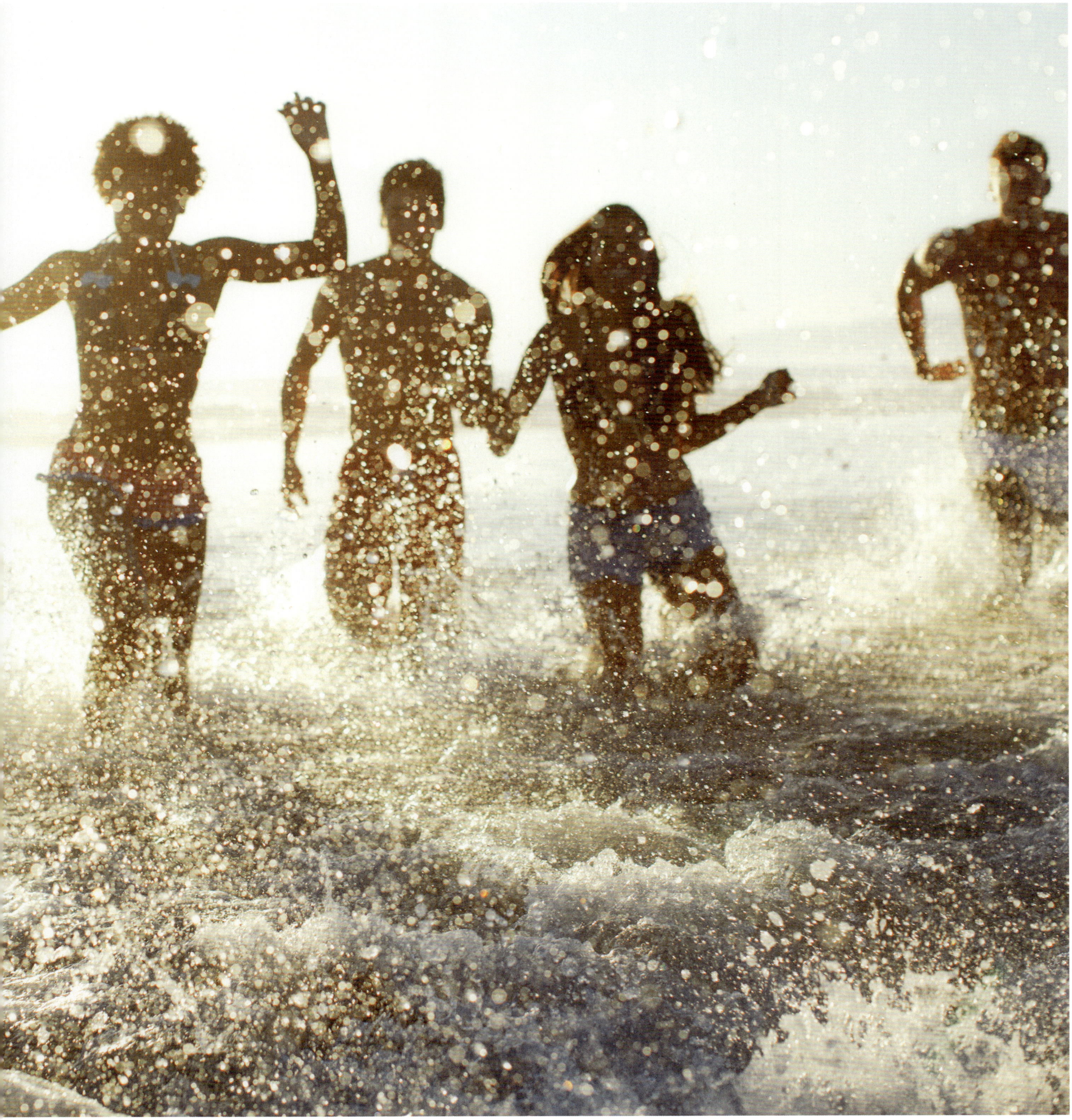

EINS MIT DEM OZEAN: SCHWIMMEN IM MEER

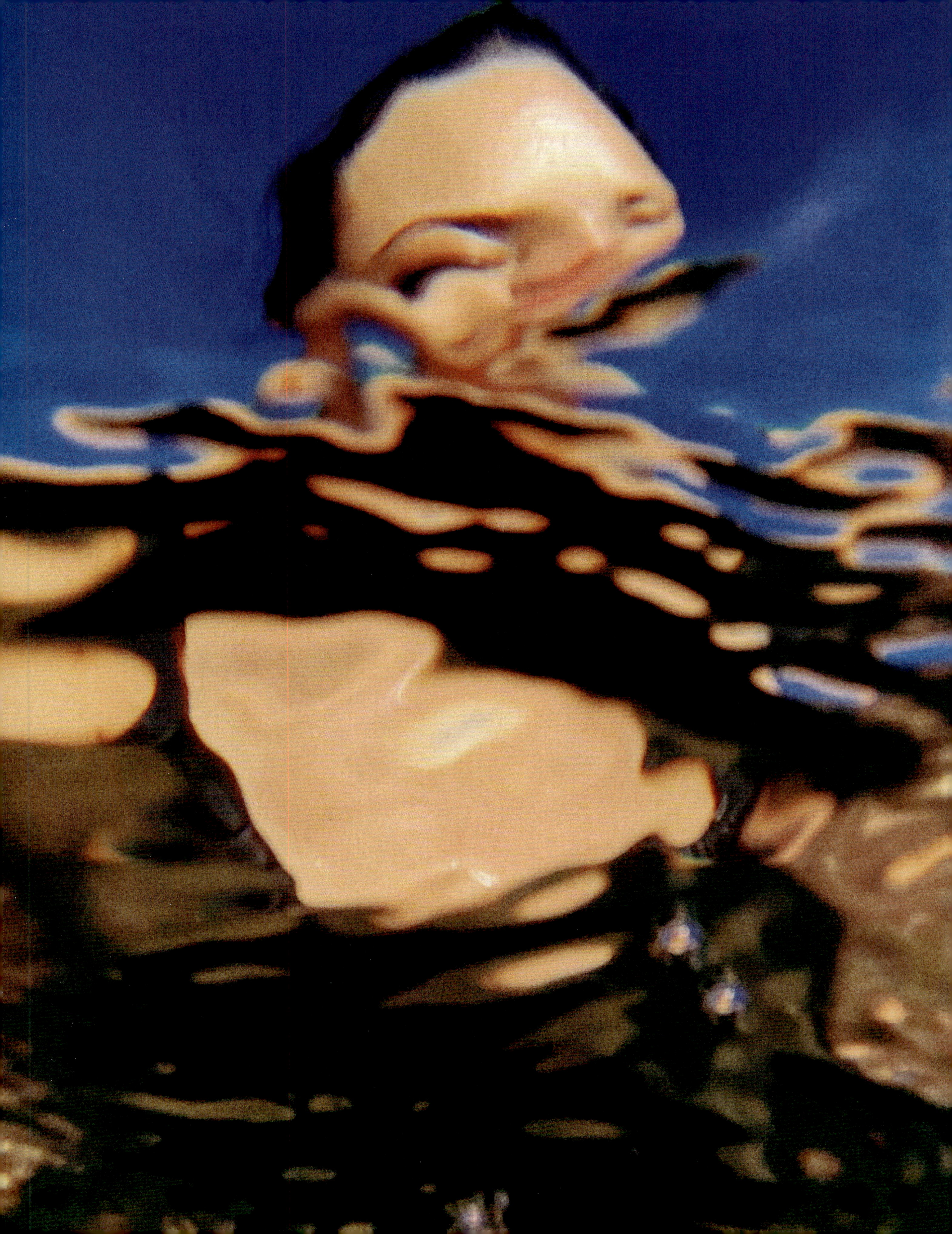

LEFT PAGE

To sea it through me, soul ecstasy. From the series *Sea it Through.* Marbella, Spain.

Durch mich hindurchsehen, Seelenekstase. Aus der Serie *Hindurchsehen.* Marbella, Spain.

Love under water. A summer moment in the Paris region, France.

Unterwasserliebe. Ein Sommermoment in der Pariser Region, Frankreich.

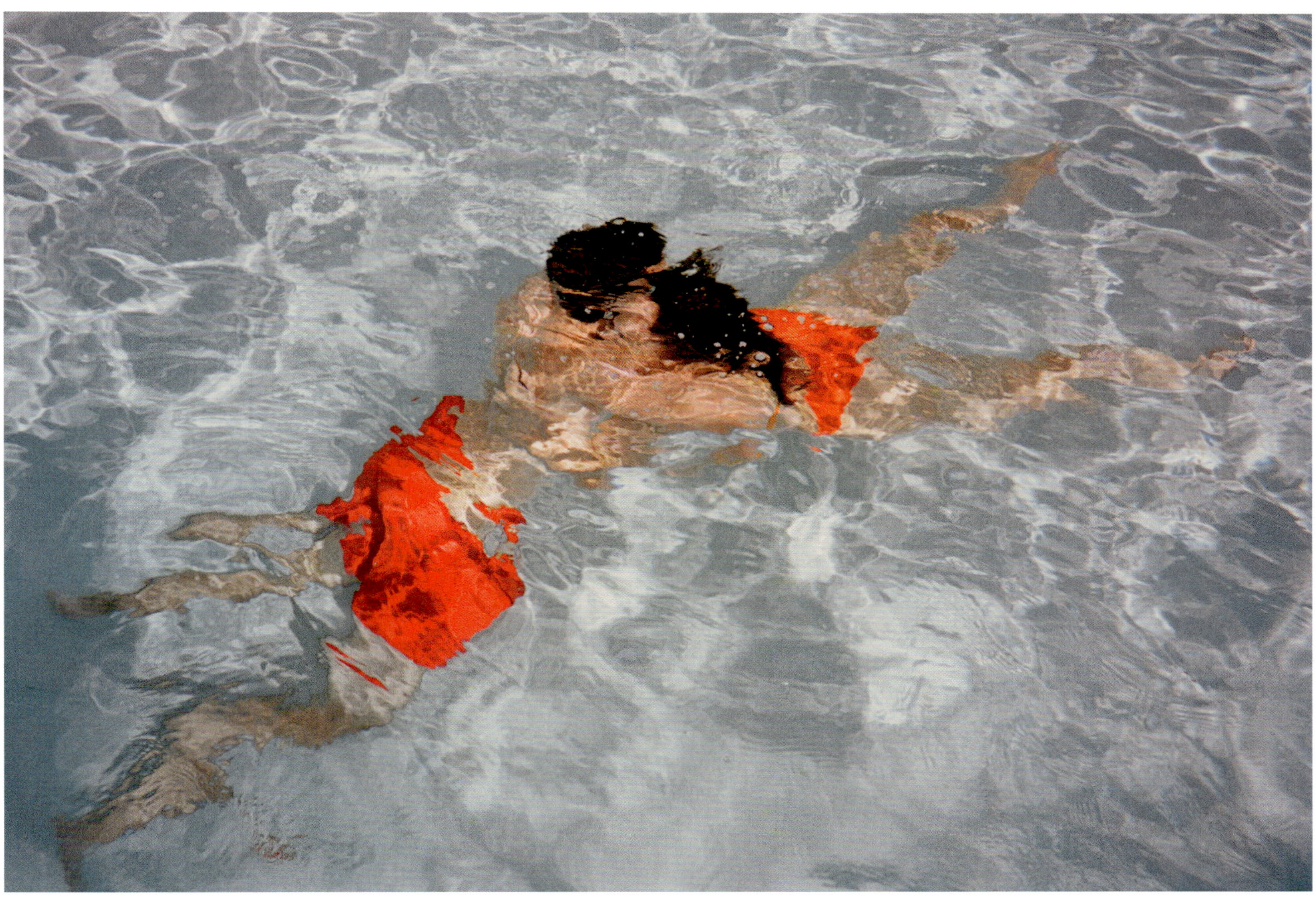

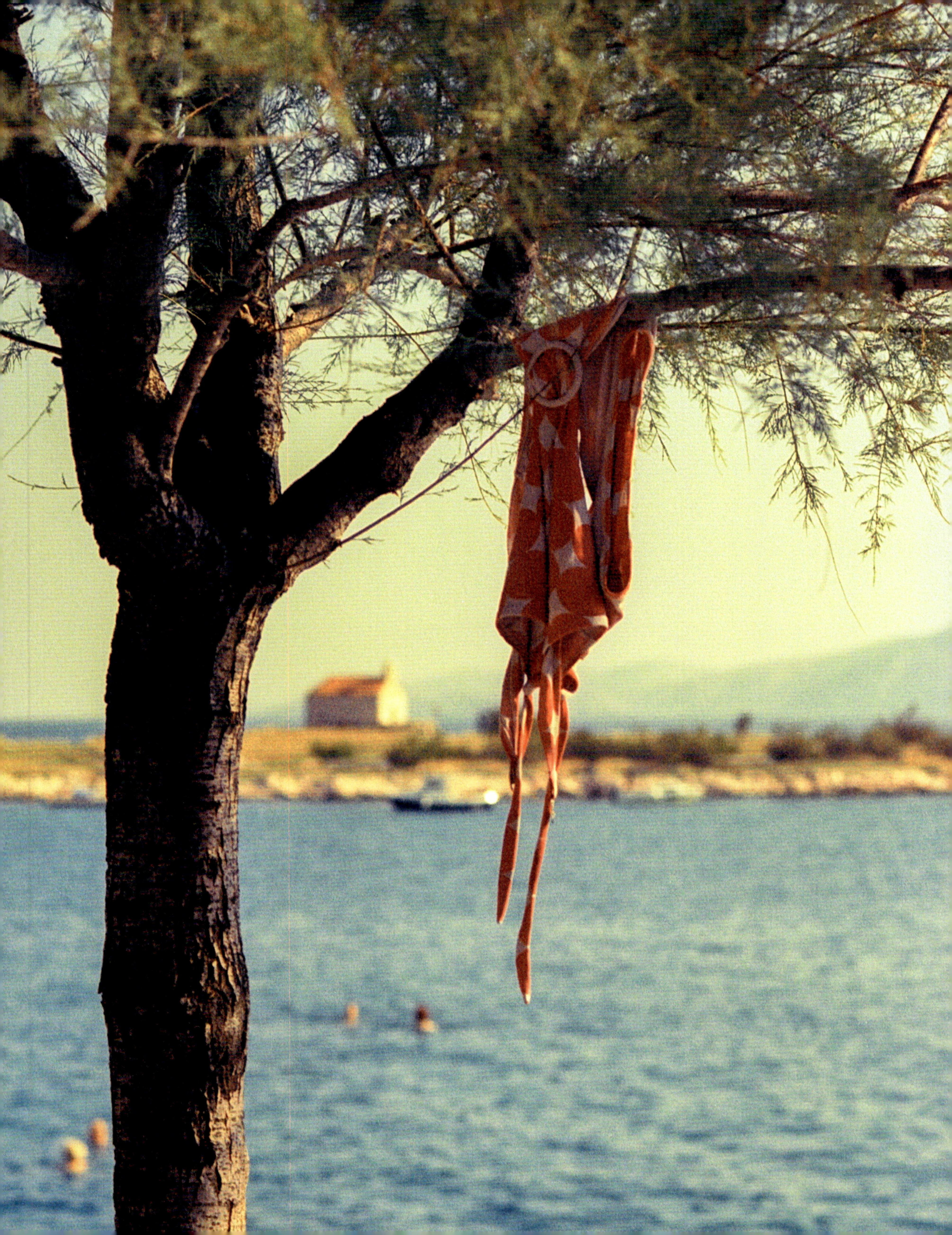

After the swim. Novi Vinodolski, Croatia.

Nach dem Schwimmen. Novi Vinodolski, Kroatien.

Volleyball was the ideal sport to relocate to the beach–because unlike other sports, the ball never touches the ground and so it can be played on any surface. As early as 1915, volleyball was documented as being played on sand, especially in the major beach life countries of the USA, Brazil, and Australia, and two-person teams soon caught on, at least among professional players. The first tournament to draw some attention was held in 1944 at Muscle Beach in Santa Monica, California, and the sport rose to global fame in the 1960s, around the same time as surfing. By 1976, the U.S. championships were being held in front of 30,000 spectators. In 1992, beach volleyball made its debut as a demonstration sport at the Olympics, with medals being awarded since 1996.

The sport is fast, exciting–and of course sexy, like life at the beach itself. The women's final in the 2016 Olympics, for instance, was the most-watched sporting event of the entire Olympics in Germany, with 8.52 million viewers, no doubt due in part to the German duo of Laura Ludwig/Kira Walkenhorst, who took the gold medal against the Brazilian team.

But it is precisely the highly marketable eroticism of beach volleyball that has caused heated controversy in recent years: While other sports have certain dress codes that prioritize concealment, in this sport, skimpy clothing is mandatory. Some female athletes did not want to bow to this directive, while others expressly welcomed the emphasis on the body and were happy to show off their body in a more revealing manner. Female track and field athletes and swimmers also wear tight-fitting clothing, they argue. Nevertheless, the International Volleyball Federation relented in 2012, permitting female players to wear wide shorts that reach down to the knee. However, very few players make use of this regulation.

Volleyball eignete sich einfach perfekt für den Umzug an den Strand – denn im Gegensatz zu anderen Sportarten darf der Ball nie den Boden berühren und kann daher auf jedem Untergrund gespielt werden. Schon 1915 wurde nachweislich Volleyball im Sand gespielt, vor allem in den großen Beachlife-Nationen USA, Brasilien und Australien, und bald setzten sich, zumindest bei den Profispielern, auch die Zweierteams durch. Das erste Turnier, das eine gewisse Aufmerksamkeit erregte, wurde 1944 am Muscle Beach von Santa Monica, Kalifornien, ausgetragen, und zu weltweitem Ruhm gelangte die Sportart in den 1960er-Jahren, etwa zeitgleich mit dem Surfen. 1976 fanden die US-Meisterschaften bereits vor 30.000 Zuschauern statt. 1992 debütierte Beachvolleyball als Demonstrationssportart bei Olympia, seit 1996 werden Medaillen verteilt.

Der Sport ist schnell, aufregend – und natürlich auch sexy, wie das Strandleben selbst. Das Damen-Finale bei Olympia 2016 war beispielsweise in Deutschland mit 8,52 Millionen Zuschauern die meistgesehene Sportveranstaltung der gesamten Spiele, was sicher auch an dem deutschen Duo Laura Ludwig/Kira Walkenhorst lag, die sich gegen das brasilianische Team die Goldmedaille holten.

Doch genau die so gut vermarktbare Erotik des Beachvolleyballs sorgte in den letzten Jahren für heftige Kontroversen: Denn während in anderen Sportarten bestimmte Kleidungsvorschriften gelten, die aufs Bedecken Wert legen, ist hier knappe Kleidung vorgeschrieben. Diesem Diktat wollten sich einige Sportlerinnen nicht beugen, während andere die Körperlichkeit ausdrücklich begrüßen und sich gern freizügig zeigen. Auch Leichtathletinnen und Schwimmerinnen würden ja knappe, enganliegende Kleidung tragen, argumentieren sie. Dennoch lenkte der Weltverband 2012 ein und erlaubte den Spielerinnen weite Shorts, die bis zum Knie reichen dürfen. Von dieser Regelung machen allerdings die wenigsten Spielerinnen Gebrauch.

Beach volleyball finds its home on beaches around the world, like here in Khao Lak, Thailand.

Beachvolleyball findet an den Stränden der ganzen Welt sein Zuhause, wie hier in Khao Lak, Thailand.

GET YOUR HANDS UP! IT'S TIME FOR BEACH VOLLEYBALL

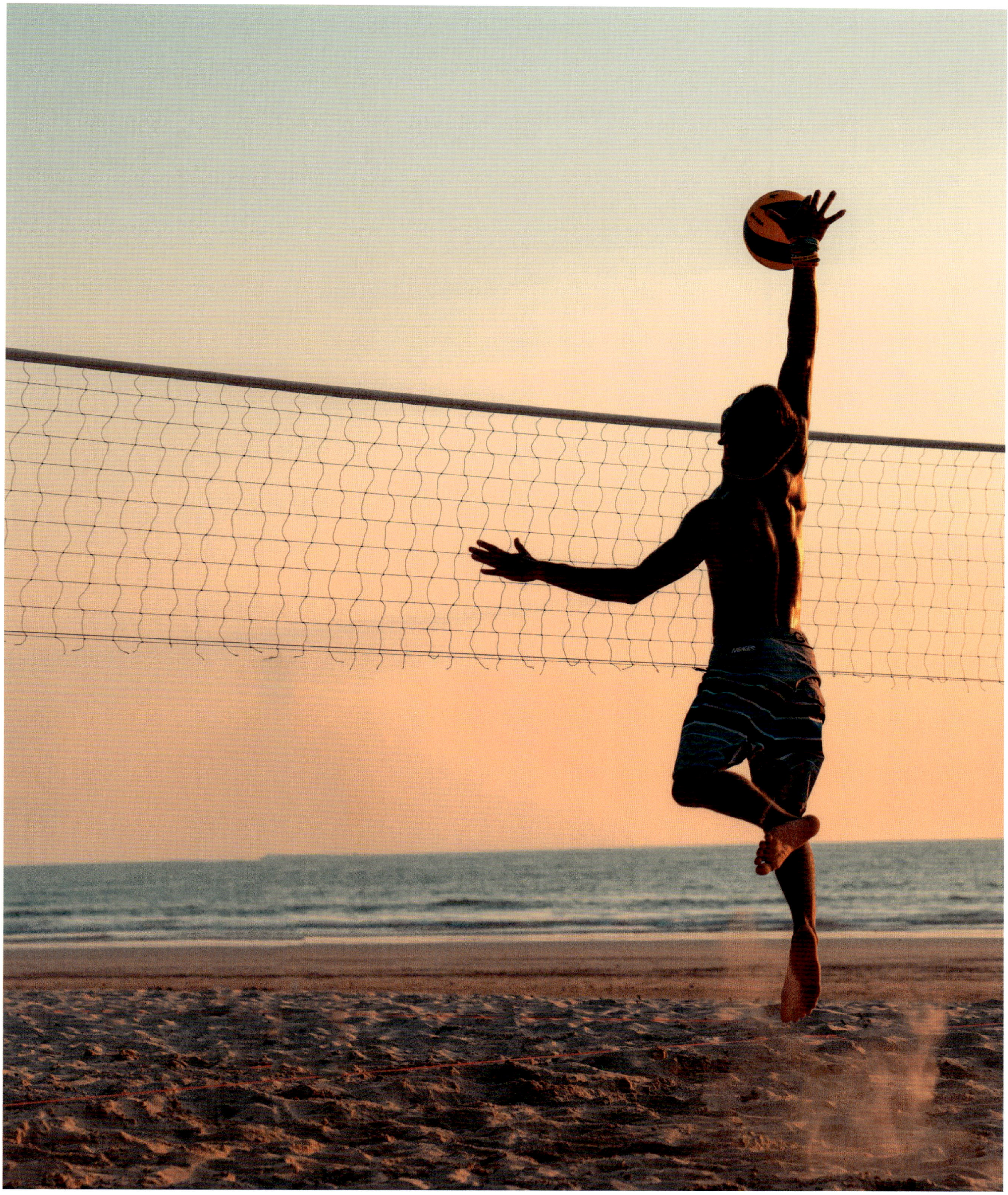

HÄNDE HOCH! BEACHVOLLEYBALL

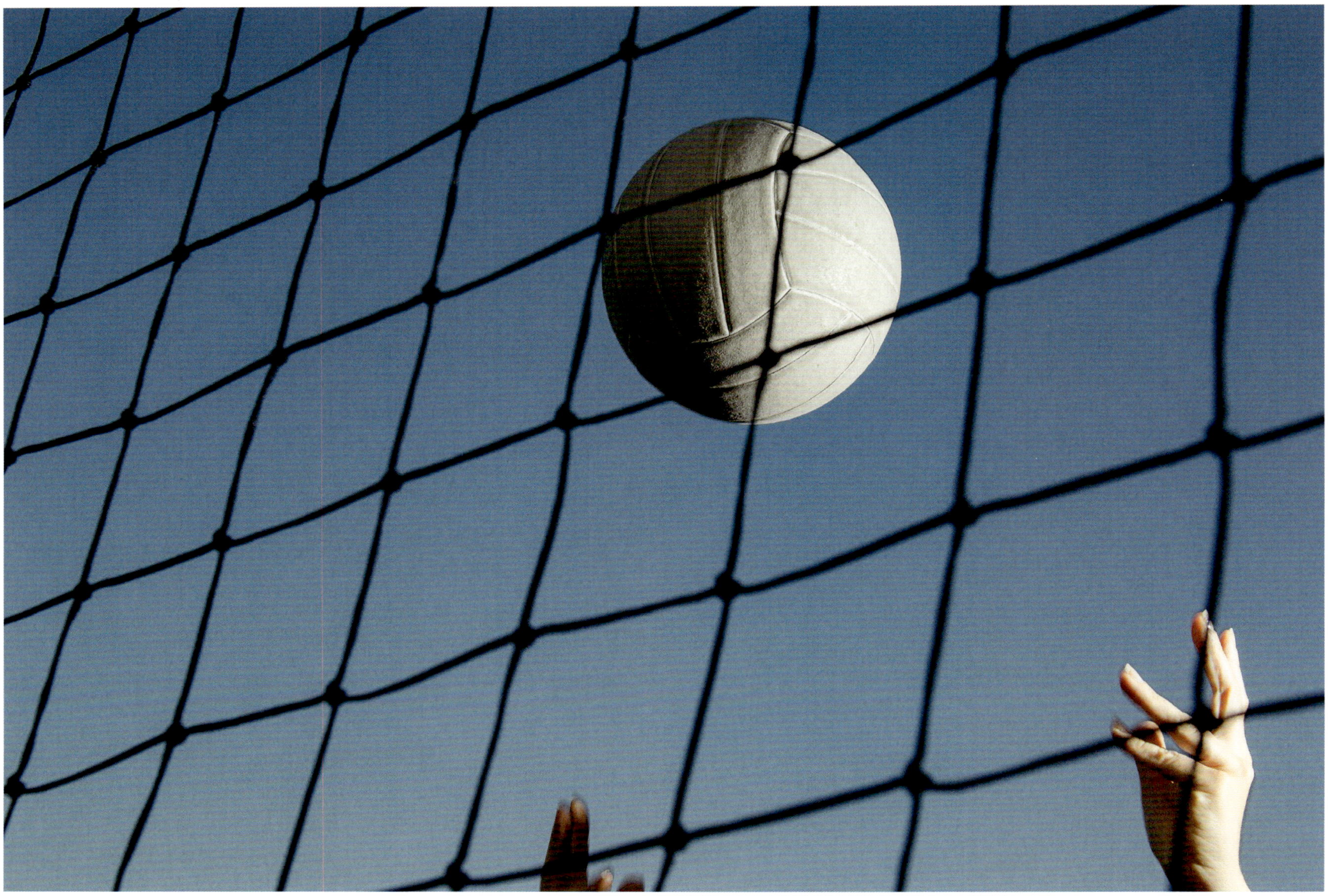

Beach volleyball is one of the most popular activities at the beach for hobby players ...

Beachvolleyball ist eine der beliebtesten Strandaktivitäten für Hobbyspieler ...

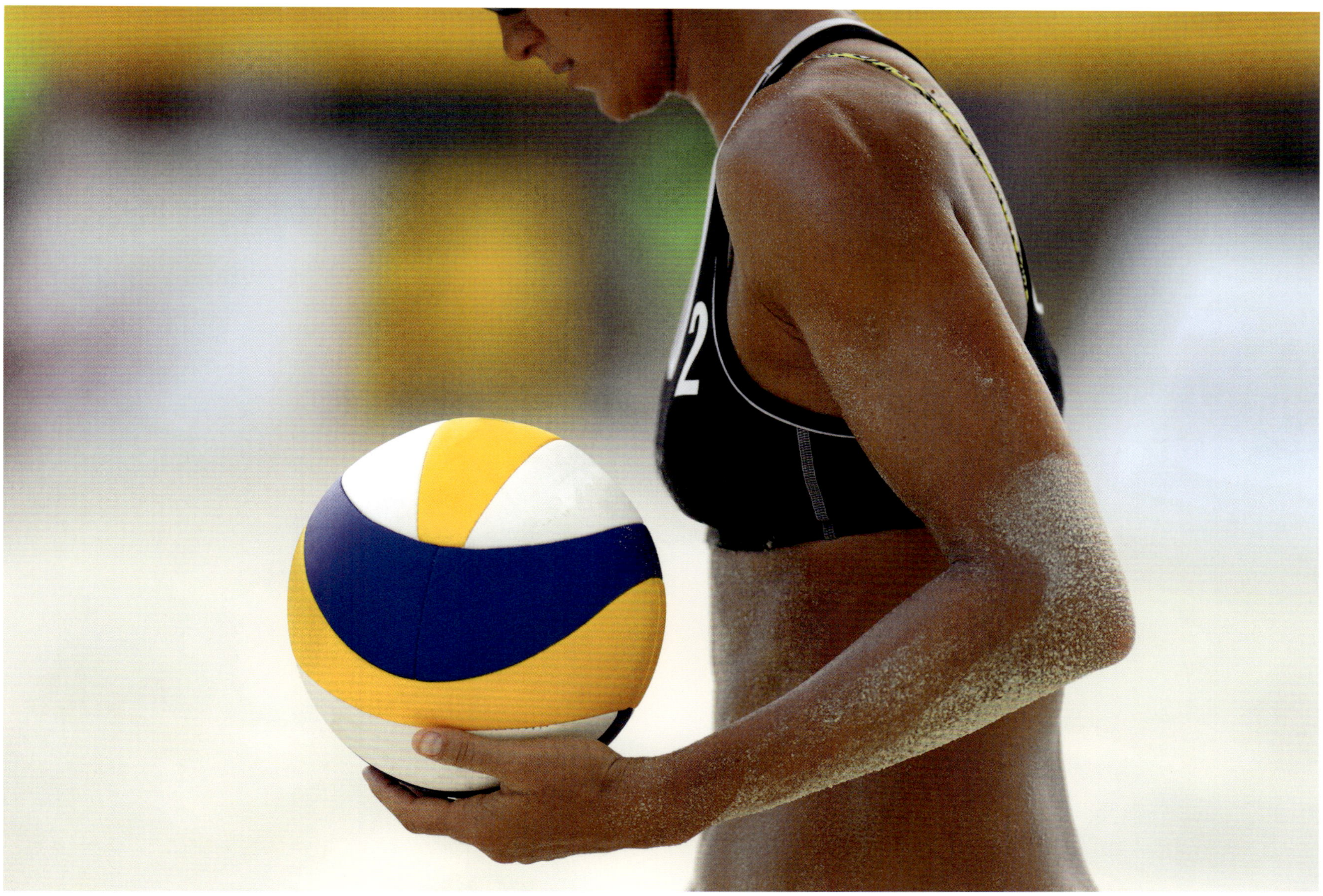

... and pros.

... und Profis.

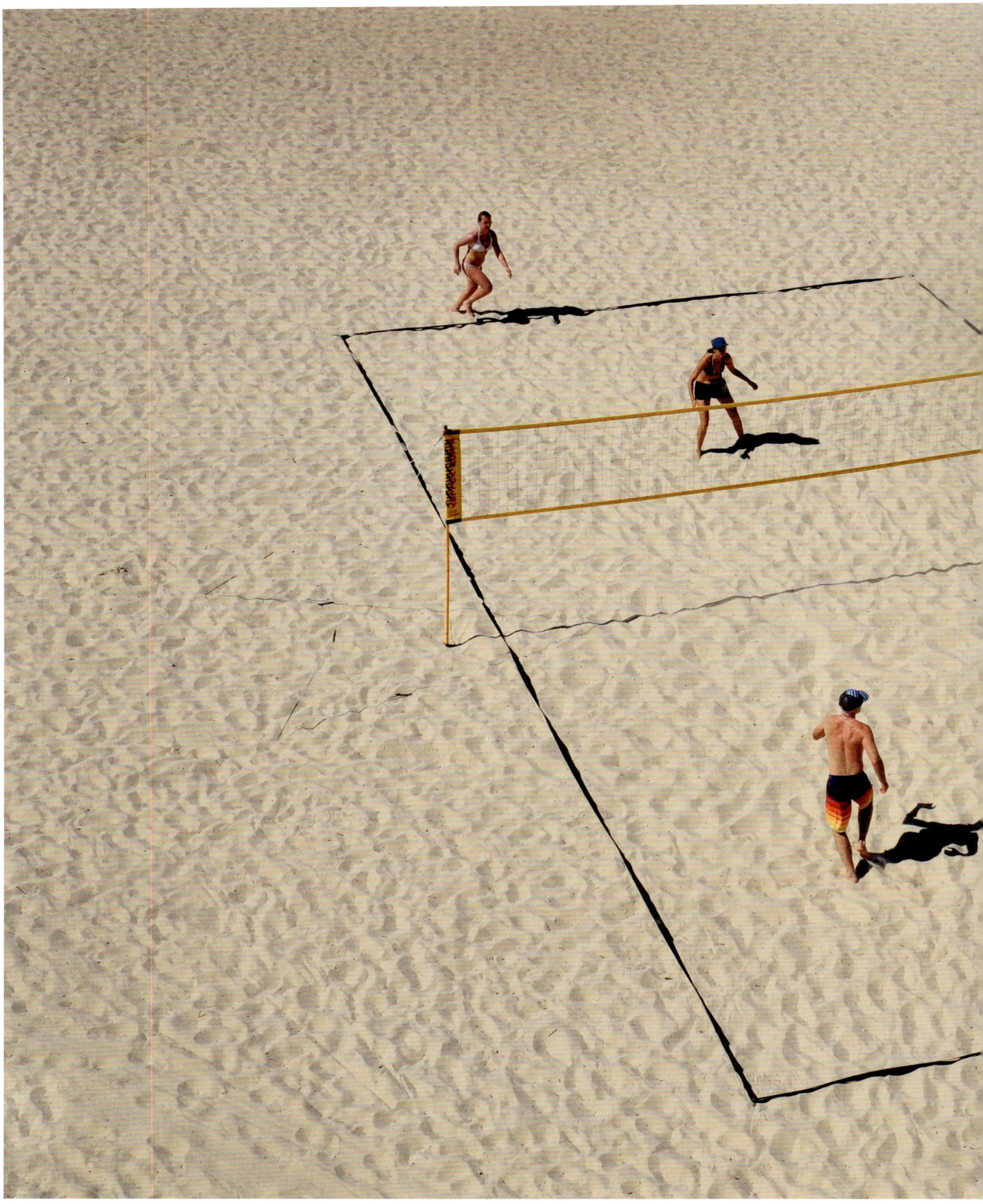

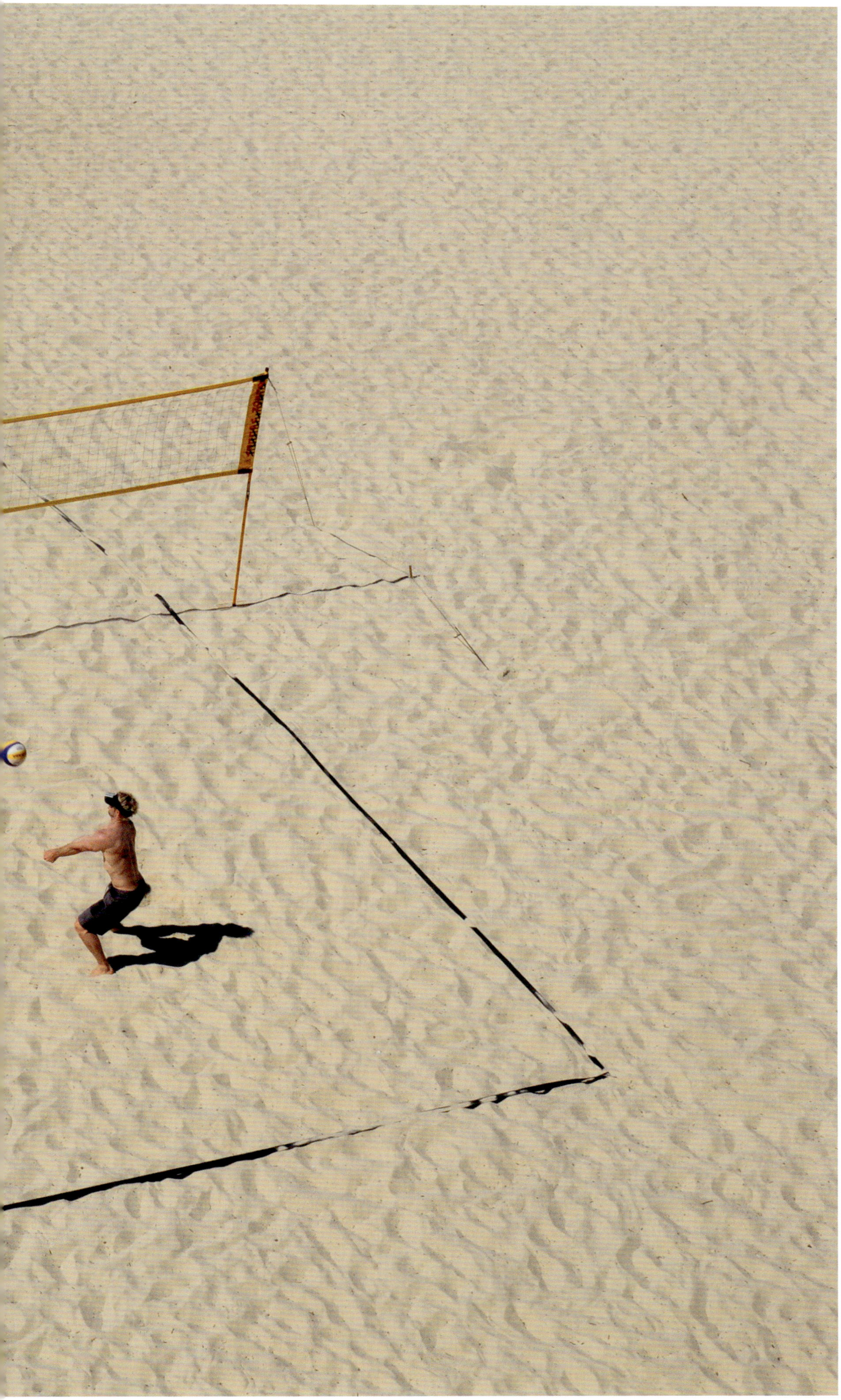

Four people on sand, a net and a ball, simple as that. Australia.

Vier Spieler auf dem Sand, ein Netz, ein Ball – so simpel wie genial. Australien.

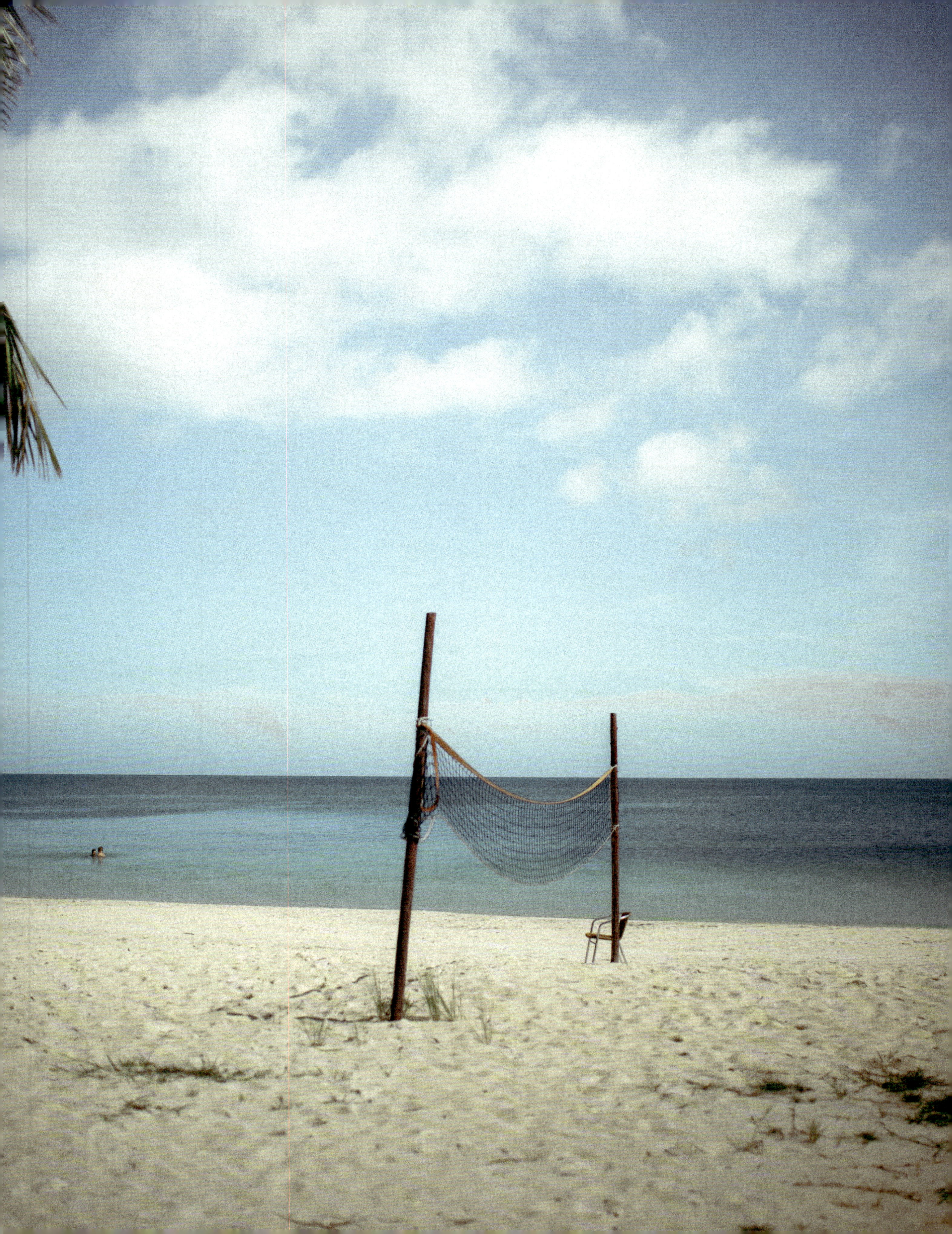

After the match.
Cayo Levisa, Cuba.

Nach dem Spiel.
Cayo Levisa, Kuba.

Beachgoers generally take it easy. Life slows down and comes to a standstill on the lounge chairs. But not everyone feels like slowing down, and fortunately things can also be fast, spectacular, and athletic at the beach, since for many people this is also part of their vacation experience. The boardwalks, often used as promenades but actually built to protect against flooding and erosion, are now firmly in the clutches of skateboarders, roller skaters, and inline skaters in some places. Halfpipes and bowls have a deeper meaning right on the beach, since an opportunity to cool off is close at hand. In addition, skaters are delighted to have as many spectators as possible, of course, which would be rather uncommon in remote industrial areas.

In California, almost every town on the beach has its own skate park, since the skateboard movement originated here. When surfers got bored with the lack of waves, they started experimenting with boards with wheels bolted underneath them on the shore. The most famous park is Venice Beach Skate Park, with the enticing postal address of 1800 Ocean Front Walk. Advice for onlookers: In the afternoon, it gets brutally crowded here. The real experts, who need a lot of space for their breathtaking feats, practice very early in the morning. The largest skate park in the world is in Guangzhou, China. The 160,000 students from the ten (!) surrounding universities have plenty of room to enjoy themselves on its almost 17,000 square meters.

It is said that everything is more fun at the beach, and that's why the mostly smooth concrete promenades also attract roller skaters and inline skaters. Skating on beachfront boulevards conveys a feeling of ease, freedom, and lightheartedness. Every year Barcelona hosts the Skate Love Festival, a huge musical event on roller skates. There are also courses and workshops on acrobatic roller skating, held by world stars on wheels. And every evening the best of the best showcase their skills on stage.

Strandmenschen lassen es im Allgemeinen gemütlich angehen. Das Leben verlangsamt sich und kommt in den Liegen zum Erliegen. Doch nicht jedem ist nach Entschleunigung zumute, und glücklicherweise kann es am Strand auch schnell, spektakulär, sportlich zugehen, denn auch das gehört für viele Menschen zum Urlaub. Die Uferpromenaden, oft zwar zum Flanieren genutzt, aber eigentlich zum Schutz vor Hochwasser und Erosion angelegt, sind in manchen Orten fest in den Händen von Skatern, Rollschuhläufern und Inlinern. Halfpipes und Bowls haben direkt am Strand einen tieferen Sinn, ist doch die Abkühlung nah. Außerdem freuen sich die Skateboarder natürlich über möglichst viele Zuschauer, die in abgelegenen Industriegebieten eher selten wären.

In Kalifornien hat so gut wie jede Stadt am Strand ihren Skatepark, schließlich stammt die Skateboard-Bewegung von hier: Als den Surfern langweilig war, weil die Wellen ausblieben, begannen sie, mit Brettern und druntergeschraubten Rollen am Ufer zu experimentieren. Der bekannteste Park ist der Venice Beach Skate Park mit der umwerfenden Postanschrift „1800 Ocean Front Walk". Tipp für Zaungäste: Am Nachmittag ist es hier brutal voll. Die echten Könner, die viel Platz für ihre atemberaubenden Kunststücke brauchen, trainieren ganz früh am Morgen. Der größte Skatepark der Welt liegt im chinesischen Guangzhou: Auf den fast 17.000 Quadratmetern können sich die 160.000 Studenten der zehn (!) umliegenden Universitäten vergnügen.

Es gilt die Maxime: Am Strand macht alles noch mehr Spaß, und deswegen ziehen die meist glattbetonierten Promenaden auch Rollschuhfahrer und Inlineskater an. Skaten auf Strandboulevards vermittelt ein Gefühl der Leichtigkeit, Freiheit und Unbeschwertheit. In Barcelona findet jedes Jahr das Skate Love Festival statt, ein riesiges Musikevent auf Rollschuhen. Dazu gibt es Kurse und Workshops im akrobatischen Rollschuhlaufen, abgehalten von Weltstars auf Rollen. Und jeden Abend zeigen die Besten der Besten auf Showbühnen ihr Können.

The legendary skatepark of
Venice Beach, Los Angeles, USA.

Der legendäre Skatepark von
Venice Beach, Los Angeles, USA.

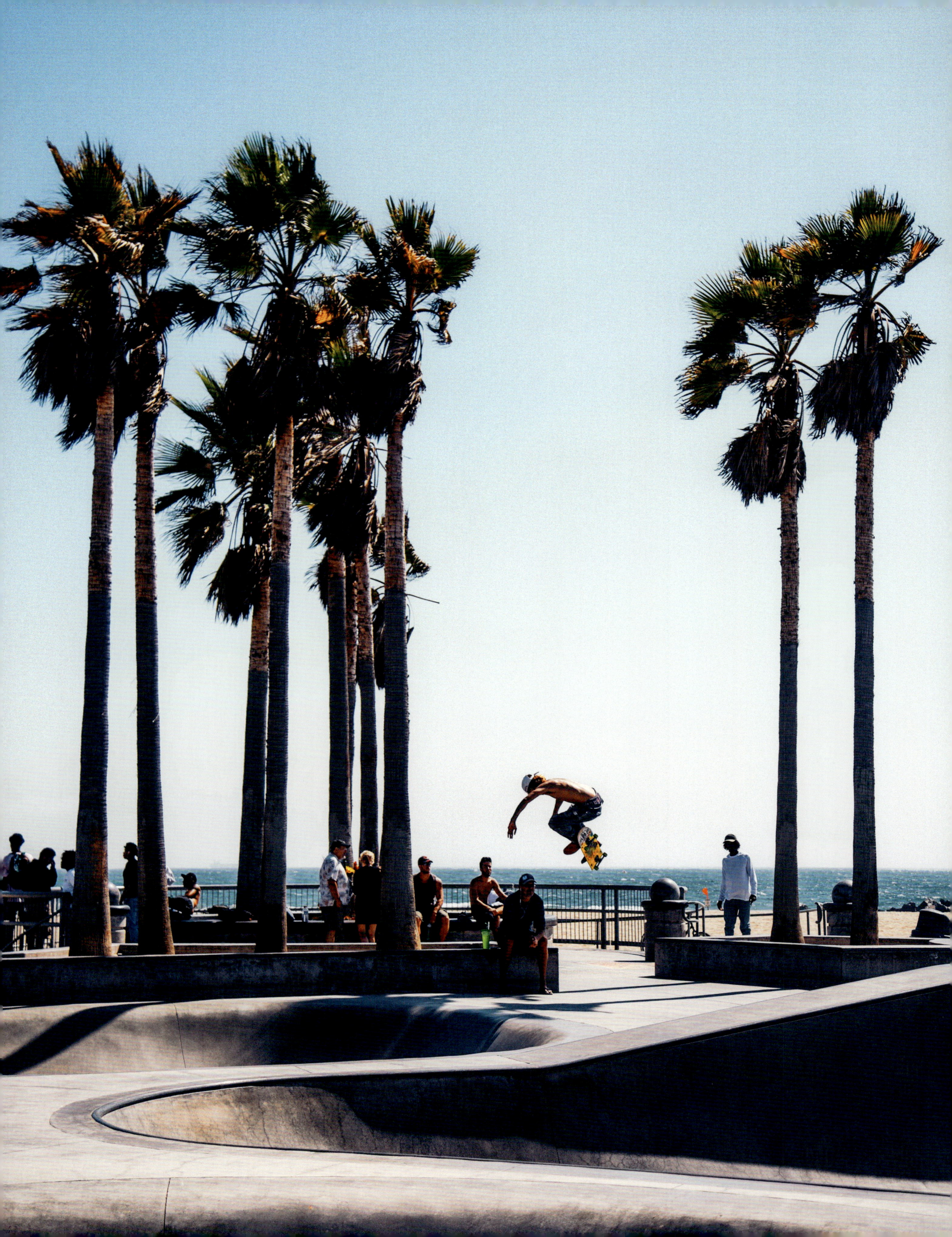

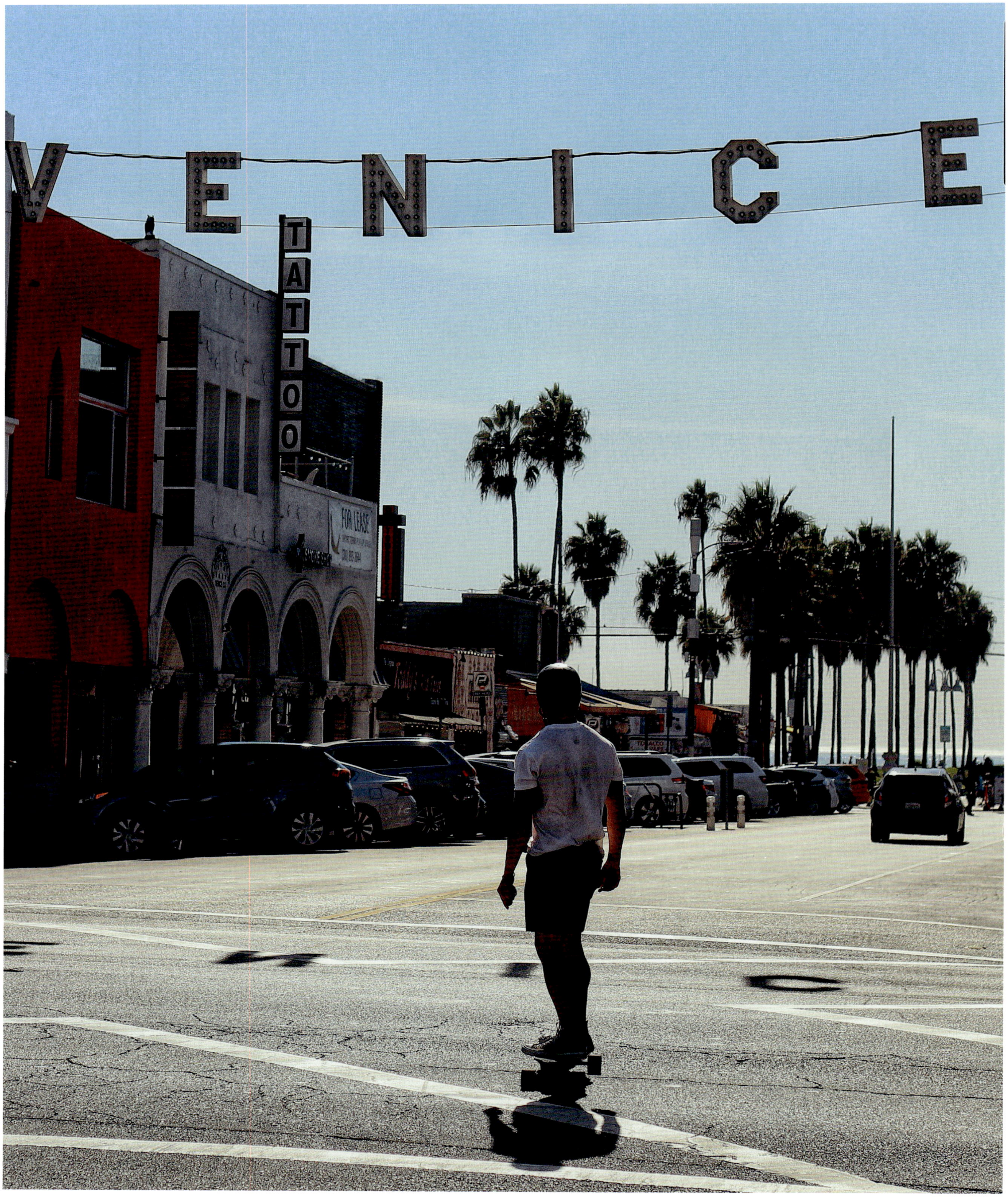
VENICE
TATTOO
FOR LEASE

On his way to a skate session on the beach.
Venice Beach, Los Angeles, USA.

Auf dem Weg zur Skatesession am Strand.
Venice Beach, Los Angeles, USA.

Roller Skaters on the seaside promenade. Location unknown.

Rollschuhläufer an der Strandpromenade. Ort unbekannt.

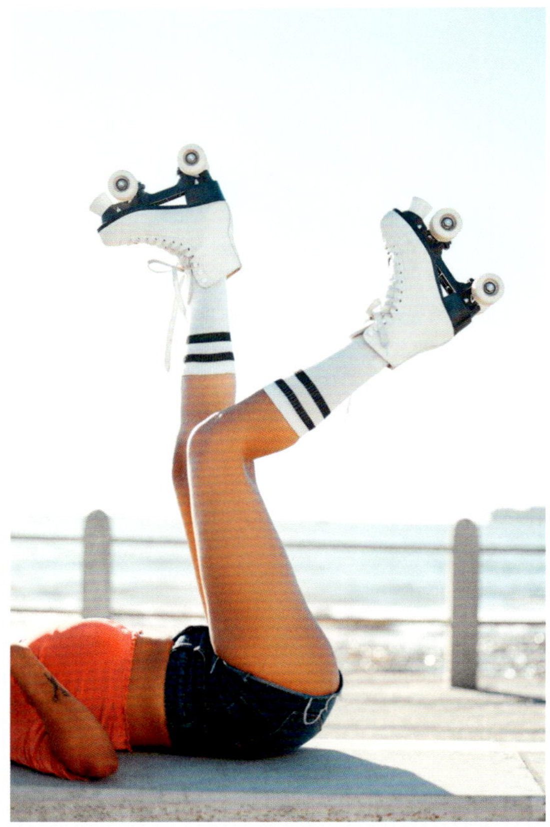

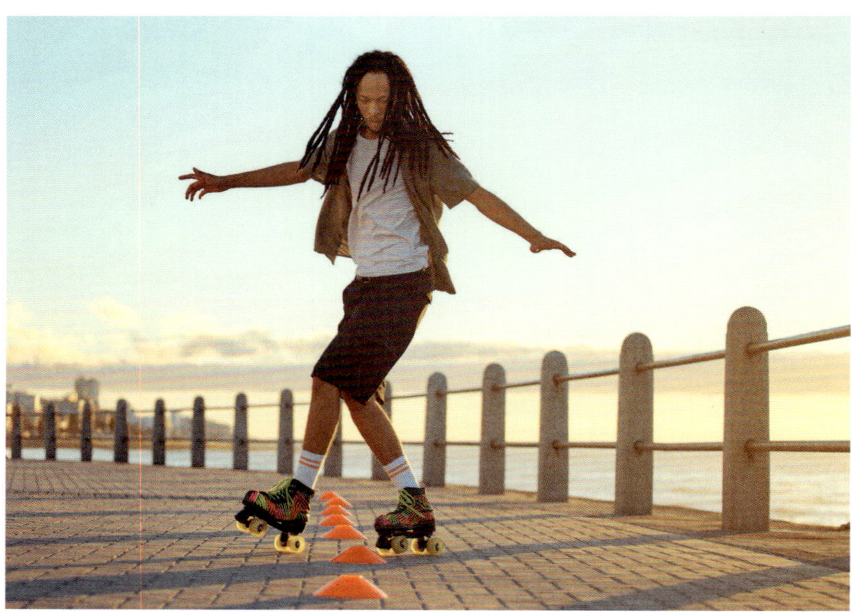

RIGHT PAGE

Roller dancer on the Skate Love Festival in Barcelona, Spain.

Rollschuhtänzerin auf dem Skate Love Festival in Barcelona, Spanien.

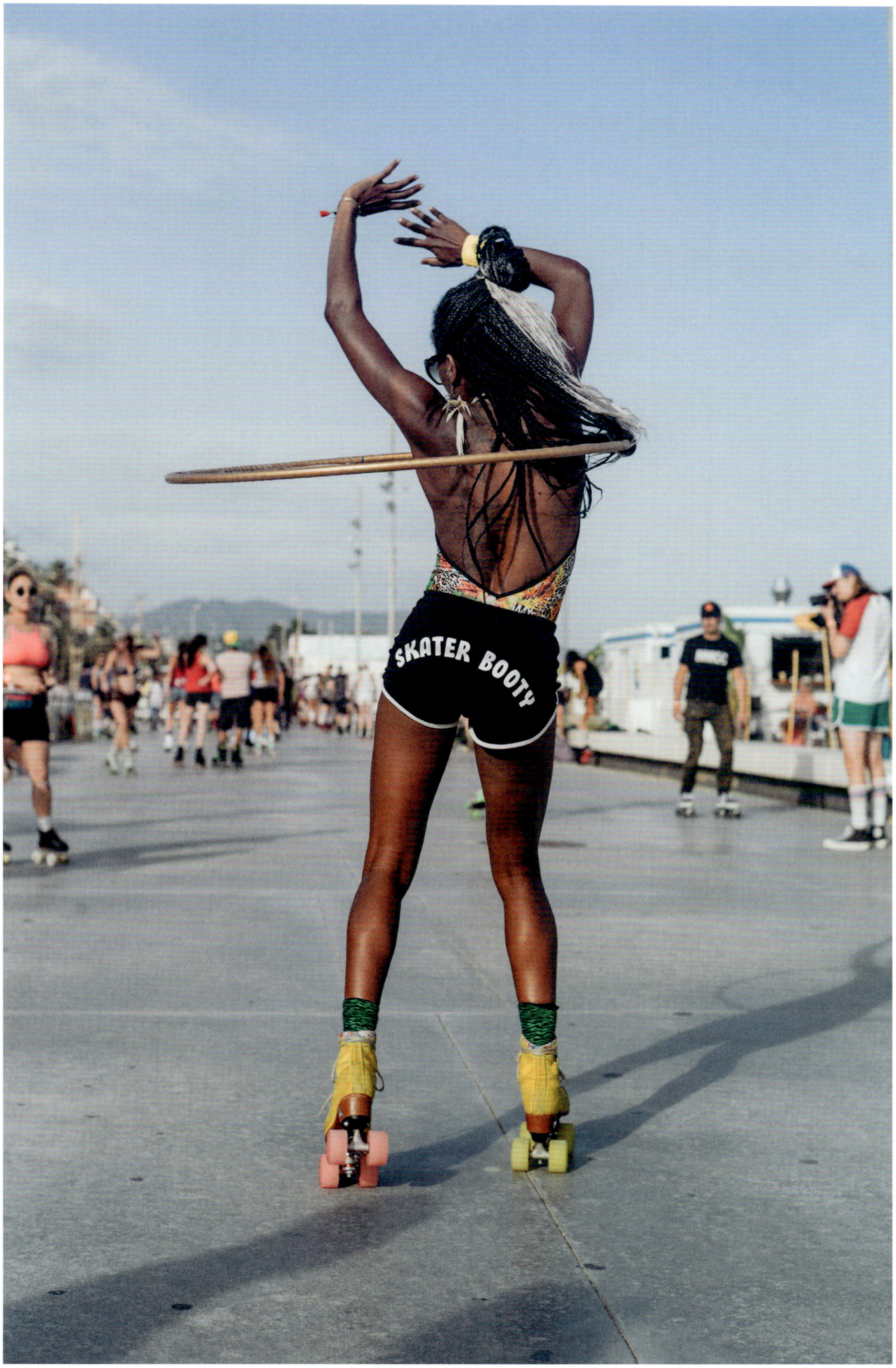
SKATER BOOTY

Skating on beachfront boulevards is the epitome of freedom and lightheartedness for many roller skaters. Location unknown.

Mit Rollschuhen auf der Strandpromenade: Für viele Skater der Inbegriff von Freiheit und Unbeschwertheit. Ort unbekannt.

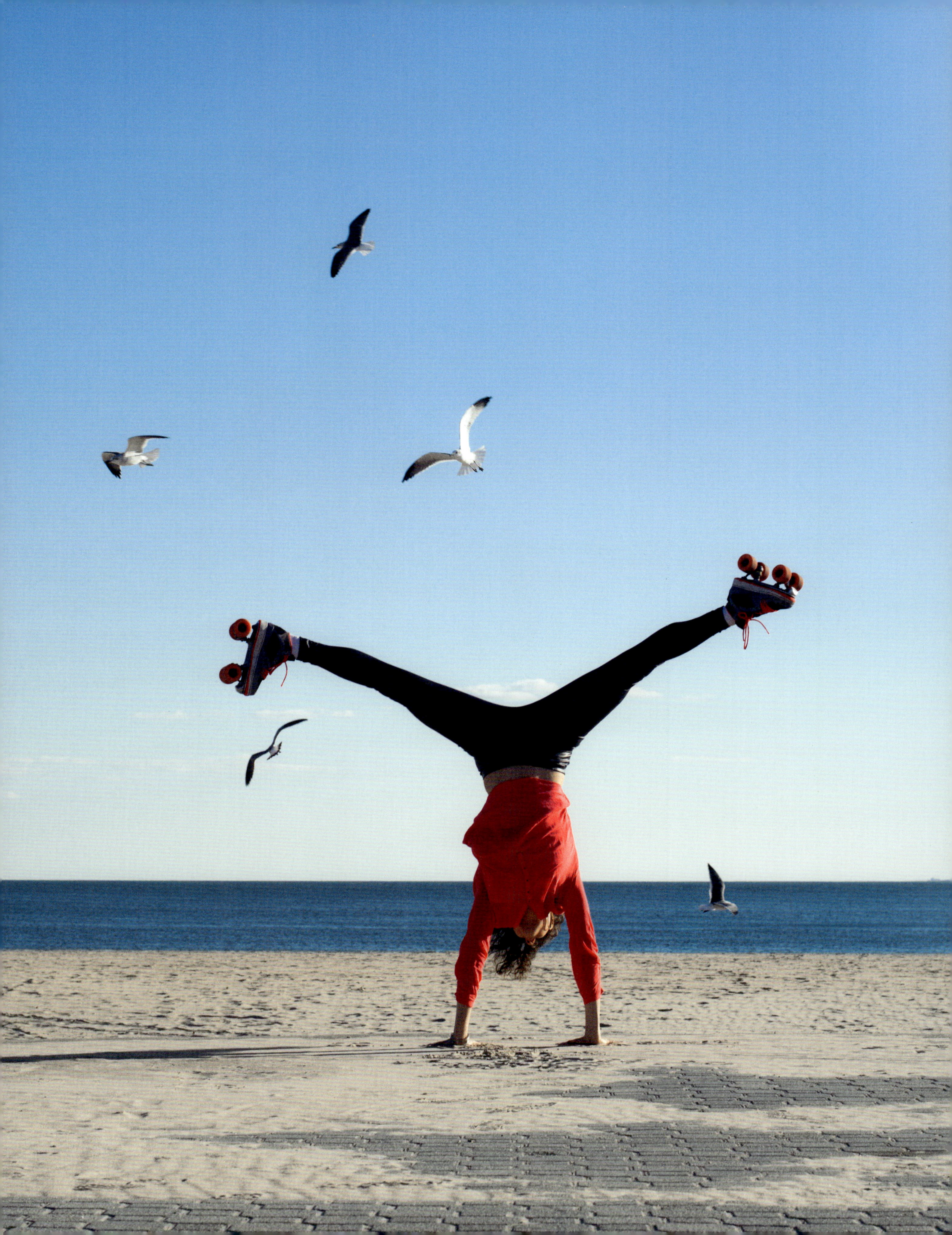

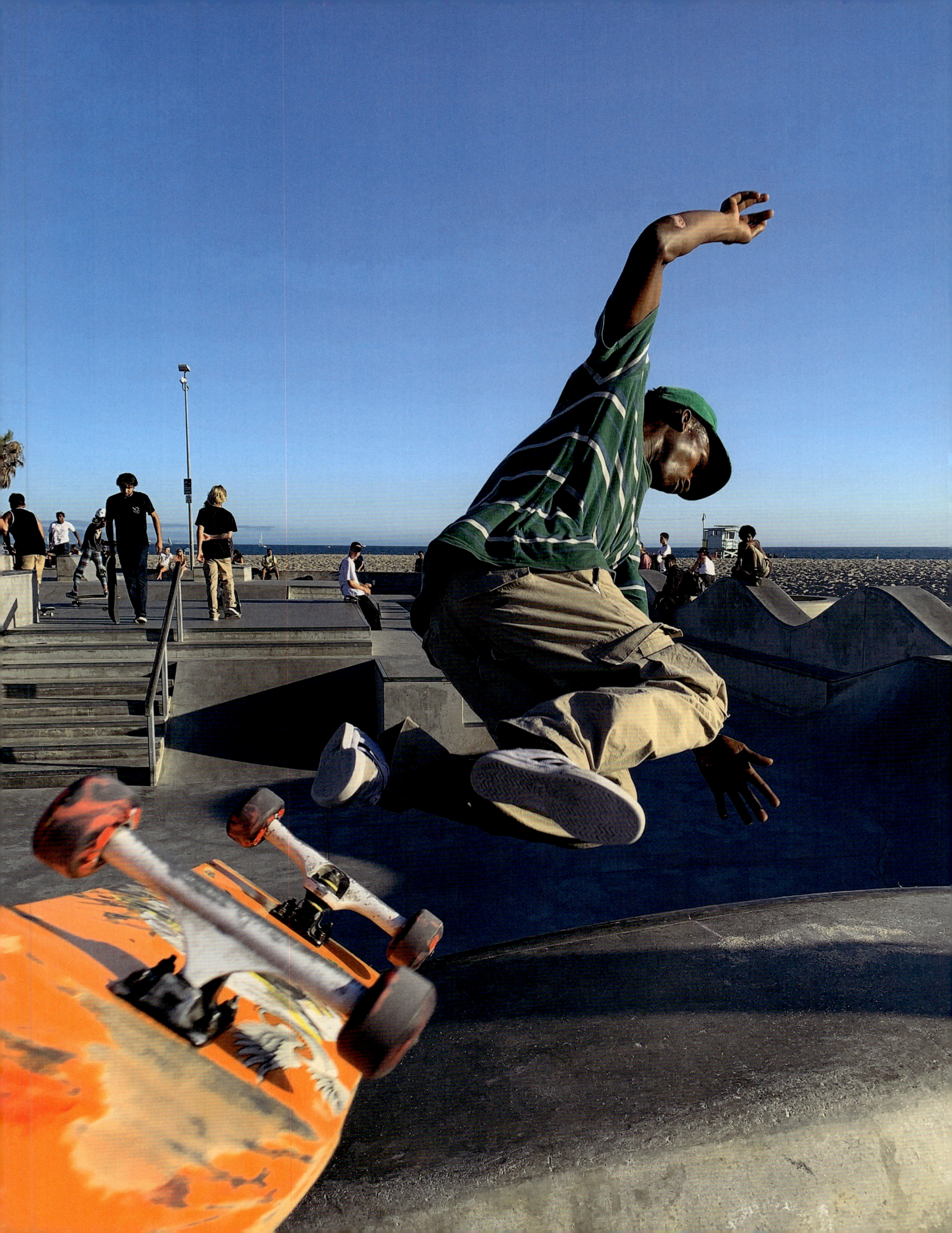

The place to be: Venice Beach, Los Angeles, USA.

Der Skate Hotspot schlechthin: Venice Beach, Los Angeles, USA.

PEOPLE

MENSCHEN

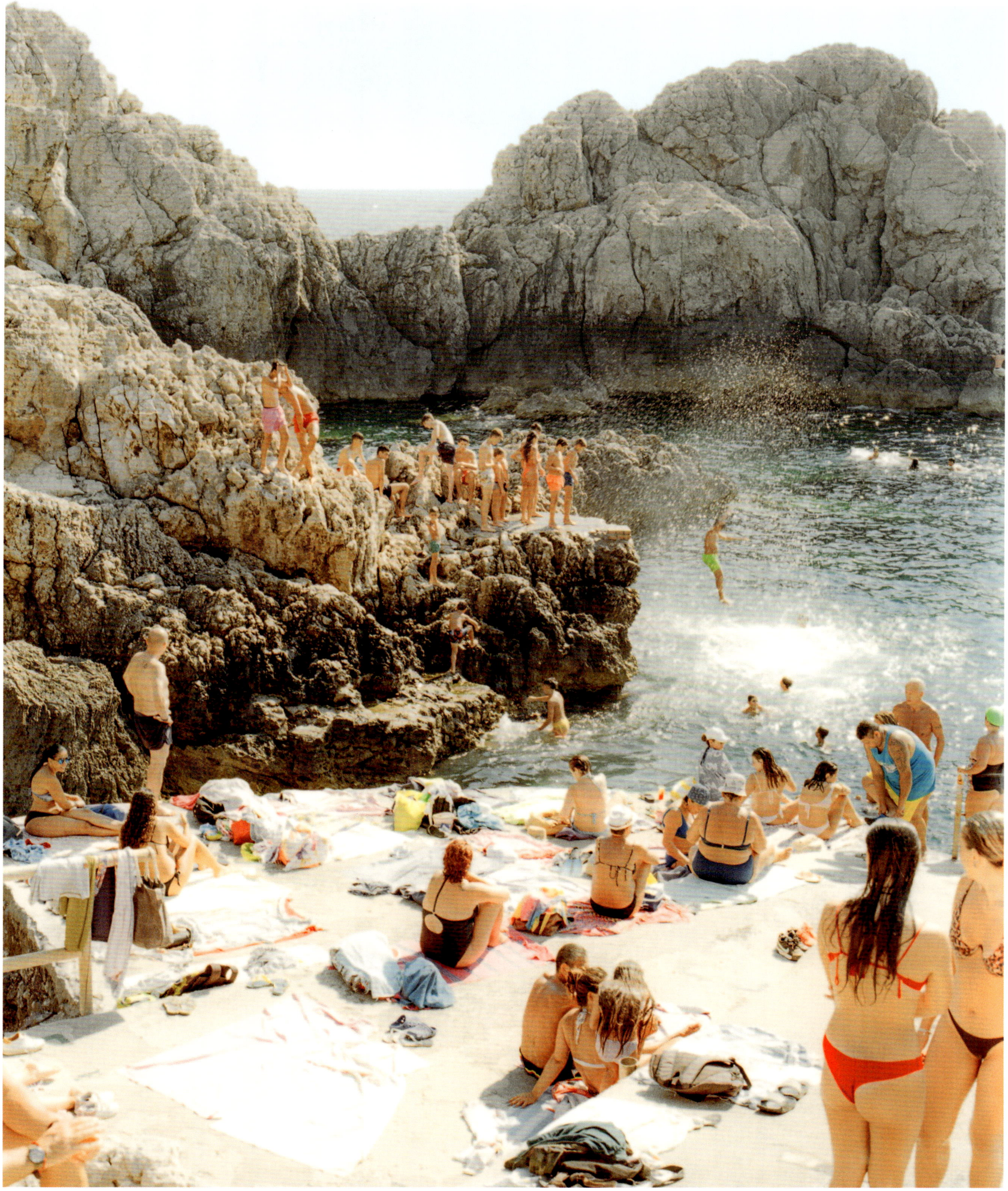

For centuries artists, intellectuals, and the nobility have been traveling to Italy, and since the 16th century it was a mandatory part of the agenda for young Englishmen from distinguished homes to take the "Grand Tour"–an extensive trip through Europe that focused primarily on Italy. Soon the custom caught on throughout Europe, and in the 17th and 18th centuries wealthy middle-class people also undertook this educational and recreational trip to the south. Goethe's trip to Italy lasted almost two years (from September 1786 to May 1788), and from the 19th century on, veritable colonies of artists evolved. Rainer Maria Rilke, Claude Debussy, Maxim Gorky, Henry James, Oscar Wilde, and Curzio Malaparte were regular visitors to Capri, as were the "savior of children" Emil von Behring, who discovered cures for tetanus and diphtheria, and the Swede Axel Munthe, who spent his life unable to decide whether he wanted to be a celebrity doctor, a psychiatrist, an animal rights activist, an impostor, or whether he in fact preferred to be a writer. Friedrich Alfred Krupp preferred Capri to his native German city of Essen, and who would have dared blame him? He even donated a famous hiking trail to the island for 25,000 lire, the Via Krupp, a narrow, spectacular serpentine path on the slopes of Monte Castiglione.
The Hamburg lawyer Friedrich Johann Lorenz Meyer wrote in 1783, "Of this rapid transition from a harsh to a mild climate (...) and of the impression of this sudden change of seasons on the soul, the most flourishing poet's imagination does not sketch an apt representation. It was the heart-rending picture of the transition from death to life."

Slim Aarons viewed it more pragmatically. "Life there was rewarding: the dollar was strong, the wine *dolce*, and the girls fantastic," recounts the U.S. photographer, who first came to Italy in 1948 on assignment from *Life* magazine to document Hollywood's rediscovery of the country–a rediscovery that culminated in the 1953 love story *Roman Holiday*, starring Gregory Peck and Audrey Hepburn. Soon, Aarons was discovering more than just Rome; he was particularly taken with the Amalfi Coast and eventually Capri. Today, with our ever-present images of Italy's dolce vita in our heads, it's hard to imagine how absolutely stunning Capri might have seemed to an American 70 years ago–or even to a native of Essen, Germany, like Krupp 150 years ago.

Seit Jahrhunderten reisen Künstler, Intellektuelle und Adlige nach Italien, und seit dem 16. Jahrhundert gehörte es zum Pflichtprogramm junger Engländer aus vornehmem Haus, die „Grand Tour" zu unternehmen – eine ausführliche Europareise mit Schwerpunkt Italien. Bald setzte sich die Mode in ganz Europa durch, und im 17. und 18. Jahrhundert unternahmen auch wohlhabende Bürgerliche die Bildungs- und Genussreise in den Süden. Goethes Italienreise dauerte fast zwei Jahre (von September 1786 bis Mai 1788), und ab dem 19. Jahrhundert entwickelten sich regelrechte Künstlerkolonien. Nach Capri kamen Rainer Maria Rilke, Claude Debussy, Maxim Gorki, Henry James, Oscar Wilde und Curzio Malaparte regelmäßig, aber auch der »Kinderretter« Emil von Behring, der Heilmittel für Tetanus und Diphterie entdeckte, und der Schwede Axel Munthe, der sich Zeit seines Lebens nie entscheiden konnte, ob er Promiarzt, Psychiater, Tierschützer, Hochstapler oder doch lieber Schriftsteller sein wollte.

Friedrich Alfred Krupp zog Capri seiner Geburtsstadt Essen vor, und wer hätte es ihm zu verdenken gewagt? Er stiftete der Insel sogar für 25.000 Lire einen berühmten Wanderweg, die Via Krupp, einen schmalen, spektakulären Serpentinenpfad am Hang des Monte Castiglione.
Der Hamburger Jurist Friedrich Johann Lorenz Meyer schrieb im Jahr 1783: „Von diesem schnellen Übergang aus einem rauen in ein mildes Klima (...) und von dem Eindruck dieses plötzlichen Wechsels der Jahreszeiten auf die Seele, entwirft die blühendste Dichterphantasie keine treffende Darstellung. Es war das herzerbebende Bild des Übergangs vom Tod zum Leben."
Slim Aarons sah es pragmatischer: „Das Leben war gut: Der Dollar war stark, der Wein war *dolce* und die Girls waren fantastisch", erzählt der US-Fotograf, der 1948 zum ersten Mal im Auftrag von *Life* nach Italien kam, um die Wiederentdeckung des Landes durch Hollywood zu dokumentieren – eine Wiederentdeckung, die 1953 in dem Liebesfilm *Ein Herz und eine Krone* mit Gregory Peck und Audrey Hepburn kulminierte. Bald entdeckte auch Aarons mehr als nur Rom; besonders die Amalfiküste und schließlich Capri hatten es ihm angetan. Heute, mit unseren ständig verfügbaren Bildern von Italiens Dolce Vita im Kopf, ist es kaum noch vorstellbar, wie absolut umwerfend Capri auf einen Amerikaner vor 70 Jahren gewirkt haben mag – oder gar vor 150 Jahren auf einen gebürtigen Essener wie Krupp.

PREVIOUS PAGE

A lively scenery on a rocky beach in Lido del Faro on Capri, Italy during a hot summer day.

Belebte Szenerie vor felsiger Strandkulisse in Lido del Faro auf Capri, Italien.

Parasols in the sun: colorful beach life at Marina Grande on Capri, Italy.

Aufgespannt und eingetaucht: Fröhlich-buntes Strandleben in Marina Grande auf Capri, Italien.

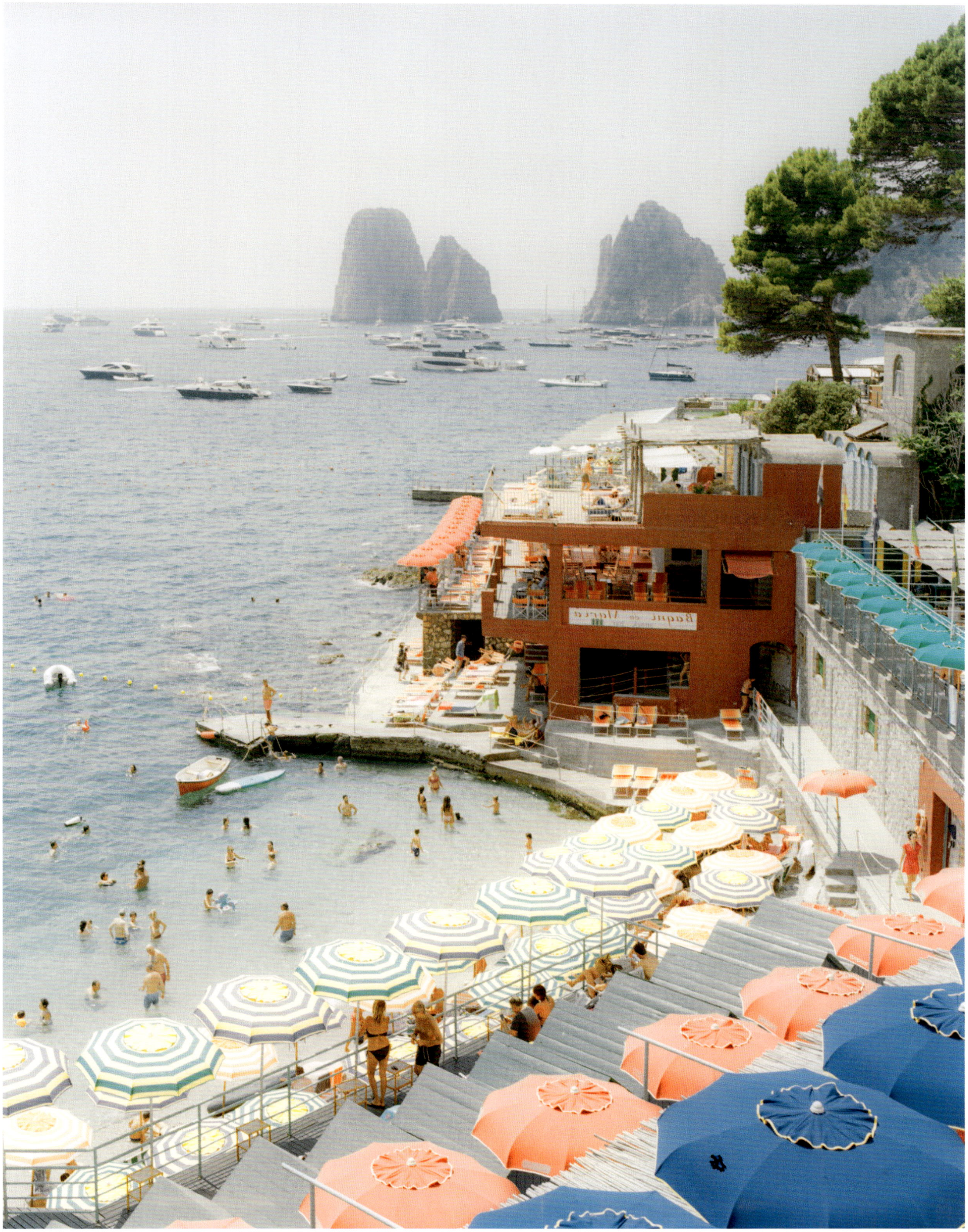

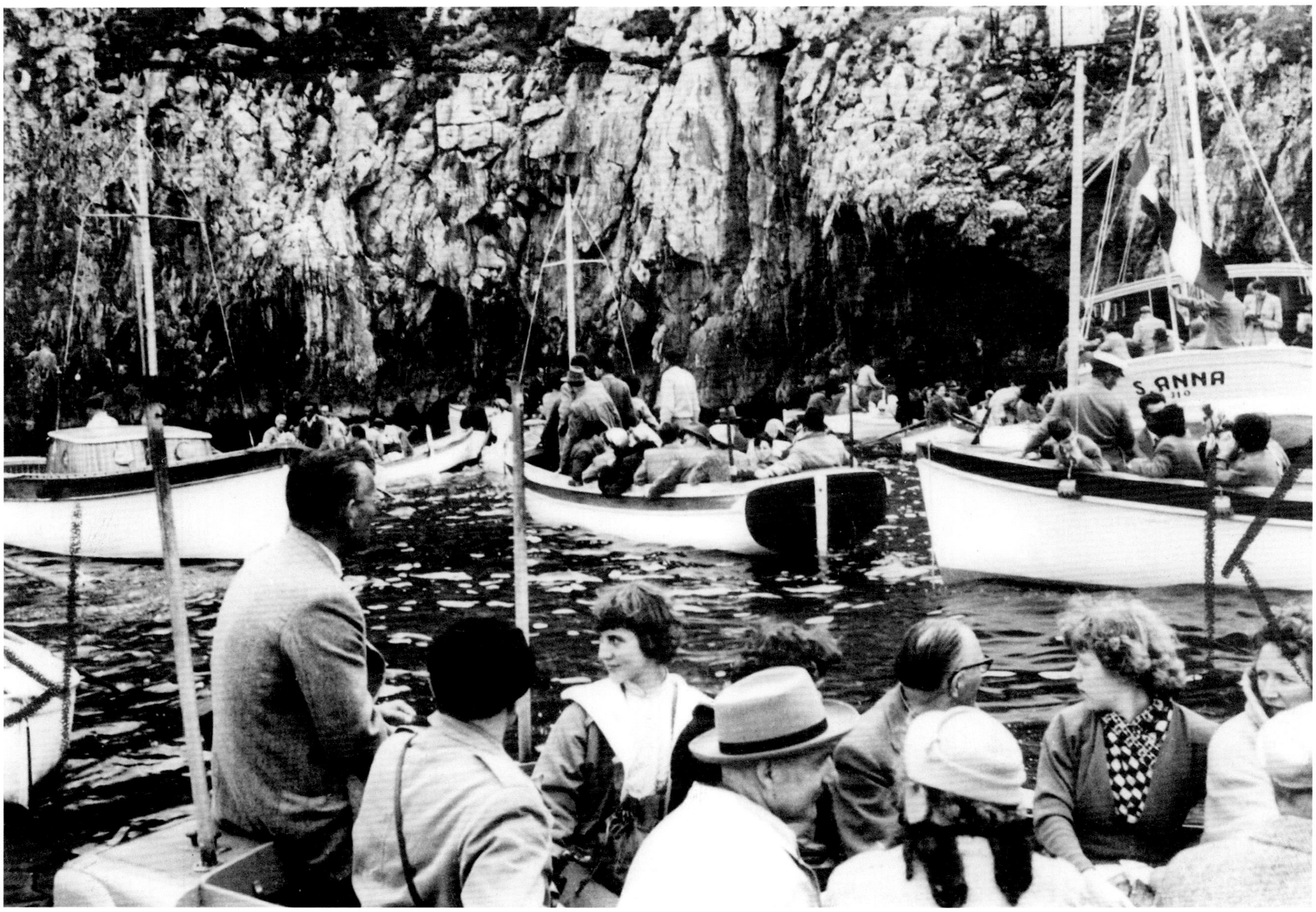

Tourist boats outside the Grotta Azzurra, Capri, Italy 1956.

Andrang von Touristenbooten vor der Grotta Azzurra, Capri, Italien 1956.

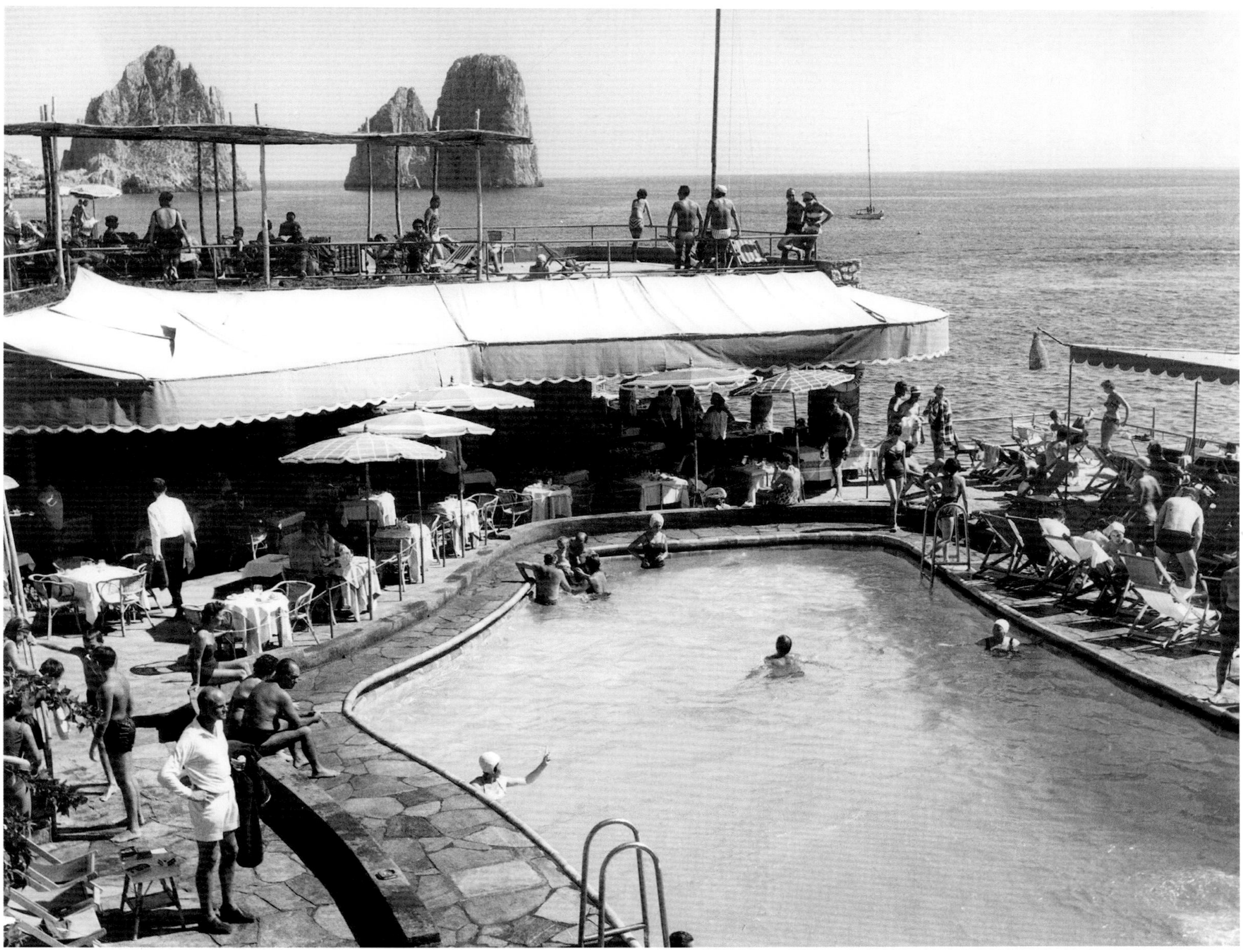

View of the luxury spa Canzone del mare and the Faraglioni rocks. Capri, Italy 1967.

Blick auf das Luxusbad Canzone del mare und die Faraglioni-Felsen. Capri, Italien 1967.

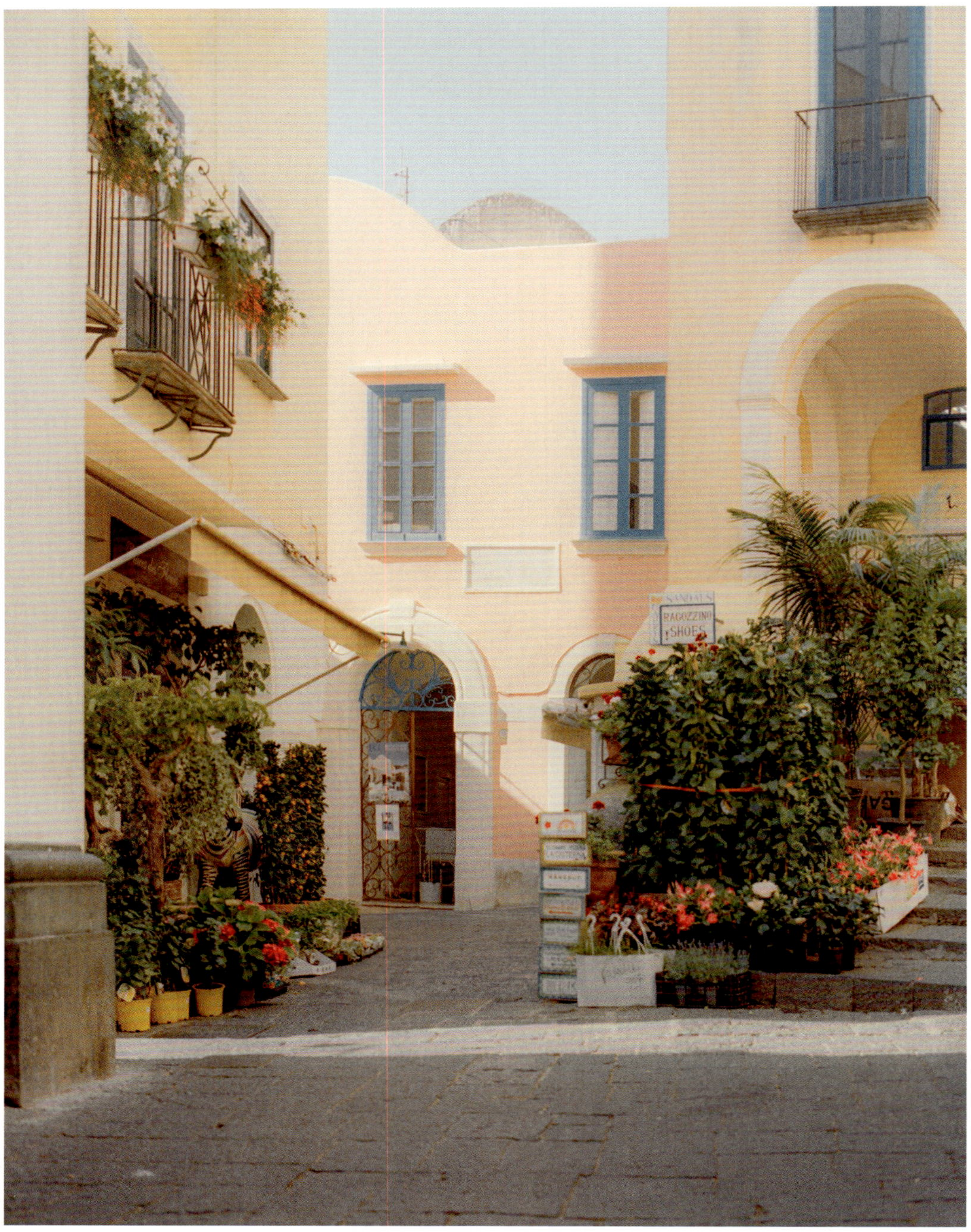

Street scenery with pastel-colored houses on Capri, Italy.

Straßenszenerie vor pastellfarbenen Häuserfassaden auf Capri, Italien.

90

RIGHT PAGE

Faraglioni, Capri, Italy: The sparkling summer sun on the clear water makes the Mediterranean idyll perfect.

Faraglioni, Capri, Italien: Die funkelnde Sommersonne auf dem klaren Wasser macht die mediterrane Idylle perfekt.

NEXT PAGE

Motorboat in front of the Faraglioni rocks, Capri, Italy.

Motorboot vor den Faraglioni-Felsen, Capri, Italien.

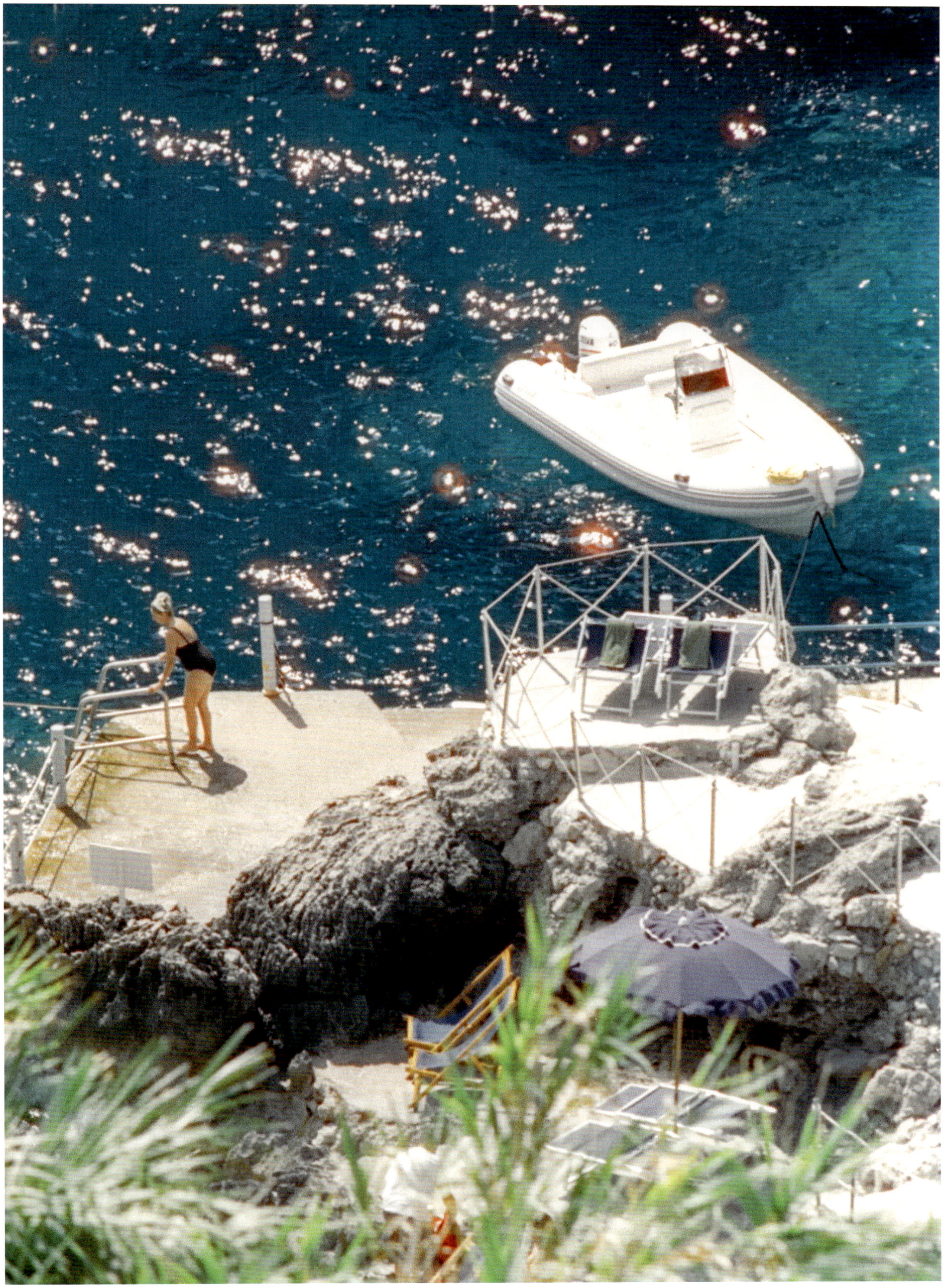

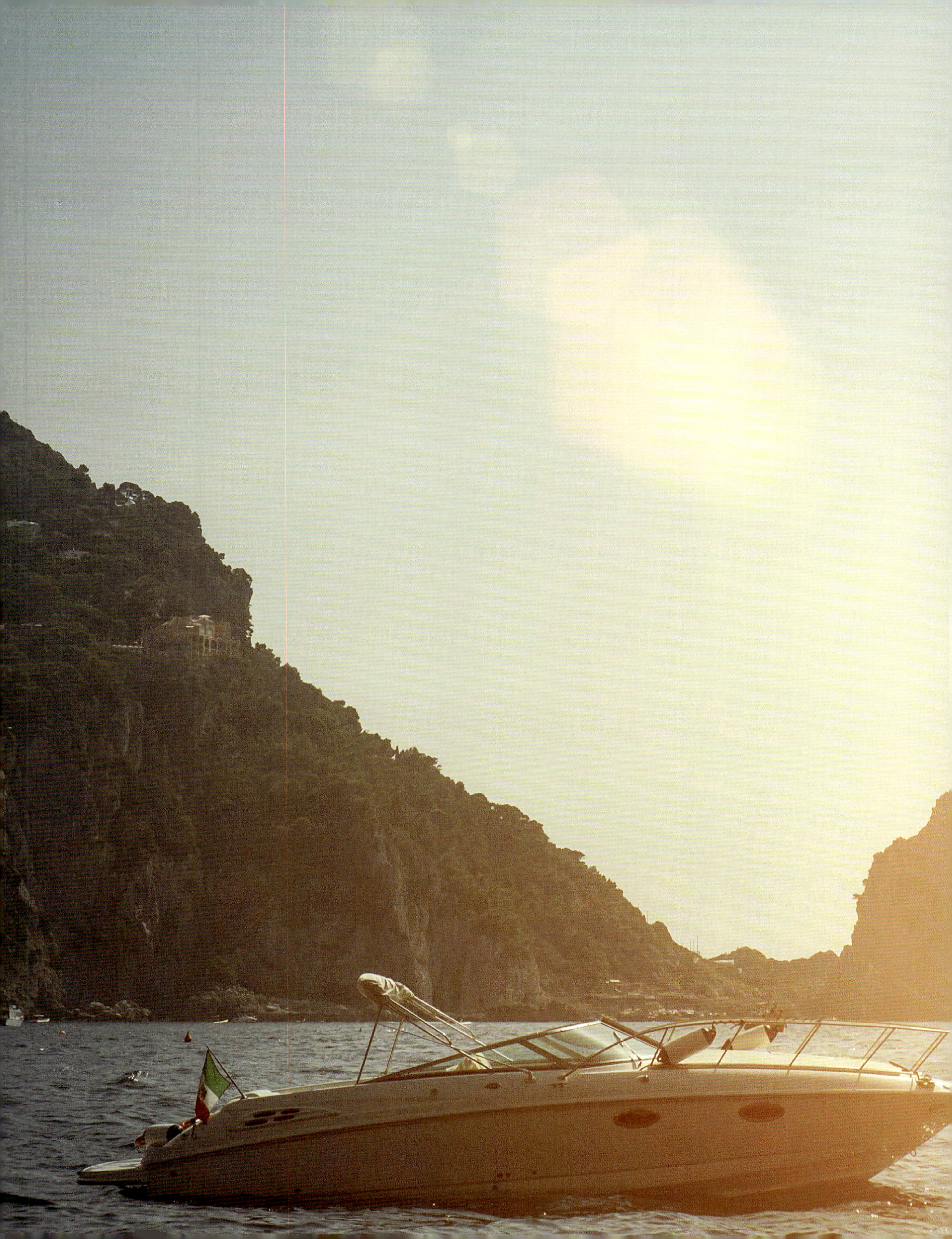

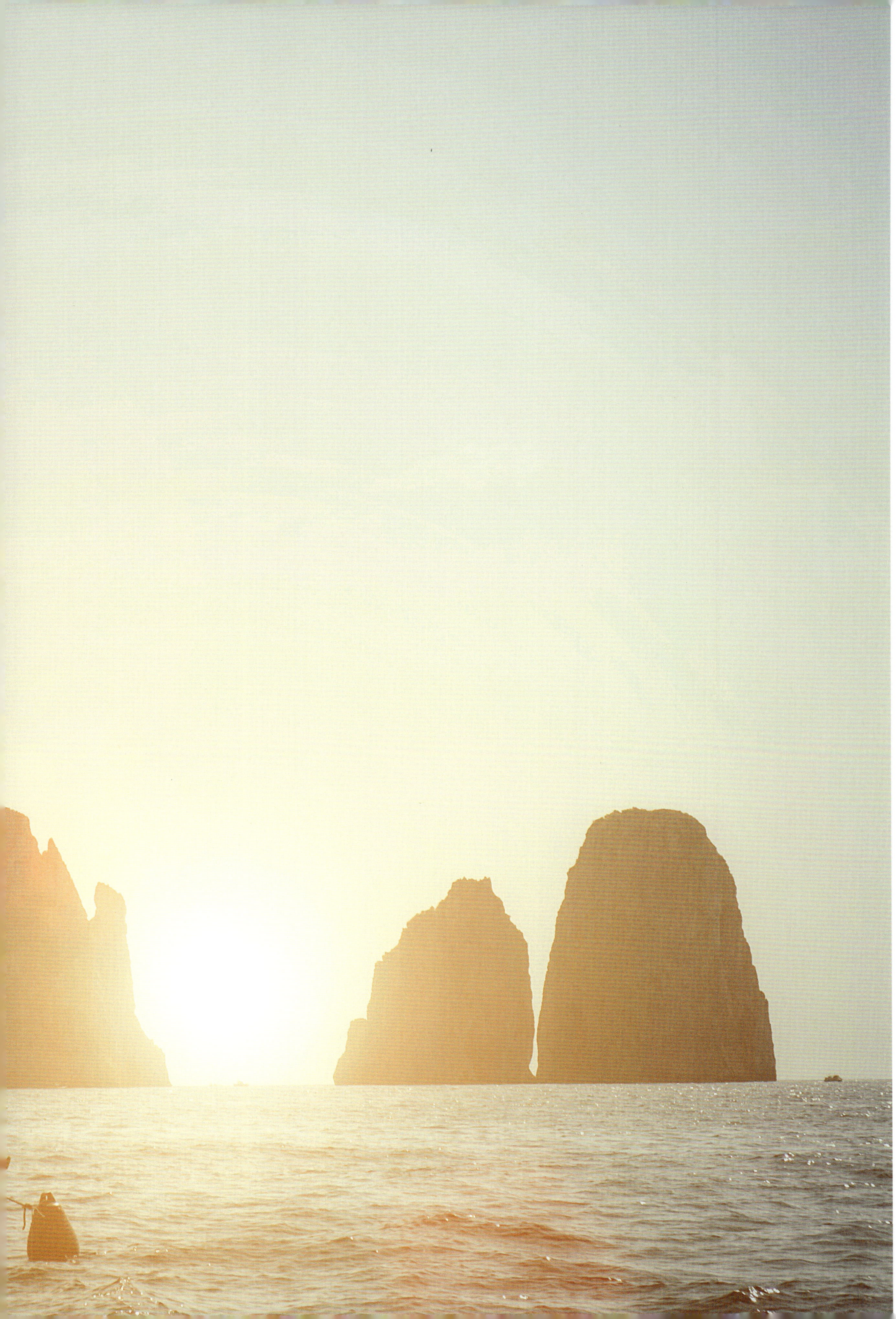

Hardly any writer was more closely linked to beach life than Thomas Mann–which surely has something to do with his home town of Lübeck, one of the most important Hanseatic cities in the history of Baltic trade. For his literary characters Hanno Buddenbrook and Tonio Kröger, the beach was an escape from crushing reality, and Mann himself also enjoyed writing copiously while ensconced a hooded wicker beach chair, whether on the Curonian Spit, in Forte dei Marmi, or later in exile in California.

At the Tyrrhenian Sea, the Mann family learned all the magnificent and quirky aspects of Mediterranean beach culture. First, they had to change hotels because an Italian noblewoman complained about the supposedly contagious coughs coming from Mann's children, then there was some trouble at the beach because his children were running around naked–still very much frowned upon in Italy today. The summer experiences of the North German family on vacation are vividly portrayed in his novella *Mario and the Magician*, although in real life the magician survived his feats.

In *The Magic Mountain*, Mann raves about walking barefoot "on the smooth, firm, water-lapped and springy sandy land at the edge of the sea," and in *Death in Venice*, the writer speaks of "carefree enjoyment of culture at the edge of the element." Tonio Kroeger, lying in the sand, observes the "mysteriously changing facial expressions that whisk over the face of the sea," and Hanno Buddenbrook, who goes through hell at school, finds salvation in life at the beach: "This gentle and dreamy playing with the soft sand that does not stain, this effortless and painless wandering by the eyes, losing themselves in the green and blue infinity, from which, free and unhindered, came with a gentle whisper a strong, fresh, wild, and wonderfully fragrant breeze that engulfed the ears and caused a pleasant dizziness, a subdued stupor in which the consciousness of time and space and everything limited sank quietly and blissfully..."

Kaum ein Schriftsteller war enger mit dem Strandleben verknüpft als Thomas Mann – was sicher mit seiner Heimat Lübeck zu tun hat, einer der wichtigsten Hansestädte im Ostseehandel. Für seine literarischen Figuren Hanno Buddenbrook und Tonio Kröger war der Strand eine Flucht vor der erdrückenden Realität, und auch Mann selbst genoss ausgiebig das Schreiben im Strandkorb, ob an der Kurischen Nehrung, in Forte dei Marmi oder später im kalifornischen Exil.

Am Tyrrhenischen Meer lernten die Manns die mediterrane Strandkultur in allen großartigen und eigenwilligen Seiten kennen. Zunächst mussten sie das Hotel wechseln, weil sich eine italienische Adlige über vermeintlich ansteckenden Husten von Manns Kindern beschwerte, dann gab es Ärger am Strand, weil die Kinder nackt herumliefen – das wird in Italien bis heute sehr ungern gesehen. In der Novelle *Mario und der Zauberer* sind die Sommererlebnisse der norddeutschen Urlauberfamilie eindrücklich geschildert, wenngleich im wahren Leben der Zauberer seine Kunststücke überlebte.

Mann schwärmt im *Zauberberg* vom Barfußlaufen „auf dem glatten, festen, gespülten und federnden Sandboden am Saume des Meeres", und im *Tod in Venedig* spricht der Schriftsteller von „sorglos genießender Kultur am Rande des Elements". Tonio Kröger beobachtet, im Sand liegend, die „geheimnisvoll wechselnden Mienenspiele, die über des Meeres Antlitz huschen", und Hanno Buddenbrook, der in der Schule durch die Hölle geht, empfindet das Strandleben als Erlösung: „Dieses zärtliche und träumerische Spielen mit dem weichen Sande, der nicht beschmutzt, dieses mühe- und schmerzlose Schweifen und Sichverlieren der Augen über die grüne und blaue Unendlichkeit hin, von welcher, frei und ohne Hindernis, mit sanftem Sausen ein starker, frisch, wild und herrlich duftender Hauch daherkam, der die Ohren umhüllte und einen angenehmen Schwindel hervorrief, eine gedämpfte Betäubung, in der das Bewusstsein von Zeit und Raum und allem Begrenzten still selig unterging ..."

Thomas Mann reading in the shade of a beach chair. Nidden, Curonian Spit, Lithuania 1932.

Thomas Mann bei der Lektüre im Schatten eines Strandkorbs. Nidden, Kurische Nehrung, Litauen 1932.

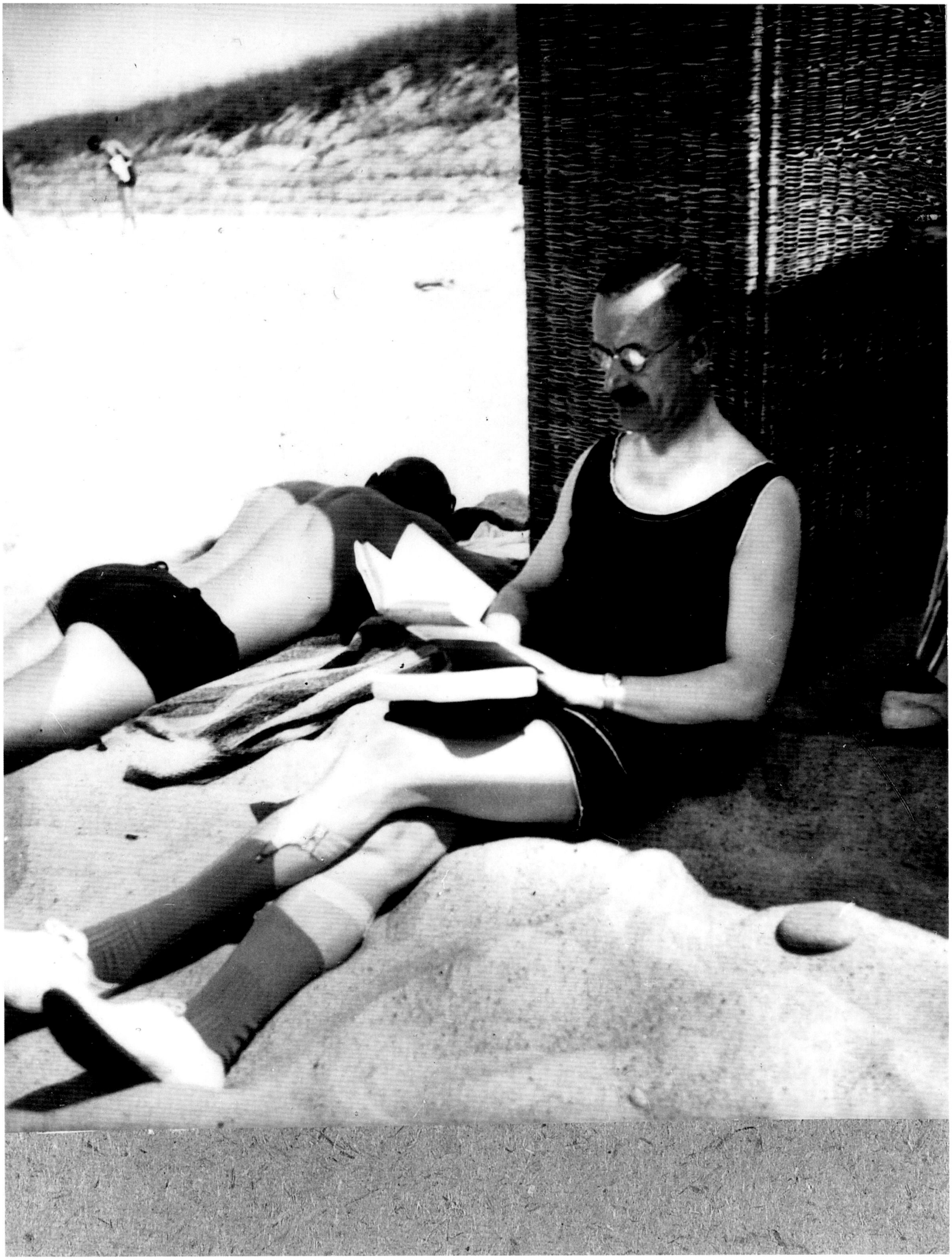

DER LITERAT AM MEER: THOMAS MANN

PEOPLE

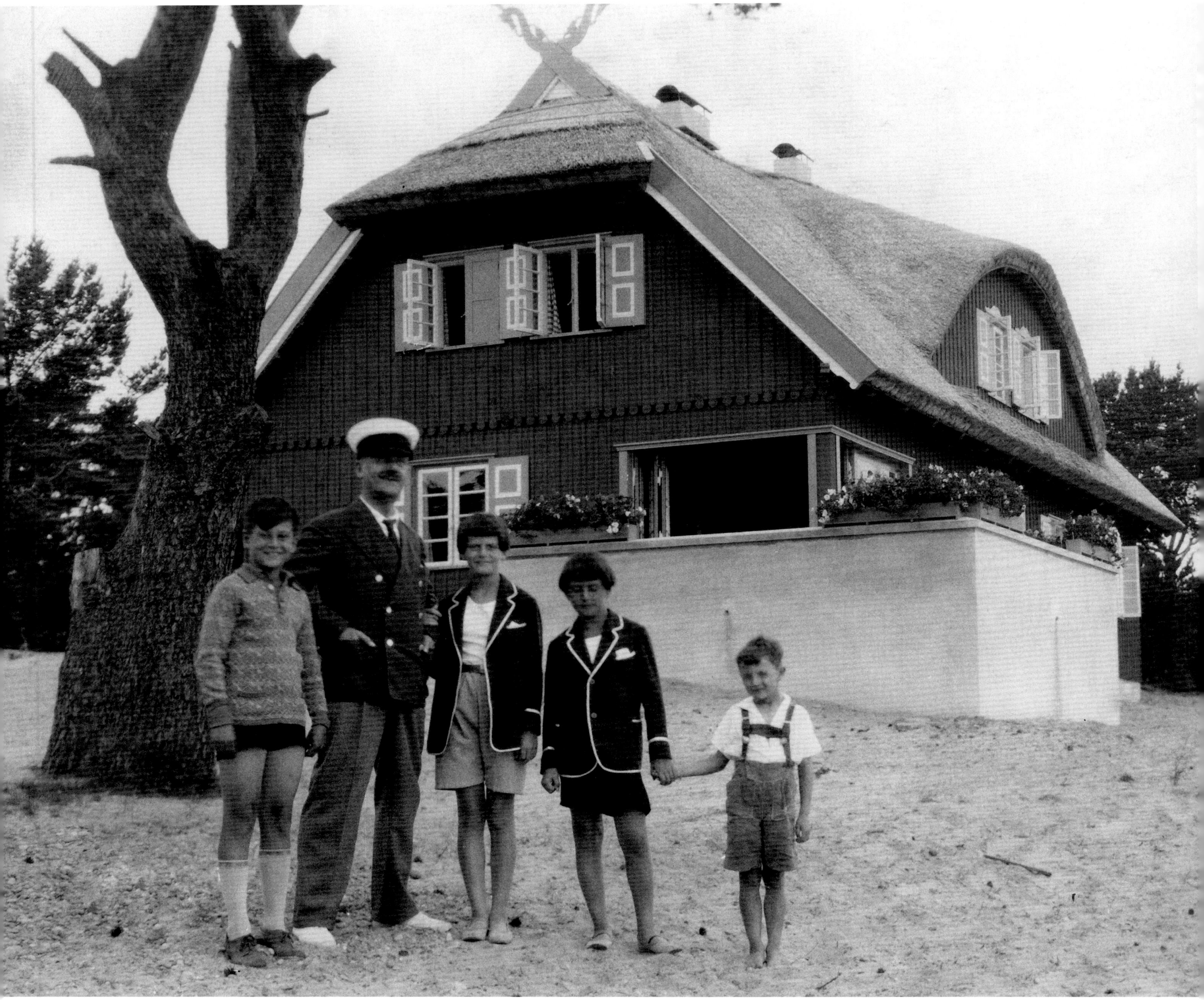

MENSCHEN

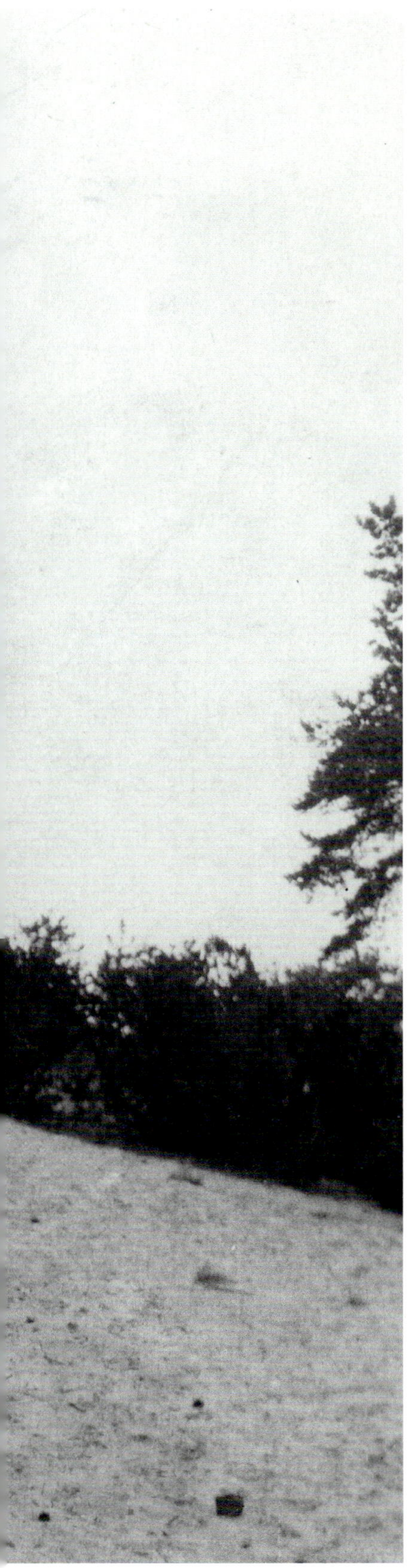

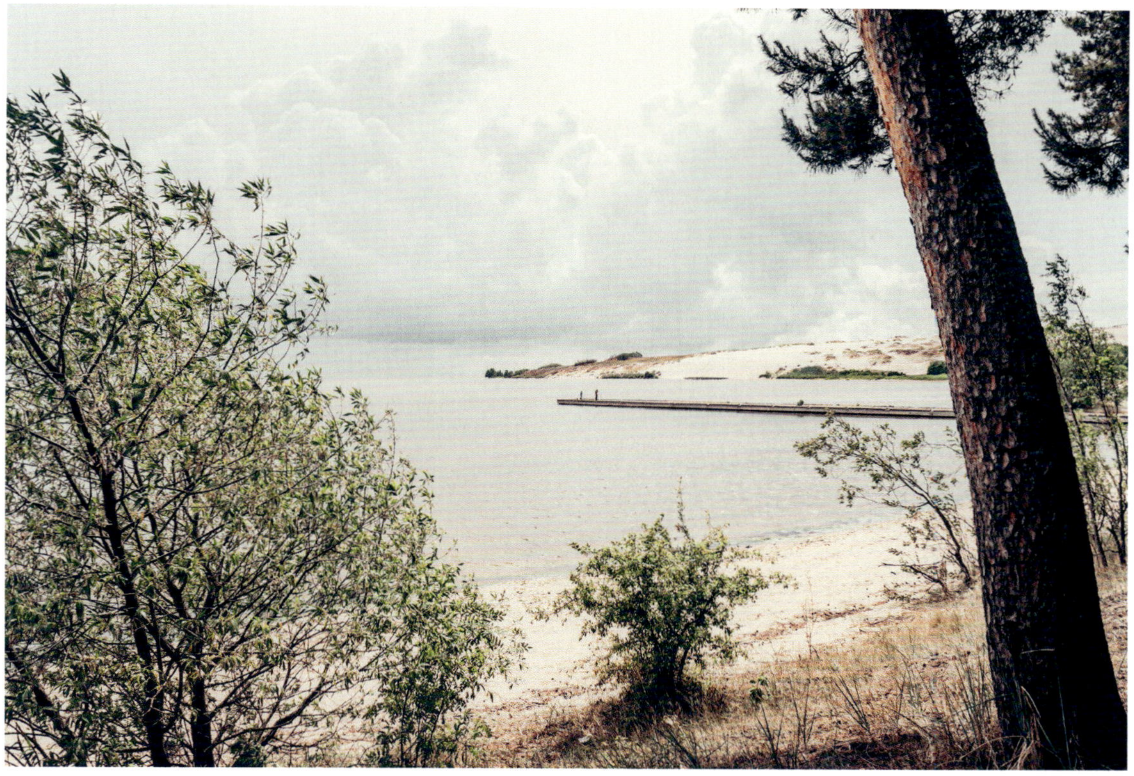

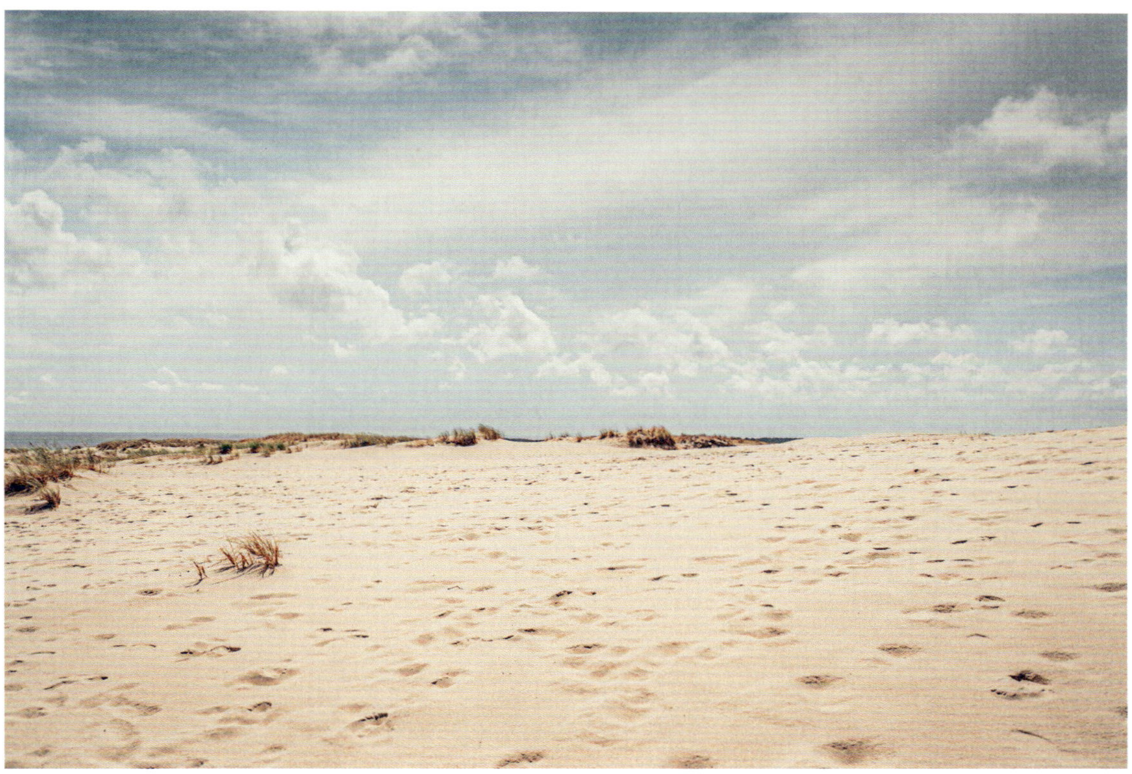

Thomas Mann with his children Elisabeth and Michael and two boys (l. and r.) on the Schwiegermutterberg in Nidden, Curonian Spit, Lithuania around 1930.

Thomas Mann mit seinen Kindern Elisabeth und Michael sowie zwei Jungen (l. und r.) auf dem Schwiegermutterberg in Nidden, Kurische Nehrung, Litauen um 1930.

Beach pines and dunes on the lagoon side of the Curonian Spit.

Strandkiefern und Dünen auf der Haffseite der Kurischen Nehrung.

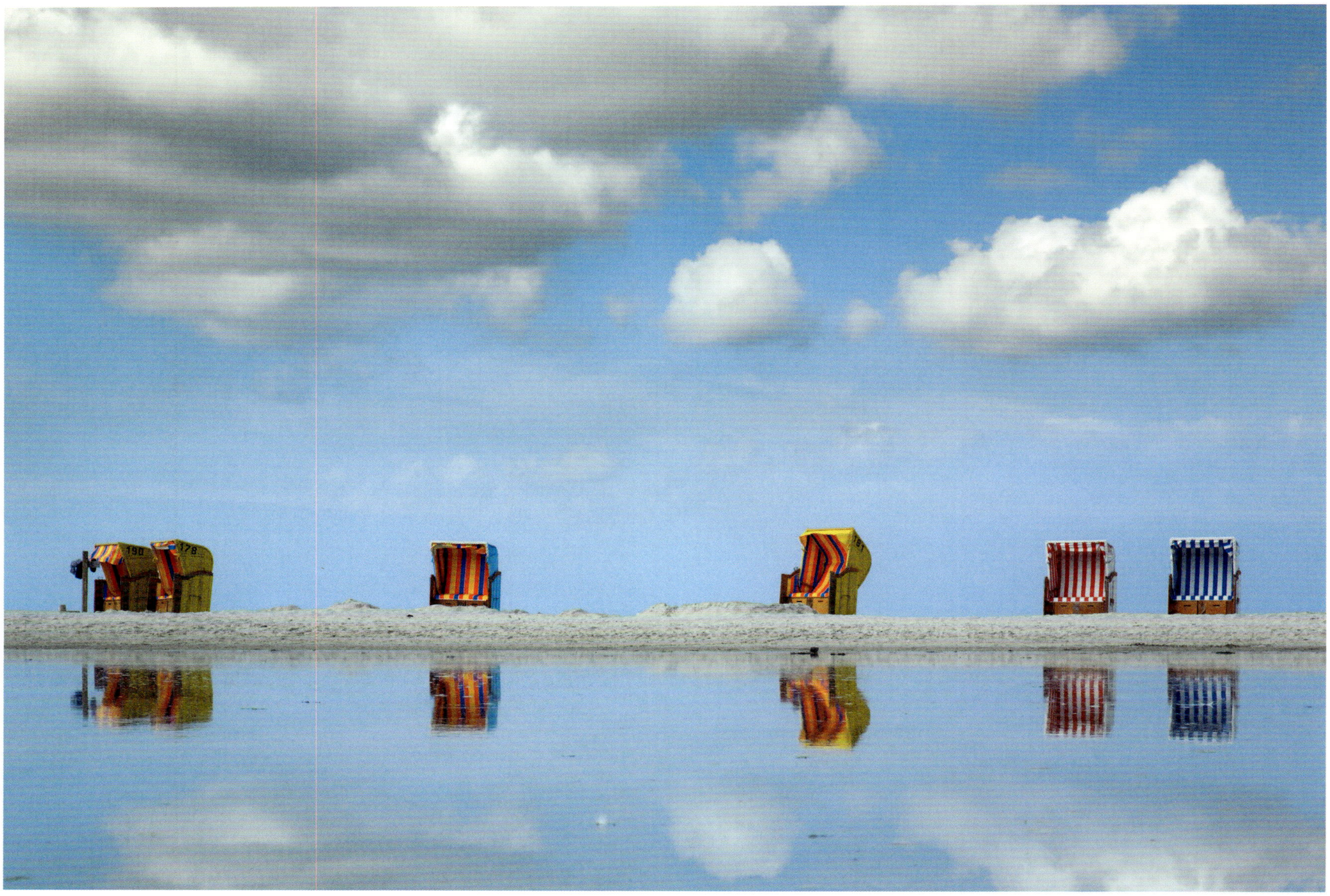

Beach chair reflections. Norddorf, Amrum, Germany.

Strandkorbreflexionen. Norddorf, Amrum, Deutschland.

RIGHT PAGE

Seagulls flying over the North Sea. England, UK.

Möwen im Flug über die Nordsee. England, UK.

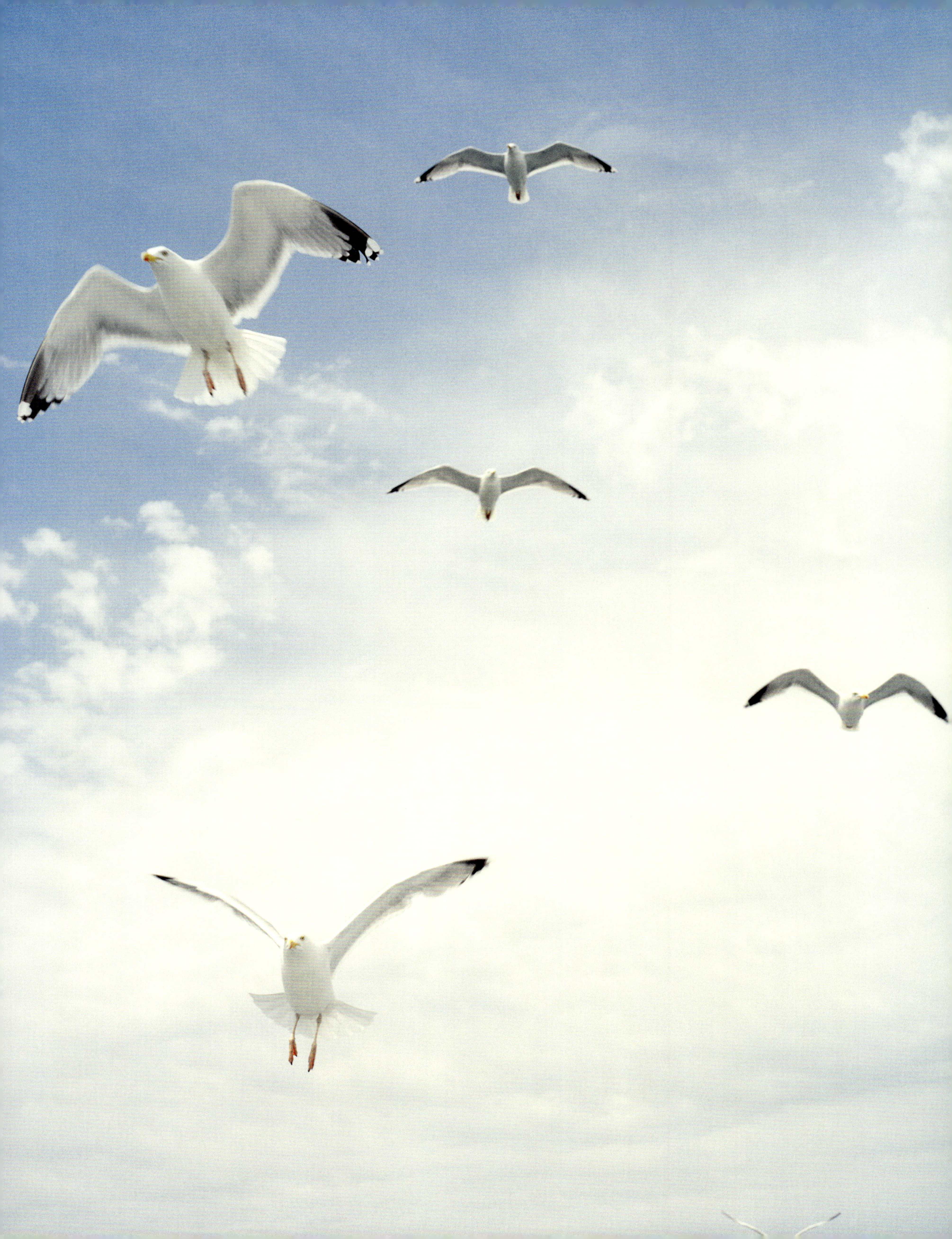

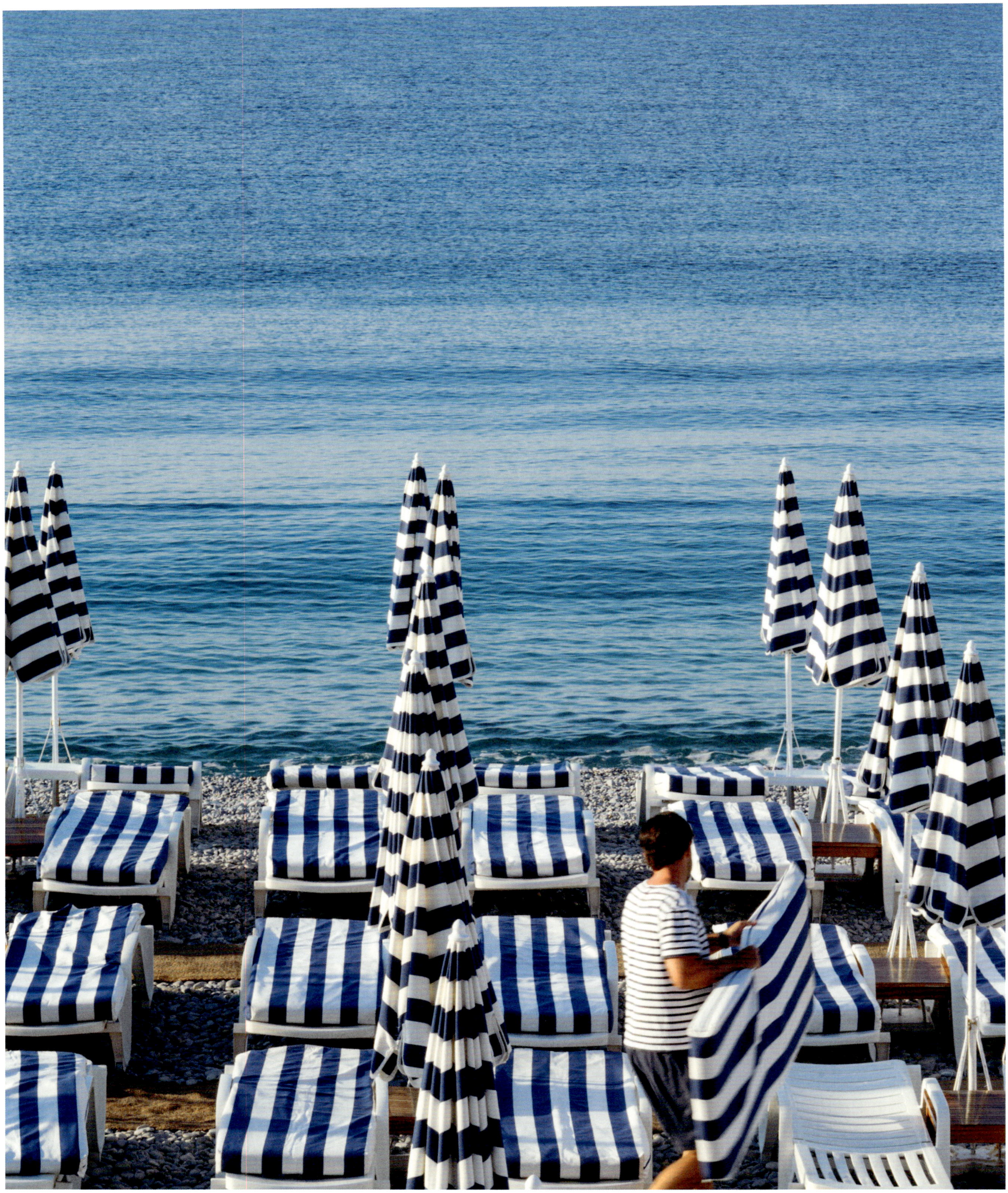

The *bagnino* is a Mediterranean archetype, combining many of the things that seem typically southern to Central Europeans: Nonchalance. Aplomb. A deep sense of calm. The blue and white striped shirt and the cigarette in the corner of his mouth.
The *bagnino* can be found at the beach clubs of Ibiza and Mykonos and Ko Samui and Orlando. Yet Italy is his natural habitat, and the international *bagnino* has to measure up to him. The profession was invented and perfected on Italian beaches. There are estimates that 90 per cent of the accessible and swimmable coastal areas in Italy are occupied by beach clubs, making it virtually impossible to imagine an Italian beach without a *bagnino*.
The *bagnino* is not a lifeguard–they are called *salvataggi*–he is a vacation facilitator. He assigns the lounge chairs to the guests, opens the umbrella (which is often not as easy as it looks), helps the guests with air pumps for their swim rings and pink flamingos, gallantly offers smokers a light, and has a first aid kit at the ready for any minor ailments. Apart from that, he is considered a reliable weather god, and is asked about 50 times a day whether it will be as hot tomorrow as it is today.
Although his services are of a rather low level and do not require lengthy training, he is highly respected, because that is what makes the *bagnino* so special: He is an integral part of the summer. Not a warden, more like a friendly neighbor. In this respect, he is much more highly regarded than a simple service provider. He also exudes the competence of an honest tradesman carrying out minor repairs immediately and free of charge. A good *bagnino* is like a good pal on a part-time basis.

That is why it is so important to get on good terms with him right from day one. How do Italians do it? At the very beginning of the vacation, they slip the *bagnino* a small tip. From then on, special requests such as an extra lounge chair are taken care of with ease.

Der Bagnino ist ein mediterraner Archetyp, weil er vieles von dem vereint, was Mitteleuropäern typisch südlich vorkommt: die Lässigkeit. Die Souveränität. Die Grundruhe. Das blau-weiß geringelte Shirt und die Zigarette im Mundwinkel.
Den Bagnino gibt es in den Beachclubs auf Ibiza und Mykonos und Ko Samui und in Orlando. Dennoch ist Italien sein natürliches Habitat, und der internationale Bagnino muss sich an ihm messen lassen. An den italienischen Stränden wurde die Profession erfunden und perfektioniert. Es gibt Schätzungen, dass 90 Prozent der in Italien zugänglichen und schwimmbaren Küstengebiete von Beach Clubs eingenommen werden. Daher ist ein italienischer Strand ohne Bagnino beinahe nicht vorstellbar.
Der Bagnino ist kein Bademeister – die heißen Salvataggi –, sondern ein Urlaubsermöglicher. Er weist den Gästen die Liege zu, klappt den Sonnenschirm auf (was oft nicht so einfach ist wie es aussieht), hilft mit Luftpumpen für die Schwimmringe und rosa Flamingos aus, gibt den Raucherinnen galant Feuer und hat auch einen Erste-Hilfe-Koffer für kleinere Wehwehchen parat. Ansonsten gilt er als verlässlicher Wettergott und wird etwa 50 Mal am Tag gefragt, ob es morgen auch so heiß sein wird.
Obwohl seine Dienstleistungen eher niederen Grades sind und keine längere Ausbildung erfordern, ist er hochgeschätzt, denn das ist das Besondere am Bagnino: Er ist ein integrativer Teil des Sommers. Kein Blockwart, sondern eher ein freundlicher Nachbar. Insofern ist er viel höher angesehen als ein einfacher Dienstleister. Dazu verströmt er noch die Kompetenz eines ehrlichen Handwerkers, der kleinere Reparaturen sofort und unentgeltlich ausführt. Ein guter Bagnino ist wie ein guter Kumpel auf Teilzeit.
Deswegen ist es wichtig, sich mit ihm vom ersten Tag an gut zu stellen. Italiener machen das folgendermaßen: Gleich zu Beginn des Urlaubs wird dem Bagnino ein kleines Trinkgeld zugesteckt. Sonderwünsche wie eine Extraliege werden fortan geschmeidig erledigt.

The *bagnino*, preparing the day at the beach. Riviera, Italy.

Der Bagnino bereitet den Strandtag vor. Riviera, Italien.

NEXT PAGE

The work is done: Everything is ready for the beach visitors.

Nach getaner Arbeit: Alles bereit für die Strandbesucher. Riviera, Italien.

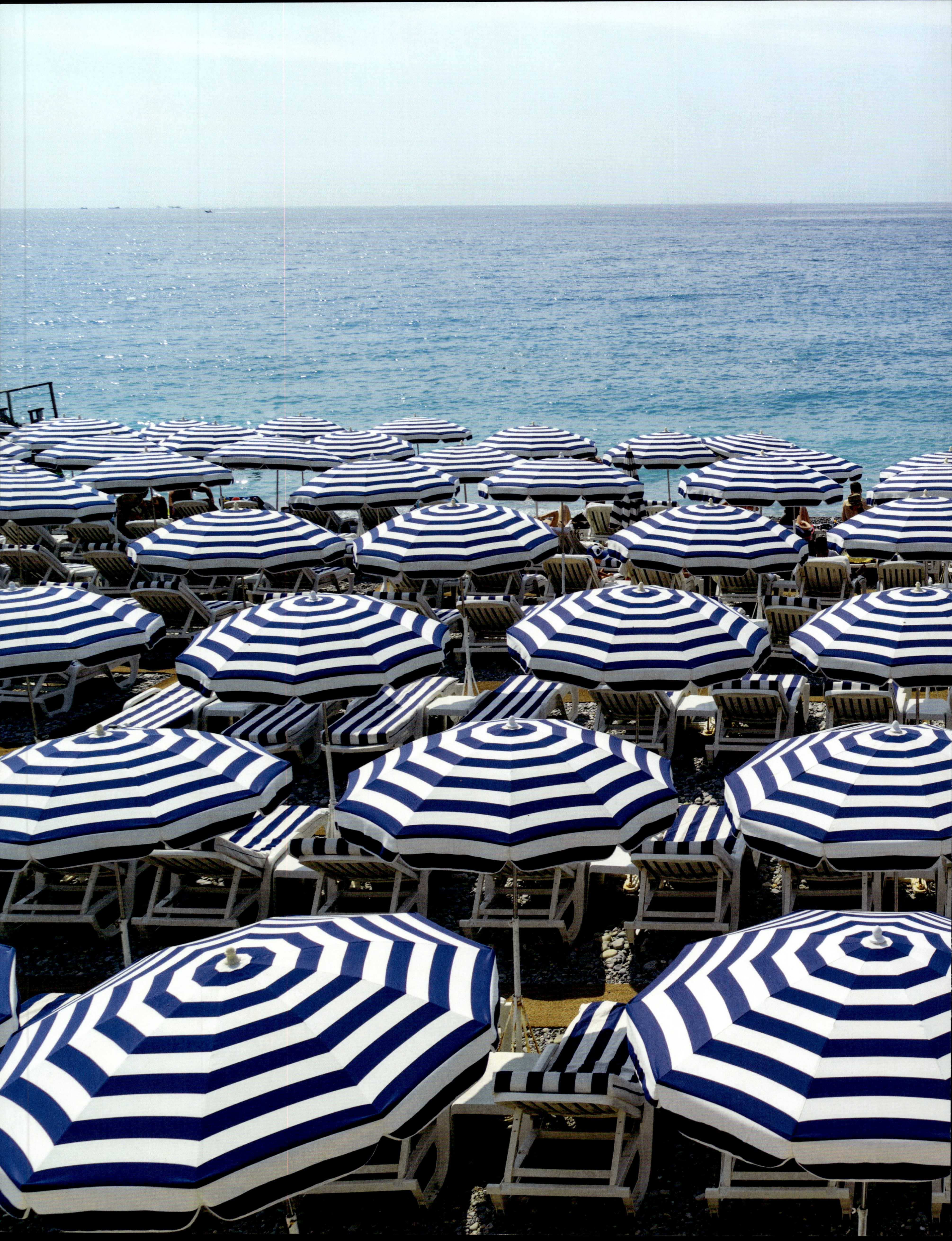

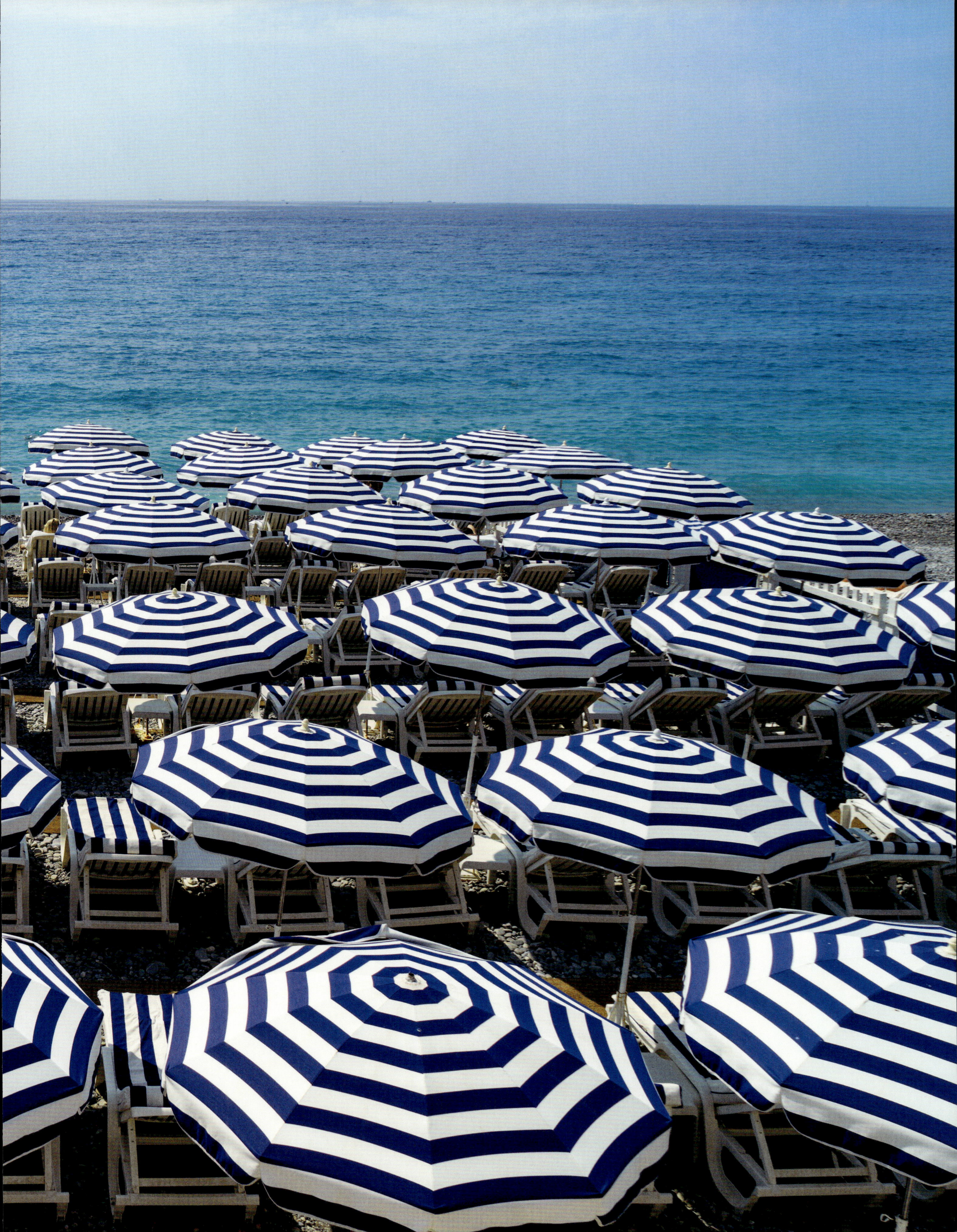

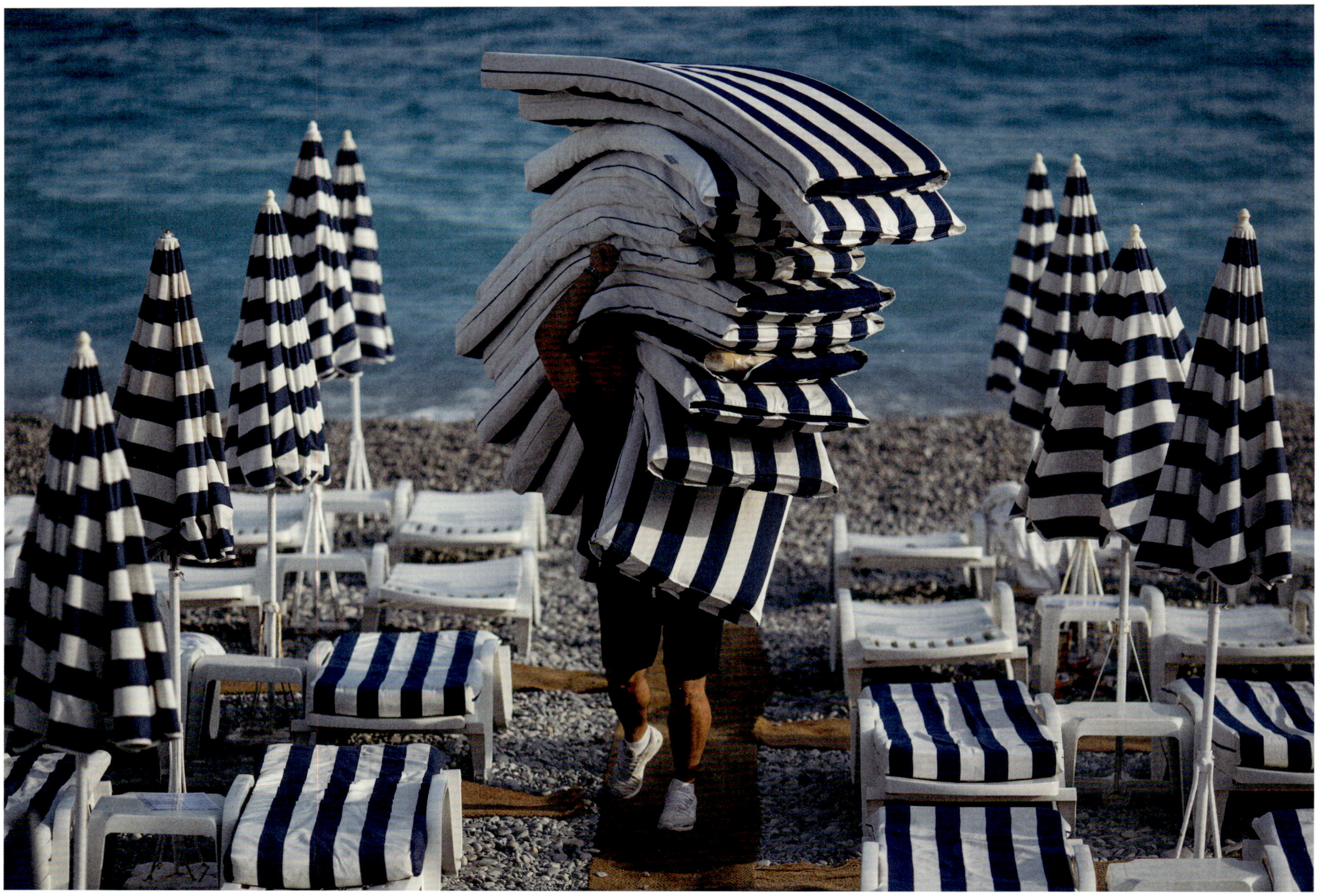

The end of the day. Riviera, Italy.

Das Ende des Strandtags. Riviera, Italien.

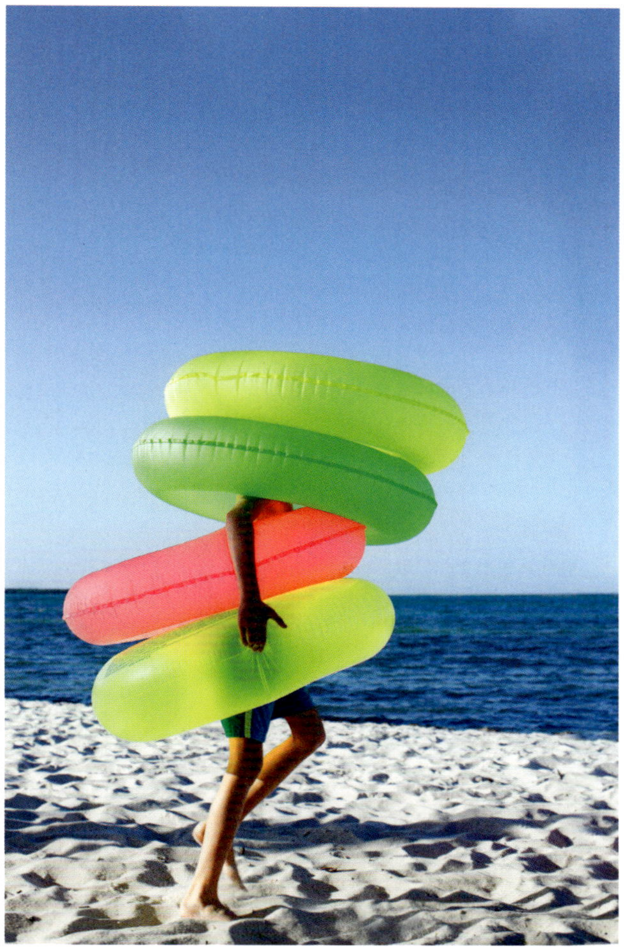

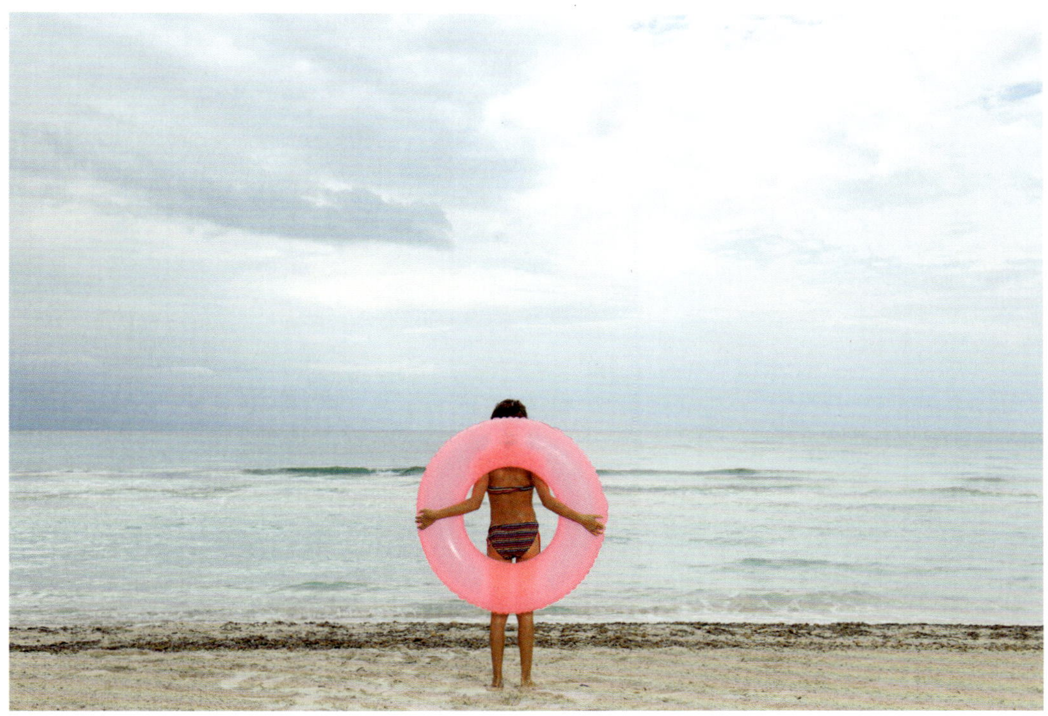

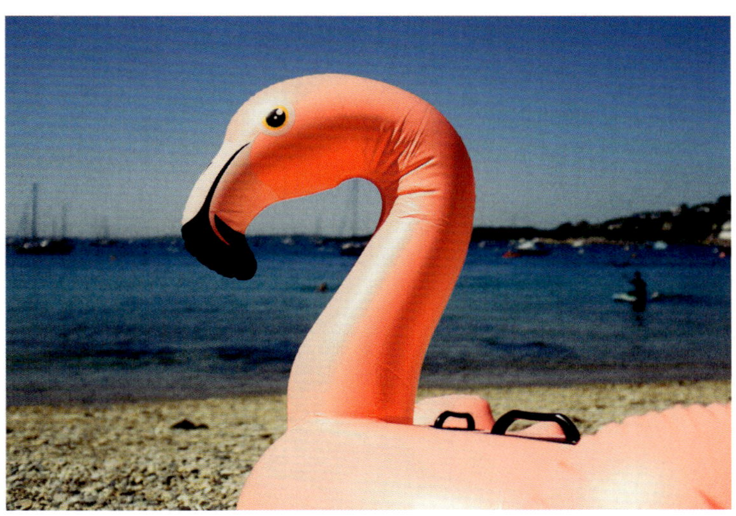

Inflatables on the beach.

Aufblasbare Schwimmreifen am Strand.

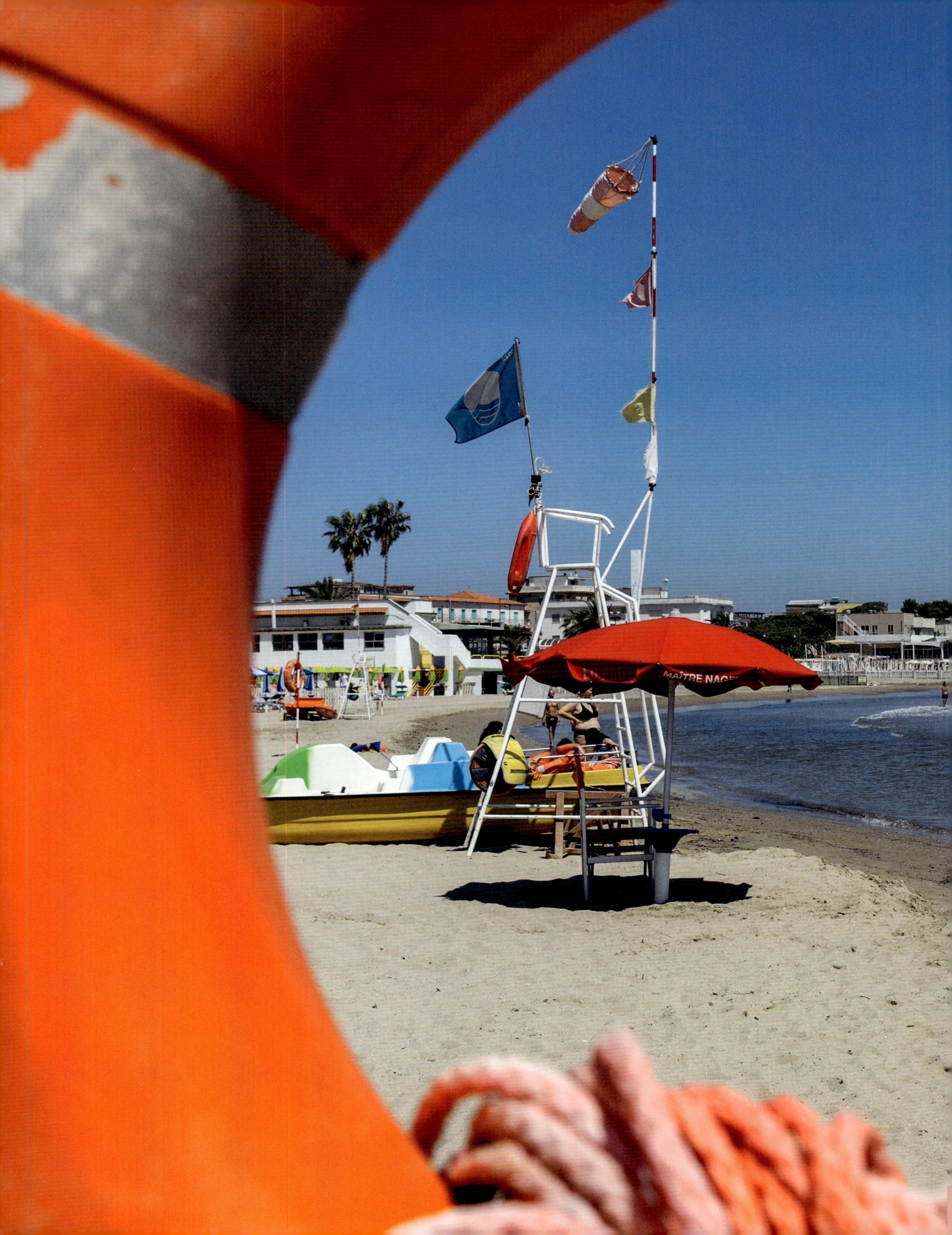
MAITRE NAGE

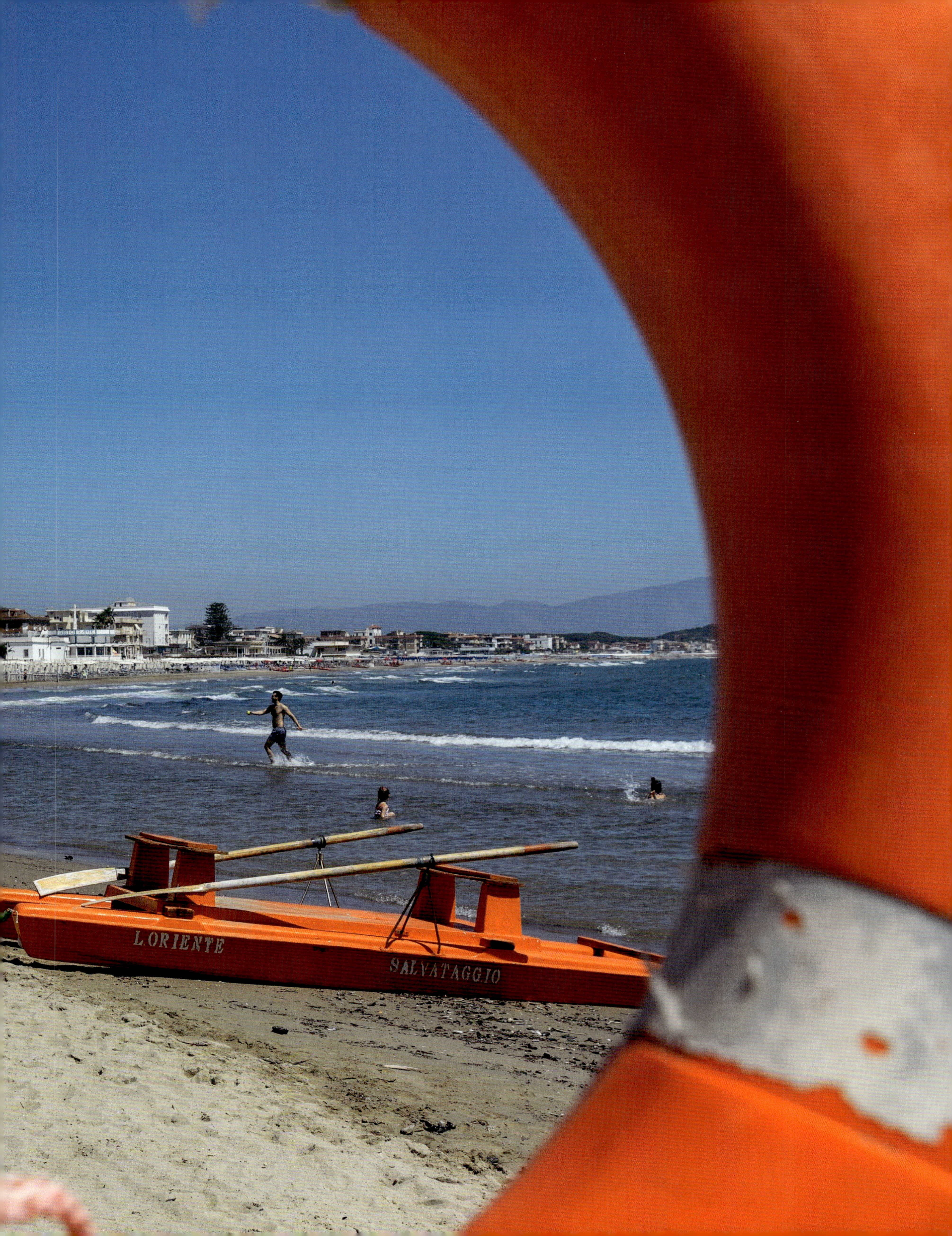
LORIENTE
SALVATAGGIO

David Hasselhoff is not only known for bringing down the Berlin Wall with his singing–he was also the leading actor (and even producer) of the most successful TV series of all time. *Baywatch* ran in 146 countries and was watched by a billion people a week; the actors used 40 tubes of sunscreen each episode. 15-year-old Leonardo DiCaprio was a heartbeat away from being cast as David Hasselhoff's son–who knows how his acting career would have developed after that? The cast included Jason Momoa and surfing world champion Kelly Slater. And let's not forget Pamela Anderson, who regularly stole the show from the intended leading man.
The world learned that when Americans have something important to say, they always look pensively out to the ocean first. The series was initially a flop; the producers were only convinced to continue after the first season by the good ratings in Great Britain and Germany–two countries that are very, very far removed climate-wise from the year-round Californian beach existence. David Hasselhoff gave up a big chunk of his salary to do it. And the iconic beach runs toward danger in slow motion? It is said they were inspired by the 1988 Olympics, when the 100-meter sprinters were shown in super slow motion for the first time, but David Hasselhoff later explained it was purely a matter of money: The slow-motion sequences saved expensive filming time.
The show's success? It wasn't the storylines, like the episode where an octopus steals surfboards, hiding them in its underwater cave until the lifeguards catch on. It was probably the women in bathing suits and the men with six-packs that generated record ratings. The producers knew that too, writing into the contracts of all the actors and actresses that they were not allowed to gain weight. And of course, it was life at the beach itself that was shown in all its glory, upbeat people, always with a chilled beverage in hand.

David Hasselhoff ist nicht nur dafür bekannt, mit seinem Gesang die Berliner Mauer zum Einsturz gebracht zu haben – nein, er war auch Hauptdarsteller (und sogar Produzent) der erfolgreichsten TV-Serie aller Zeiten. *Baywatch* lief in 146 Ländern und wurde von einer Milliarde Menschen wöchentlich gesehen; pro Folge verbrauchten die Schauspieler 40 Tuben Sonnencreme. Der 15-jährige Leonardo DiCaprio wäre um ein Haar als David Hasselhoffs Sohn gecastet worden – wer weiß, wie sich seine Schauspielkarriere danach entwickelt hätte? Tatsächlich dabei waren Jason Momoa und Surf-Weltmeister Kelly Slater. Und natürlich Pamela Anderson, die dem eigentlichen Hauptakteur regelmäßig die Show stahl.
Die Welt lernte, dass Amerikaner, wenn sie etwas Wichtiges zu sagen haben, immer erst nachdenklich aufs Meer hinaus blicken. Dabei war die Serie zunächst ein Flop; nur die guten Quoten in Großbritannien und Deutschland überzeugten die Produzenten, nach der ersten Staffel weiterzumachen – zwei Länder, die klimatisch vom ganzjährigen kalifornischen Stranddasein sehr, sehr weit entfernt liegen. David Hasselhoff verzichtete dafür auf einen Großteil seines Gehalts. Und die ikonischen Strandläufe in Richtung Gefahr in Zeitlupe? Angeblich ließ man sich von den Olympischen Spielen 1988 inspirieren, als die 100-Meter-Sprinter erstmals in Superzeitlupe gezeigt wurden, doch David Hasselhoff erklärte später, pure Geldnot habe dahinter gestanden: Mit den Slow-Motion-Sequenzen verringerte man die teure Drehzeit.
Der Erfolg der Show? An den Handlungen lag es nicht, etwa jener Folge, in der ein Oktopus Surfbretter stiehlt und sie in seiner Unterwasserhöhle versteckt, bis ihm die Rettungsschwimmer auf die Schliche kommen. Es waren wohl eher die Frauen im Badeanzug und die Männer mit Sixpack, die für Rekordquoten sorgten. Das wussten auch die Produzenten: Sie schrieben allen Darstellerinnen und Darstellern in den Vertrag, dass sie nicht an Gewicht zulegen durften. Und natürlich war es das Leben am Strand selbst, das in aller Ausführlichkeit gezeigt wurde, mit gutgelaunten Menschen, stets einen kühlen Drink in der Hand.

PREVIOUS PAGE

Lifeguard station in Caserta, Italy.

Rettungsschwimmerstation in Caserta, Italien.

Rescue is at hand: David Hasselhoff during the famous slow motion run in *Baywatch*. Malibu Beach, USA 1989.

Die Rettung naht: David Hasselhoff beim berühmten Slow-Motion-Run in *Baywatch*. Malibu Beach, USA 1989.

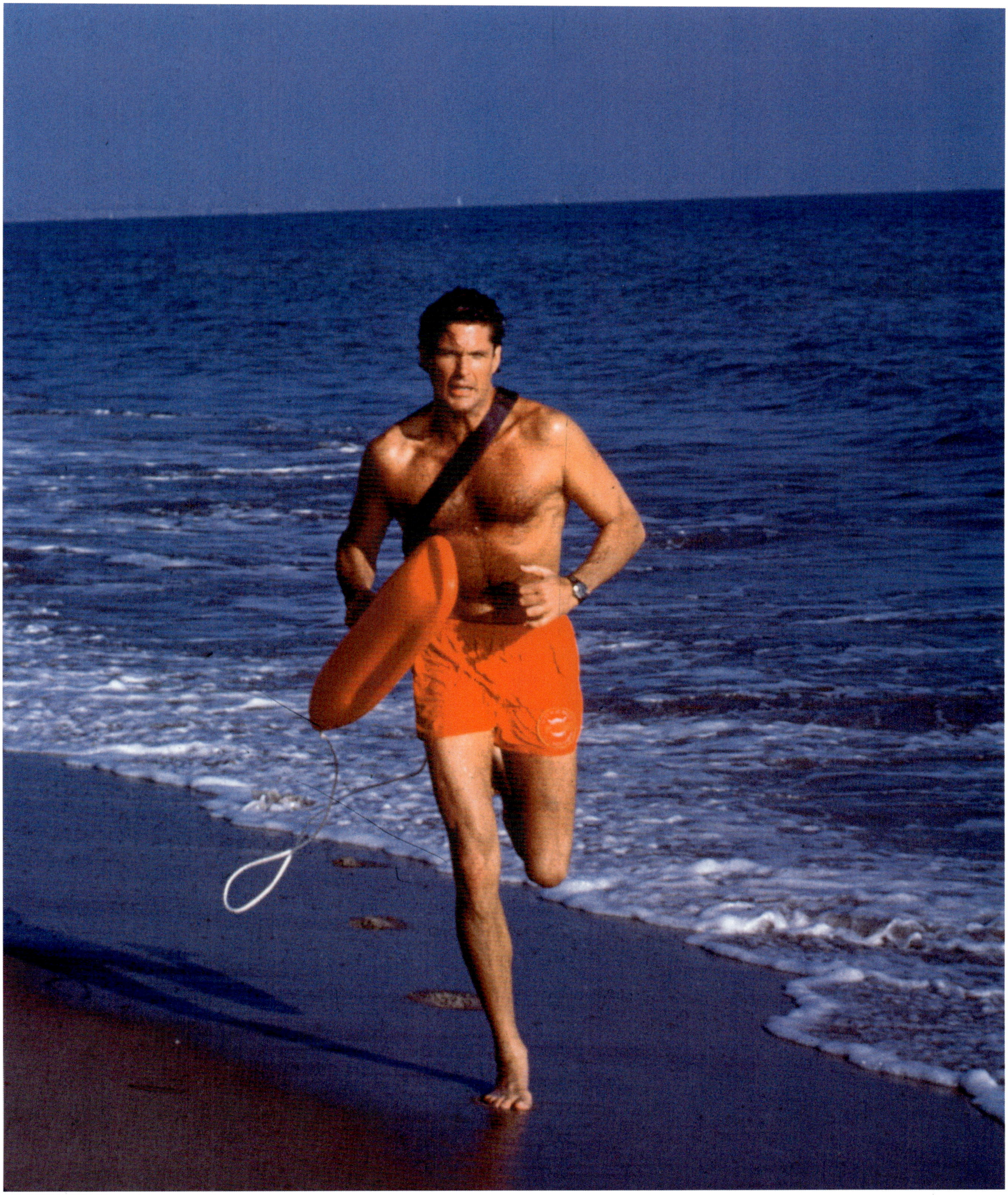

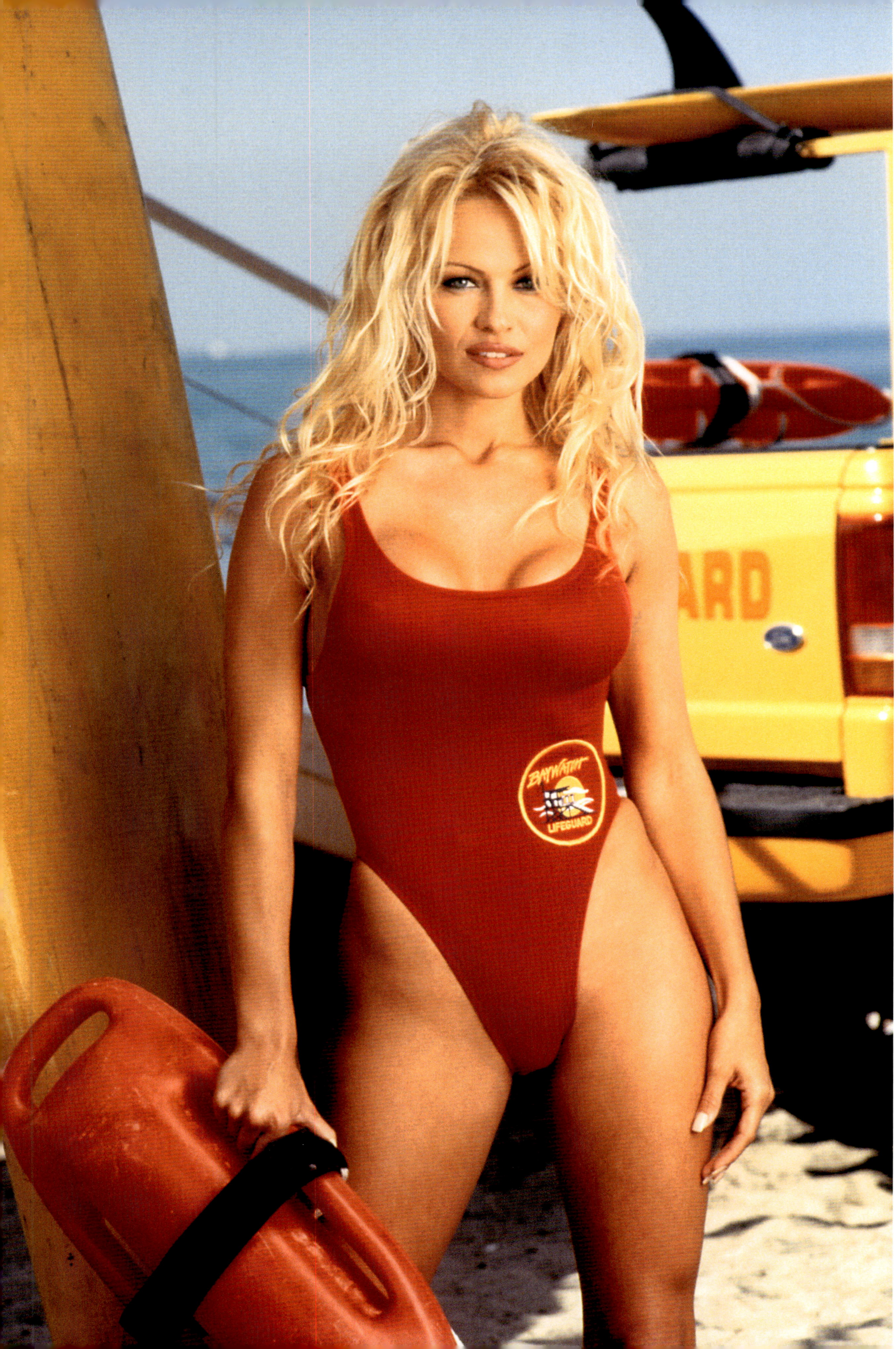
BAYWATCH
LIFEGUARD

Pamela Anderson in *Baywatch*, Malibu Beach, USA 1995.

Pamela Anderson, David Charvet, Yasmine Bleeth, Jason Simmons in *Baywatch*, Malibu Beach, 1994.

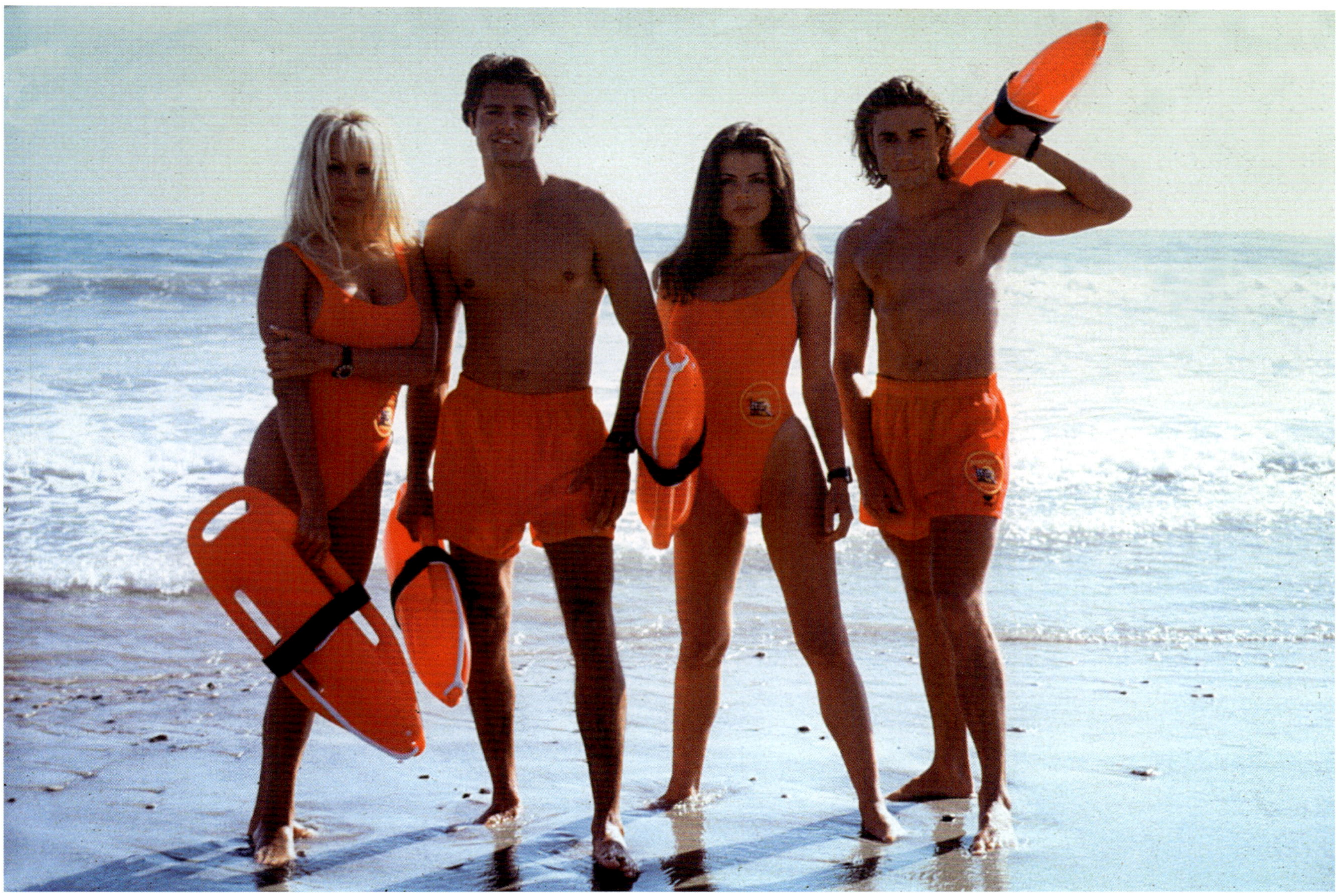

NEXT PAGE

Today's baywatch at Santa Monica Beach, USA.

Die moderne Baywatch am Strand von Santa Monica, USA.

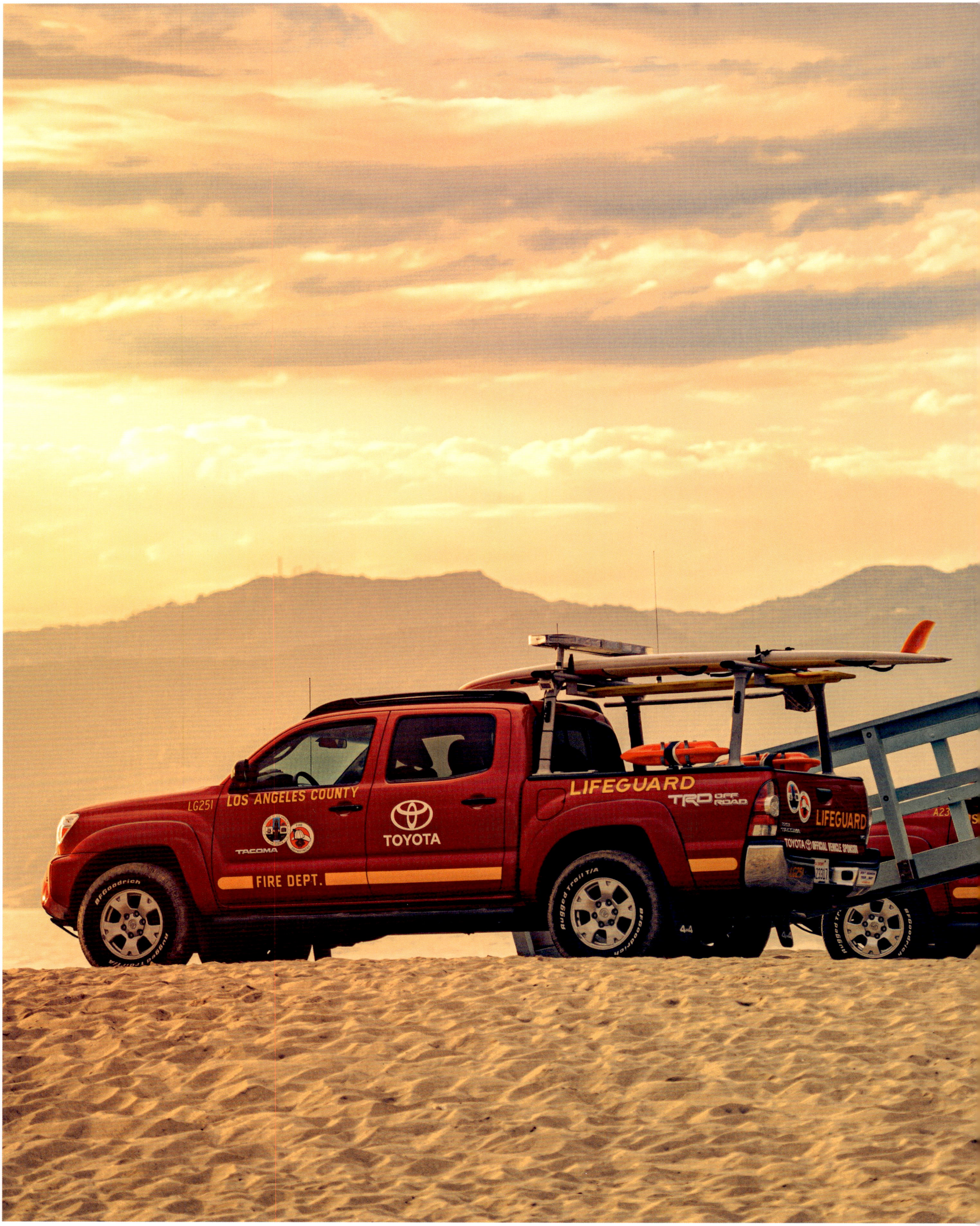
LOS ANGELES COUNTY
LG251
TOYOTA
TACOMA
FIRE DEPT.
LIFEGUARD
TRD OFF ROAD
LIFEGUARD
TOYOTA OFFICIAL VEHICLE SPONSOR

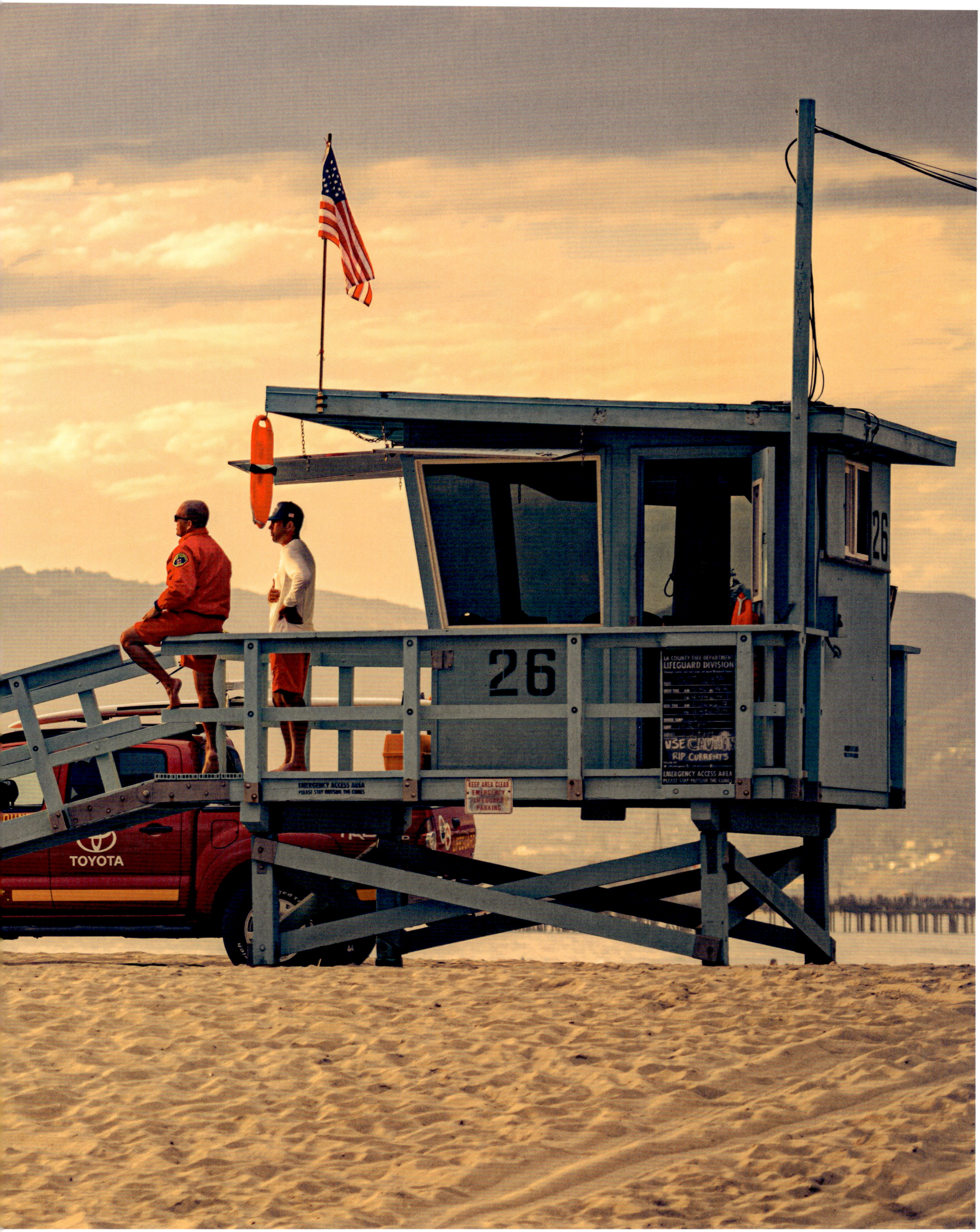
26
26
LA COUNTY FIRE DEPARTMENT
LIFEGUARD DIVISION
RIP CURRENTS
EMERGENCY ACCESS AREA
EMERGENCY ACCESS AREA
TOYOTA

NO PEOPLE

MENSCHENLEER

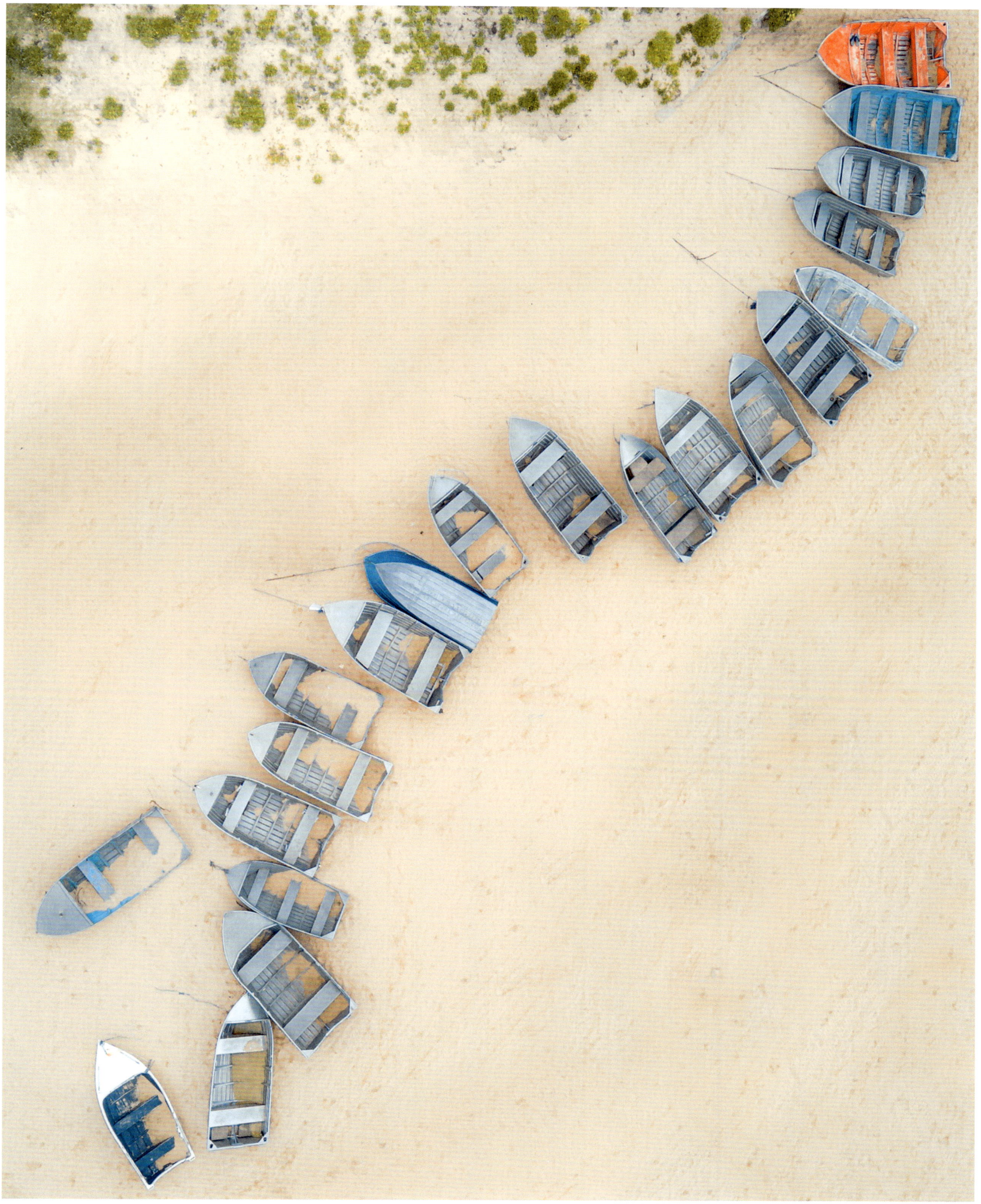

And at some point, the time comes. The lounge chairs are folded up and stacked on top of one another. The umbrellas are removed from their holders, the mounts from the sand. The swept-out cabins are loaded onto small trucks. The lifeguards tidy up their towers. The lifeboats are anchored on shore. Everyone's tan begins to fade, the first rain clouds gather, sweaters are dragged out. The *bagnini* watch the goings-on from their own cabins, not doing much more than usual.
In Italy, it's called *la stagione*–which is translated somewhat inadequately as "season." Instead, it is *the summer* that defies any kind of classification by the calendar, *the life*. And just many societies count in years and Indigenous Americans count in moons, beach people count in *stagioni*. When a season is over, it is unmistakable proof that we have grown older once again.
A cycle comes to an end, autumn sets in with all its decay. And this triggers a tiny depression. Nothing bad, not a deep black hole, just a little gloominess in the state of mind, a bit of sadness that doesn't last long but is still noticeable, a wet bath towel wrapping itself around the soul. The feeling doesn't last forever, a day or two at most. Because beach people are sure of one thing, and that is that the next *stagione* is sure to come.
Of course, the end of a season also has financial consequences. If you haven't taken advantage of it now, you'll not have the chance again. Many seasonal workers receive unemployment benefits and have far too much time on their hands, squandering their money in bars or spending their time being internet trolls. Others move to the mountains for the winter season and take jobs as cooks, waiters, and bartenders. Overall, they earn decent money, much more than at the beach, because winter vacationers spend more than summer guests. The job is tough on the bones, of course. And snowy. And overall characterized by rather cool charm and decidedly veiled eroticism.

Und irgendwann ist es soweit. Die Liegestühle werden zusammengeklappt und übereinander gestapelt. Die Sonnenschirme werden aus ihren Halterungen entfernt, die Halterungen aus dem Sand. Die ausgefegten Kabinen werden auf Kleinlaster geladen. Die Rettungsschwimmer räumen ihre Türmchen auf. Die Rettungsboote werden an Land verankert. Die Bräune aller Beteiligten verblasst zusehends, erste Regenwolken ziehen auf, Pullover werden hervorgeholt. Die Bagnini schauen sich das Treiben von ihren Kabinen aus an, machen also nicht viel mehr als sonst auch.
In Italien heißt es *la stagione* – das ist mit „Saison" nur unzureichend übersetzt. Es ist vielmehr *der* Sommer, der sich jeder kalendarischen Einordnung entzieht, *das* Leben. Und so, wie gewöhnliche Menschen in Jahren zählen und Native Americans in Monden, zählen Strandmenschen in *stagioni*. Ist eine Saison vorbei, ist der untrügliche Beweis erbracht, dass wir wieder älter geworden sind.
Ein Zyklus geht zu Ende, der Herbst setzt ein mit all seinem Verfall. Und das löst eine klitzekleine Depression aus. Nichts Schlimmes, kein tiefes schwarzes Loch, nur eine kleine Eintrübung der Befindlichkeit, etwas Traurigkeit, die nicht lange anhält, aber doch spürbar ist, ein nasses Badehandtuch, das sich um die Seele wickelt. Das Gefühl bleibt nicht ewig bestehen, höchstens ein oder zwei Tage. Denn Strandmenschen haben eine Gewissheit: Die nächste *stagione* kommt bestimmt.
Natürlich hat das Ende einer Saison auch ökonomische Folgen. Wer jetzt nicht geerntet hat, erntet nimmermehr. Viele Saisonarbeiter bekommen Arbeitslosengeld, haben deutlich zu viel Zeit und vertändeln sie in den Bars oder als Internettrolle. Andere ziehen für die Wintersaison in die Berge und übernehmen dort Jobs als Koch, Kellner und Barmann. Insgesamt verdienen sie ordentliches Geld, viel mehr als am Strand, denn Winterurlauber geben mehr aus als Sommergäste. Der Job ist natürlich knöchern. Und schneereich. Und insgesamt von eher kühlem Charme und dezidiert verhüllter Erotik.

PREVIOUS PAGE

Stranded: Boats at Coogee Beach in Sydney, Australia.

Gestrandet: Boote am Coogee Beach in Sydney, Australien.

Imagine doors. From the series *Marbelland*. Marbella, Spain.

Mögliche Türen. Aus der Serie *Marbelland*. Marbella, Spanien.

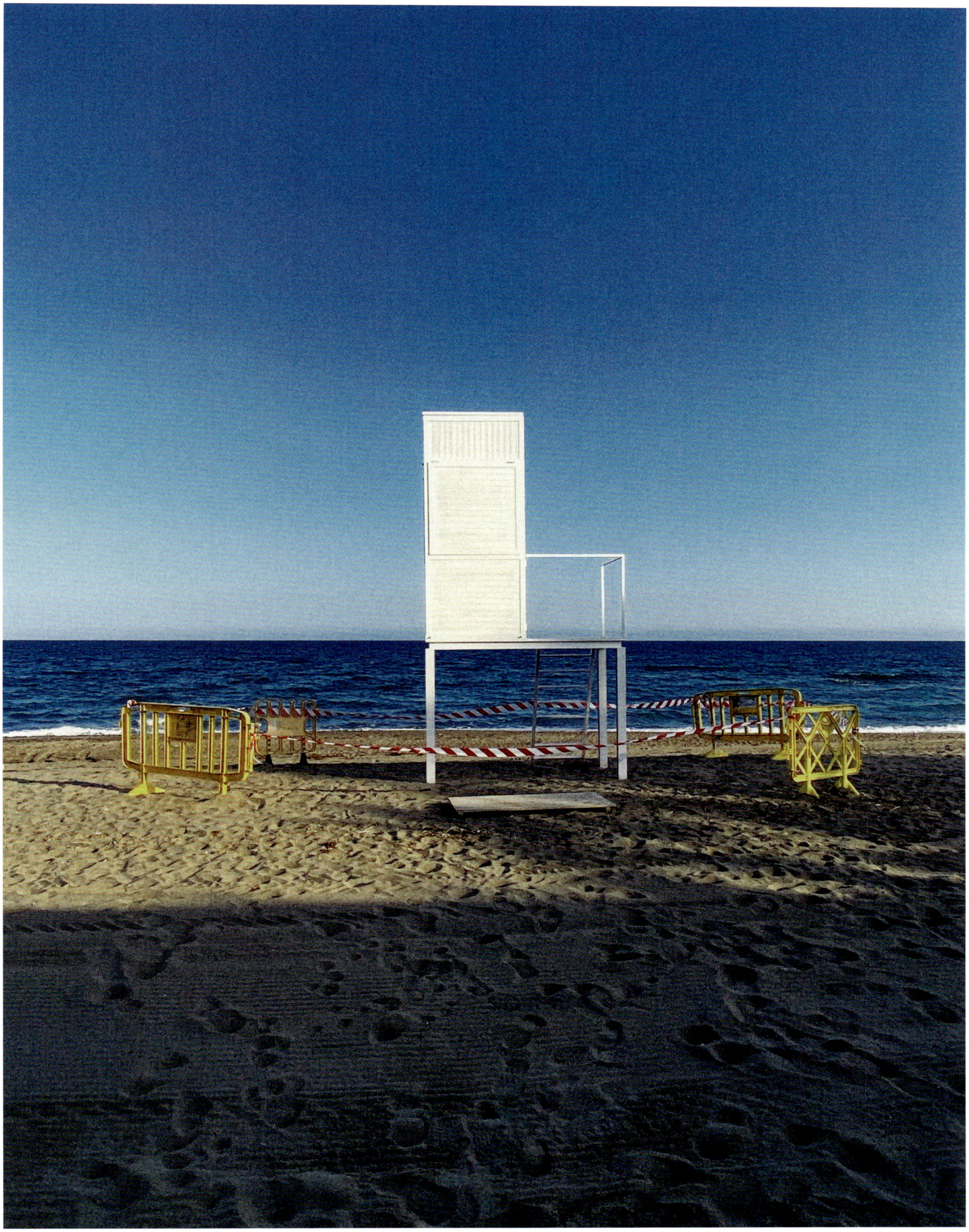

POST-SOMMER-BLUES: DAS ENDE DER SAISON

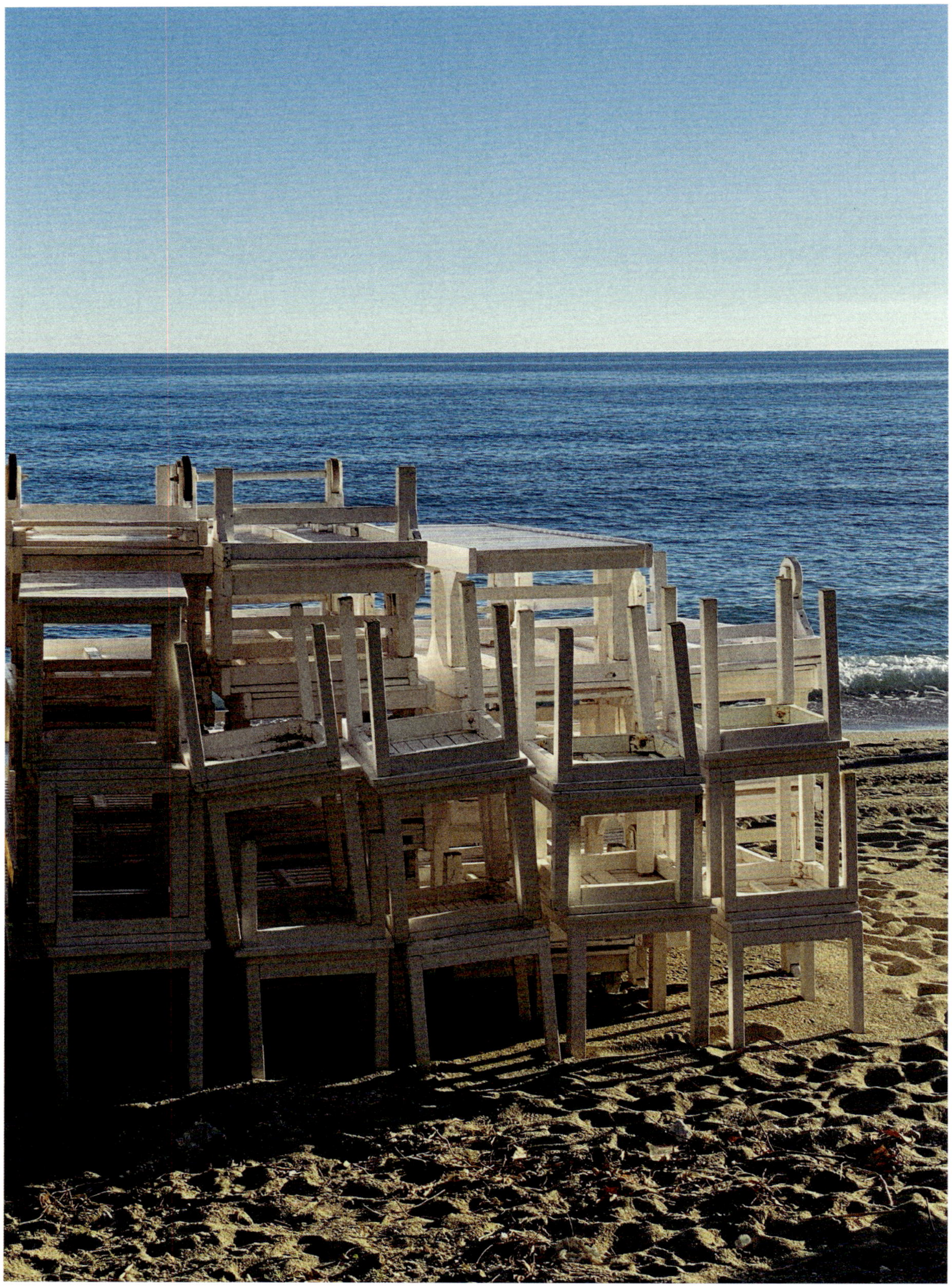

Costa del Sol LeWitt. From the series *Marbelland,* Marbella, Spain.

Costa del Sol LeWitt. Aus der Serie *Marbelland.* Marbella, Spanien.

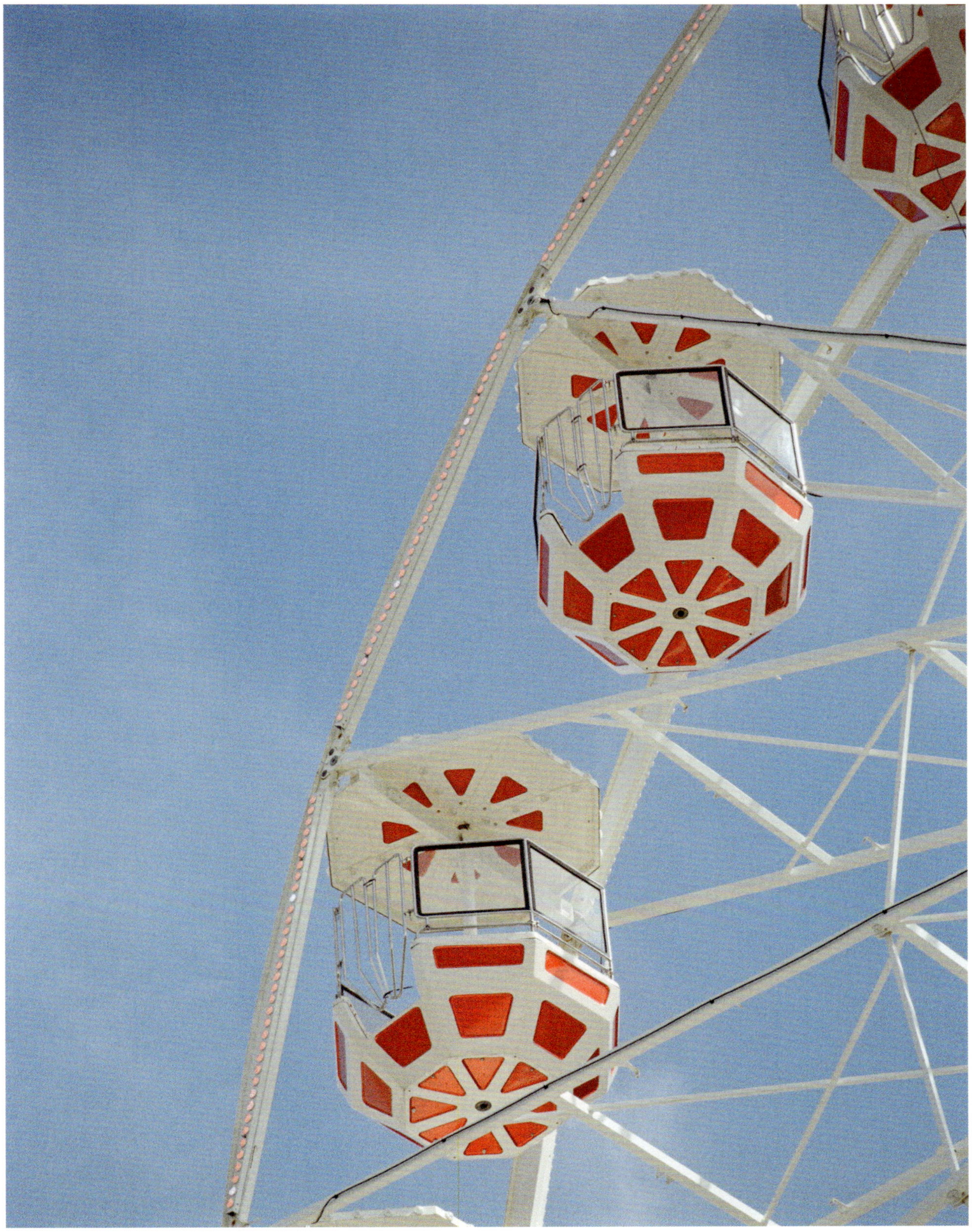

The Holiday Wheel.
Deauville, France.

Das Riesenrad.
Deauville, Frankreich.

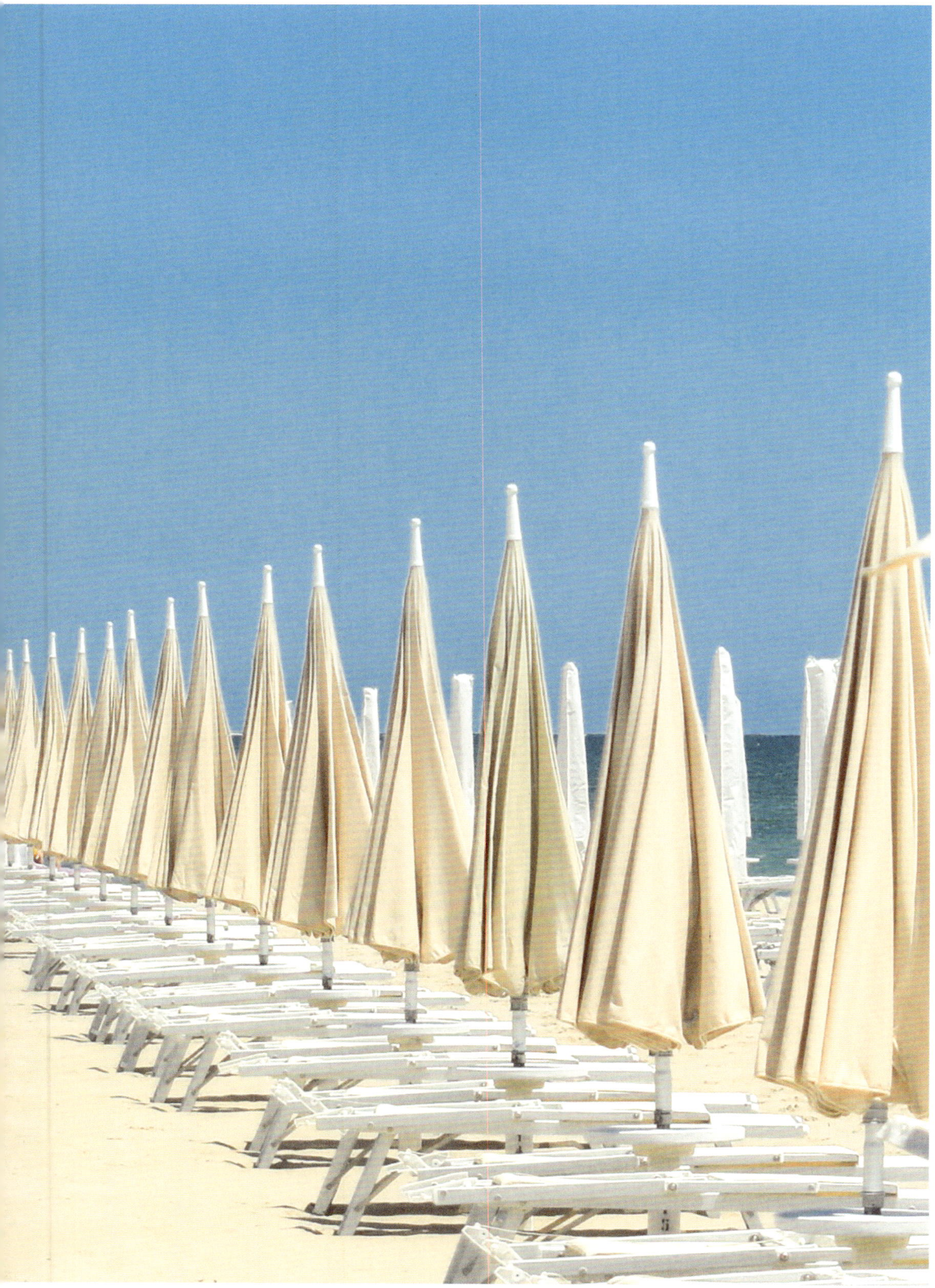

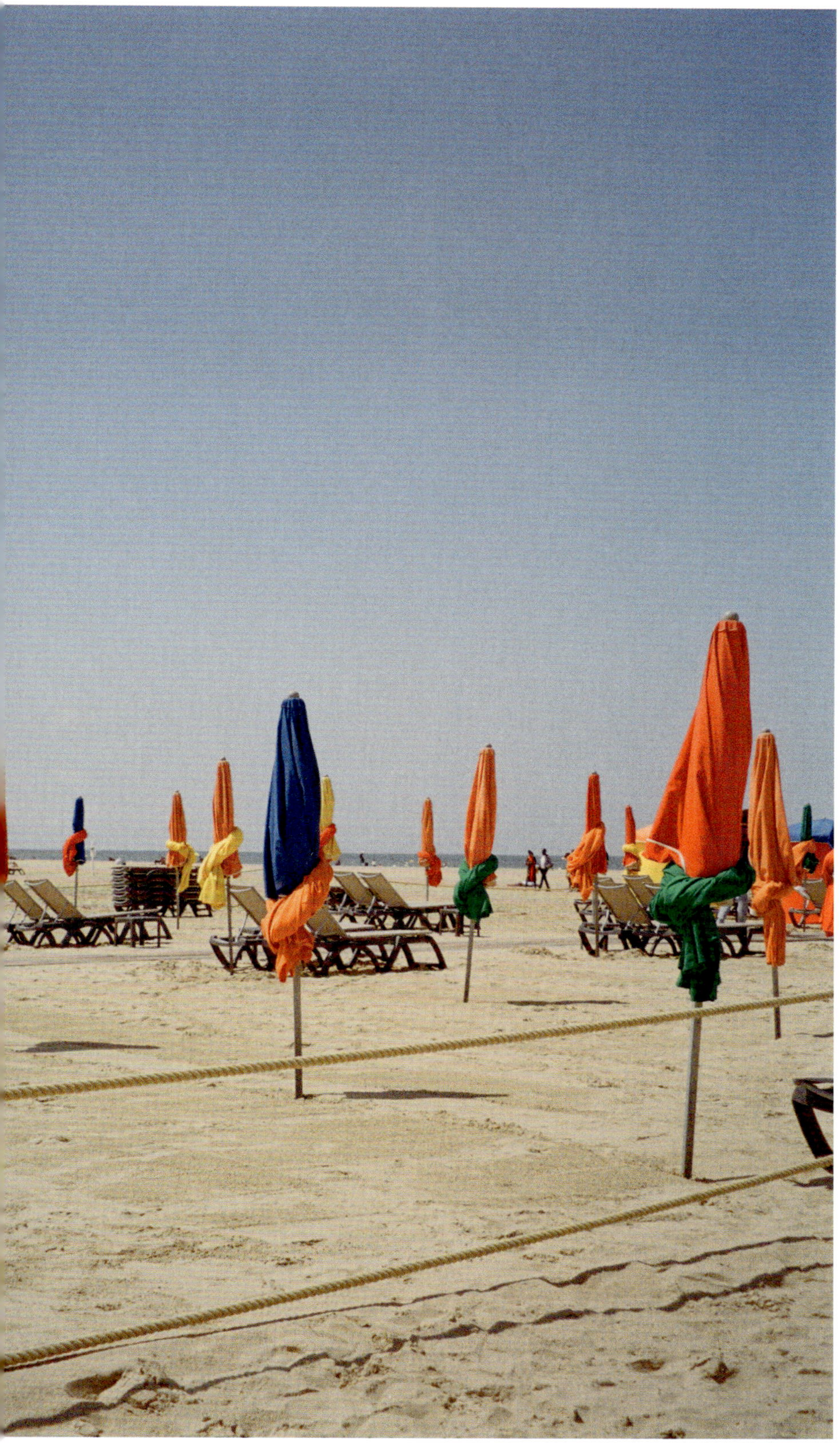

LEFT PAGE

Closed for today. Rimini, Italy.

Feierabend. Rimini, Italien

Sun for free. The characteristic colorful parasols at the beach of Deauville, France.

Sonne kostenlos. Die charakteristischen bunten Sonnenschirme am Strand von Deauville, Frankreich.

Off duty. Watch tower in Austinmer, Wollongong, Australia.

Außer Dienst. Turm der Strandwache in Austinmer, Wollongong, Australien.

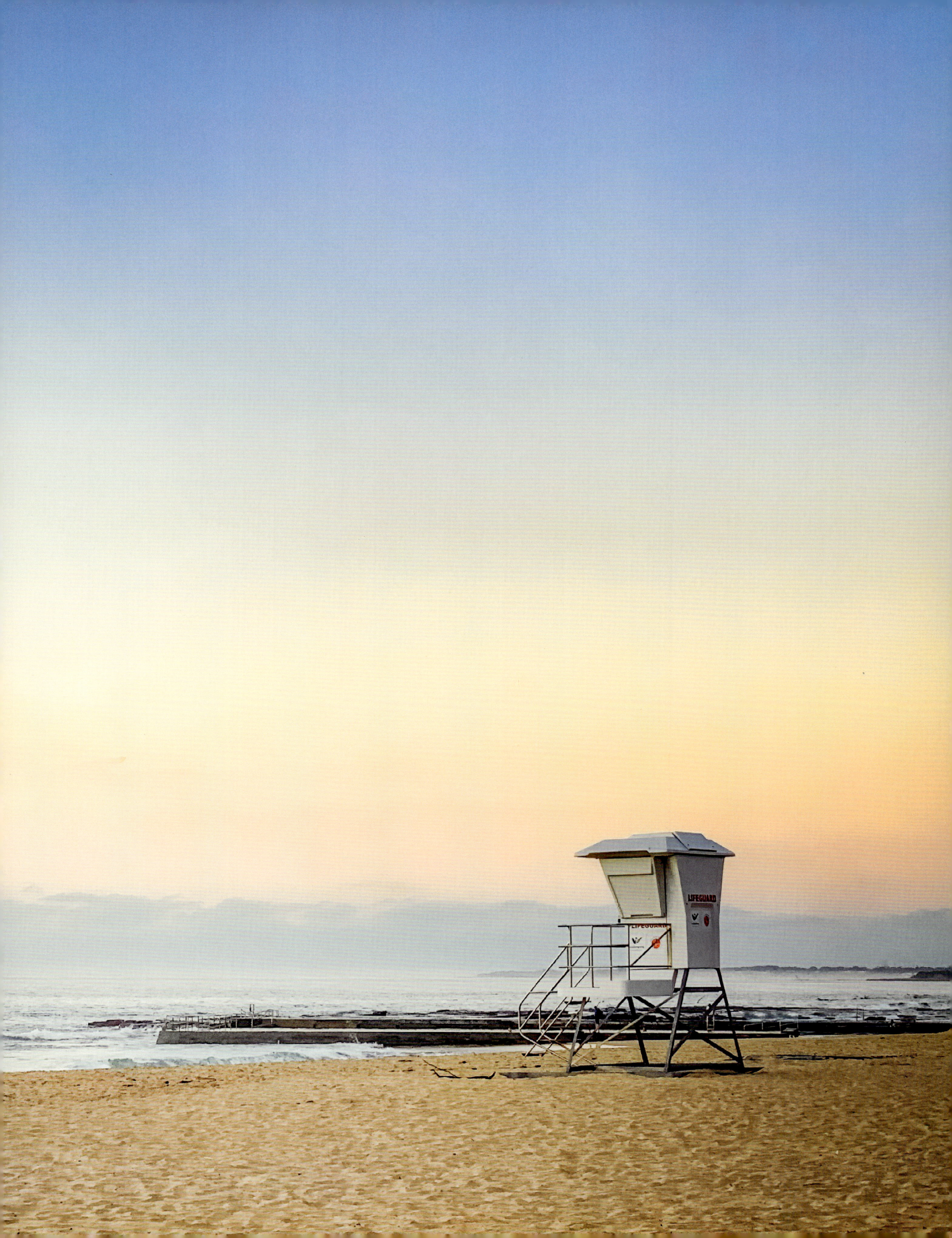
LIFEGUARD
LIFEGUARD

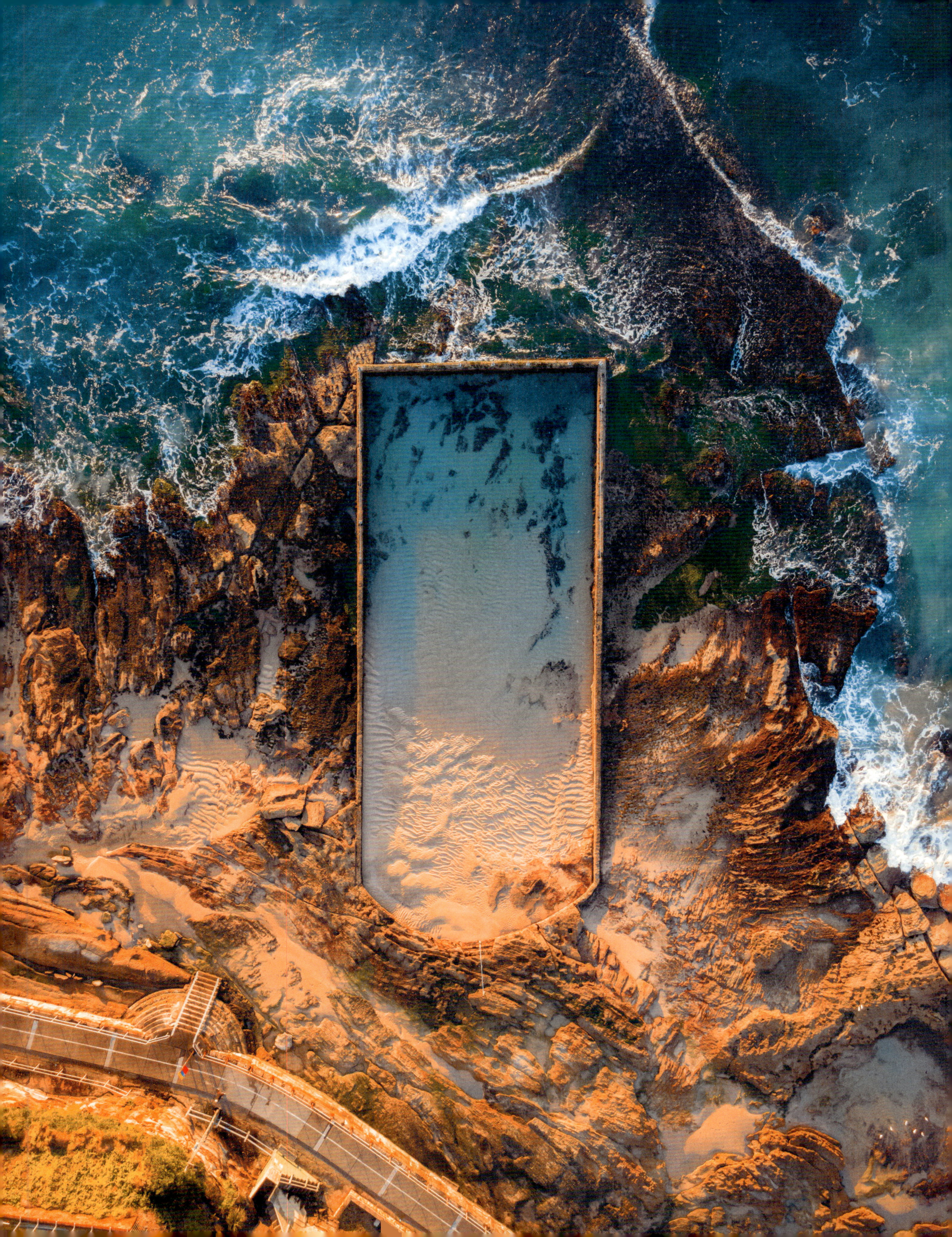

There's much to see, but everything that meets the eye has an almost meditative vastness and pleasant repetition: the still damp sand. The steady, gentle crash of the waves. The fishing boat on the horizon that barely seems to be making any progress. The footprints we leave behind, transient much like ourselves. The view spreads out in every direction. The wind is naturally much stronger on the shore, the smell of salt and seaweed is deliciously intense; our lips taste like the ocean. We are so close to our primal element that we become one with it.
But a solitary beach is not only a meditation, it is an adventure. Walking on the beach as the tide recedes, you feel like an explorer. You take a closer look, finding exotic shells, colorful little stones, maybe a small, brightly colored float that has come loose from a fishing net; once made of clay, they are now made of plastic. And naturally we dream of making a grand discovery. Of amber at the Baltic Sea, Roman coins at the Mediterranean, Spanish gold doubloons in the Caribbean. Such finds are rare, but they do happen. On many a beach, professionals walk up and down with their metal detectors during the off-season, grumpy fellows who don't like to talk about their spoils. On the shores, children play and marvel, searching for shells and secrets.
For many centuries, combing the beach for items that washed up was a lucrative occupation for those who lived on the dangerous coasts near busy sea routes. The whole village rejoiced especially after storms and when ships ran aground. And if any sailors survived the accident and made it to shore, they were beaten to death as a preventive measure, so they could not lay claim to any finds.
Among the most amazing modern-day discoveries is an ancient gold drinking vessel found by a snorkeler on the beach of Key West, Florida (valued at $1 million). Hunters and collectors of sea glass even travel from the United States to island of Helgoland in the North Sea in search of it. These shards of glass that have been rounded by salt water can be more valuable than amber in certain colors.

Es gibt zwar viel zu sehen, aber alles, was das Auge erblickt, ist von einer beinahe meditativen Weite und angenehmen Repetition: der noch feuchte Sand. Der stete, sanfte Wellenschlag. Das Fischerboot am Horizont, das kaum voranzukommen scheint.
Die Fußspuren, die wir hinterlassen, vergänglich wie wir selbst. Der Blick weitet sich in jede Richtung. Der Wind ist am Ufer naturgemäß viel stärker zu spüren, der Salz- und Algengeruch von einer köstlichen Intensität; unsere Lippen schmecken nach Meer. Wir sind so nah an unserem Urelement, dass wir eins mit ihm werden.
Aber ein einsamer Strand ist nicht nur eine Meditation, sondern auch ein Abenteuer. Wer bei zurückweichender Flut am Strand spazieren geht, fühlt sich als Entdecker und schaut ganz genau hin. Findet exotische Muschelschalen, farbenfrohe Kiesel, vielleicht einen kleinen, bunten Schwimmkörper, der sich vom Fischernetz gelöst hat; früher waren sie aus Ton, heute sind sie aus Kunststoff. Und natürlich träumen wir von den ganz großen Entdeckungen. Von Bernstein an der Ostsee, Römermünzen am Mittelmeer, spanischen Golddublonen in der Karibik. Solche Funde sind selten. Aber sie kommen vor. An so manchen Stränden gehen Profis in der Nebensaison mit Metalldetektoren auf und ab, mürrische Burschen, die ungern über ihre Ausbeute reden. An den Ufern spielen und staunen Kinder, Muscheln und Geheimnissen auf der Spur.
Viele Jahrhunderte lang war die Suche nach Strandgut eine einträgliche Beschäftigung für all jene, die an gefährlichen Küsten vielbefahrener Seewege lebten. Besonders nach Stürmen und auf Grund gelaufenen Schiffen freute sich das ganze Dorf. Und falls Seeleute die Havarie überlebten und es ebenfalls ans Ufer schafften, wurden sie vorsichtshalber erschlagen, um keine Ansprüche auf die Funde geltend machen zu können. Zu den erstaunlichsten Fundstücken der heutigen Zeit gehört ein antikes Trinkgefäß aus Gold, das ein Schnorchler am Strand von Key West, Florida fand (Wert: eine Million Dollar). Und für Meerglas reisen Jäger und Sammler sogar aus den USA auf die Nordseeinsel Helgoland. Meerglas, vom Salzwasser rundgeschliffene Glasscherben, können in bestimmten Farben wertvoller sein als Bernstein.

One of the ocean rock pools of Cronulla Beach, Sydney, with a view from above. Along the east coast of Australia many of these pools can be admired.

Einer der Ocean Rock Pools von Cronulla Beach, Sydney, mit Blick von oben. Entlang der Ostküste Australiens kann man viele dieser Pools bewundern.

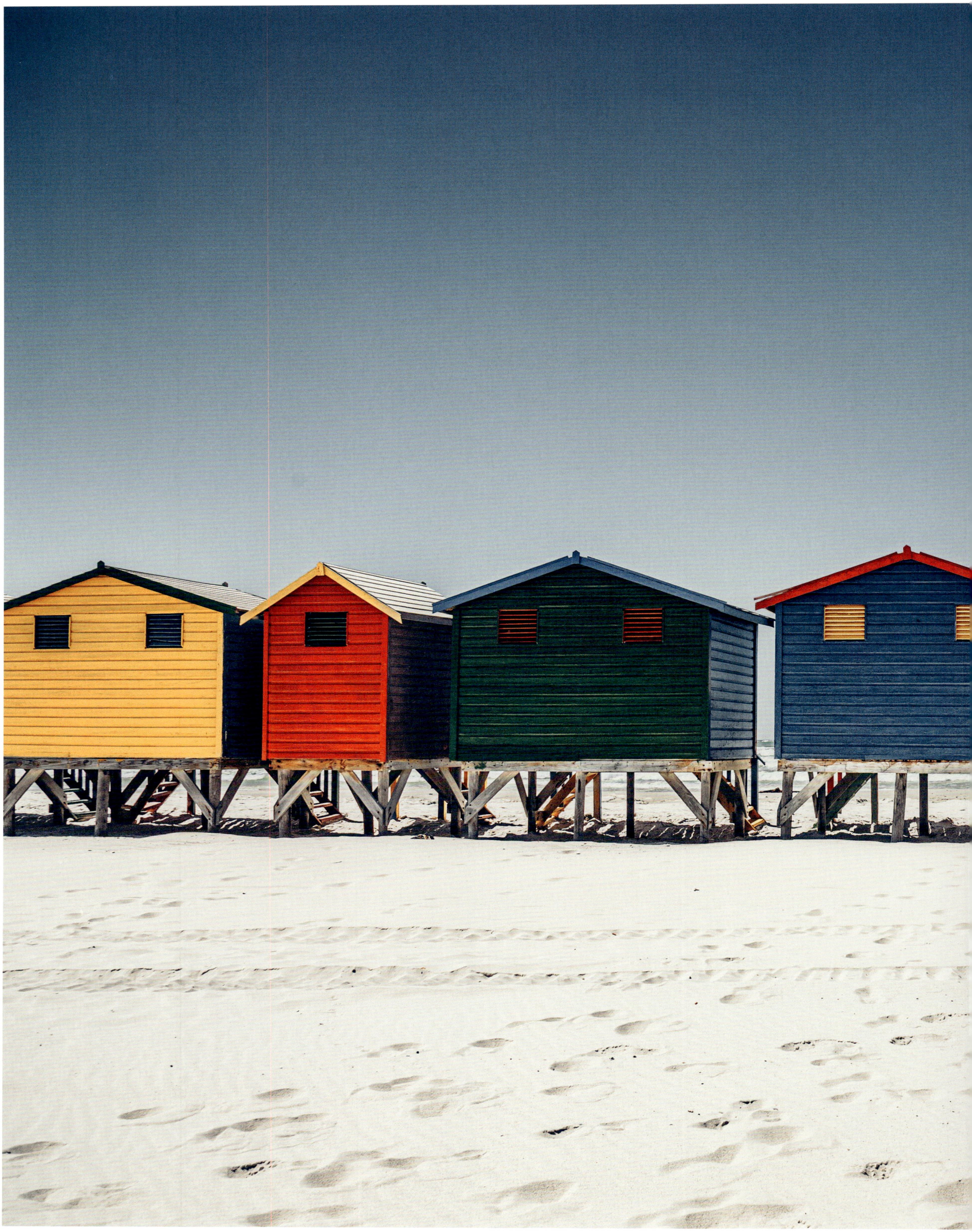

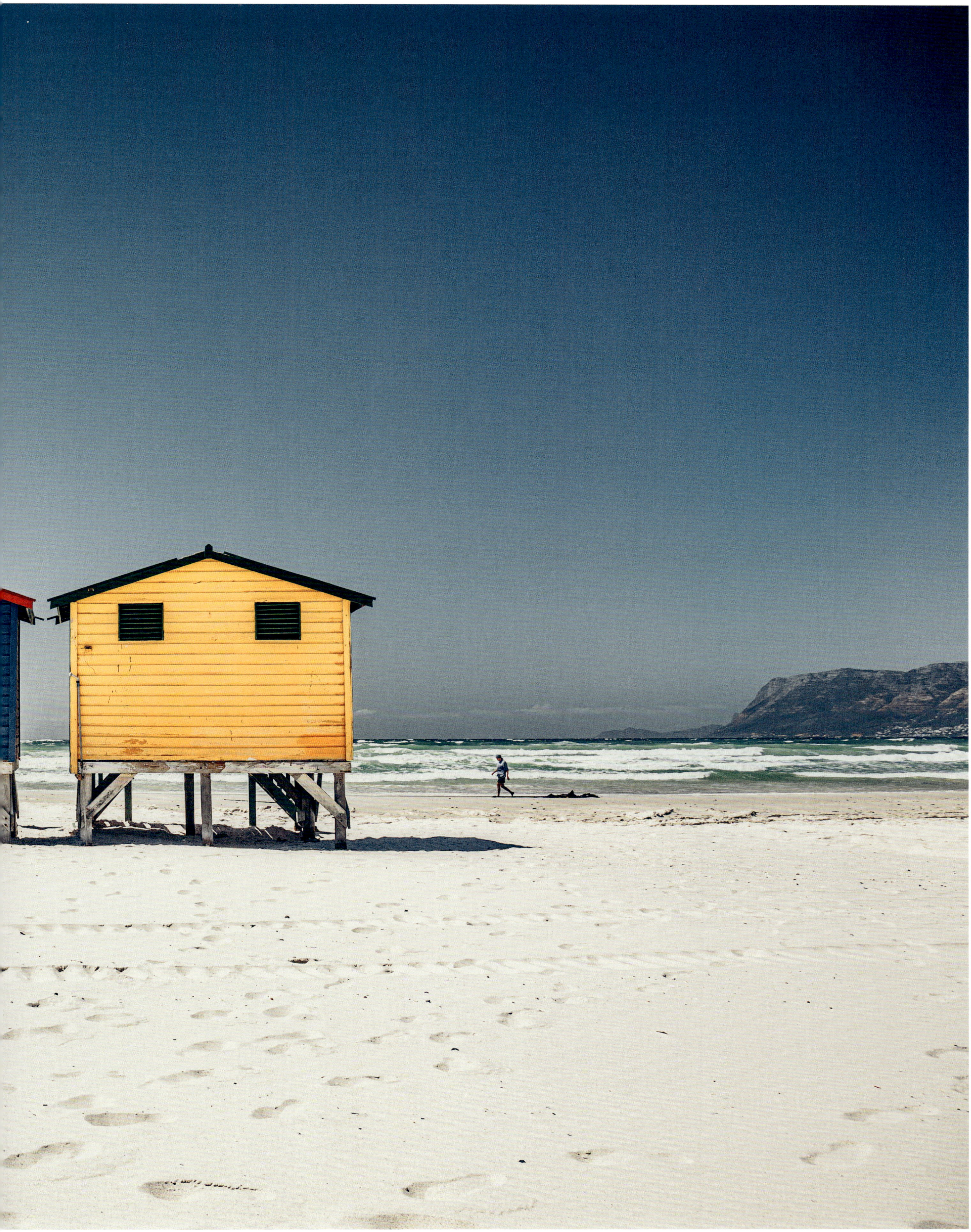

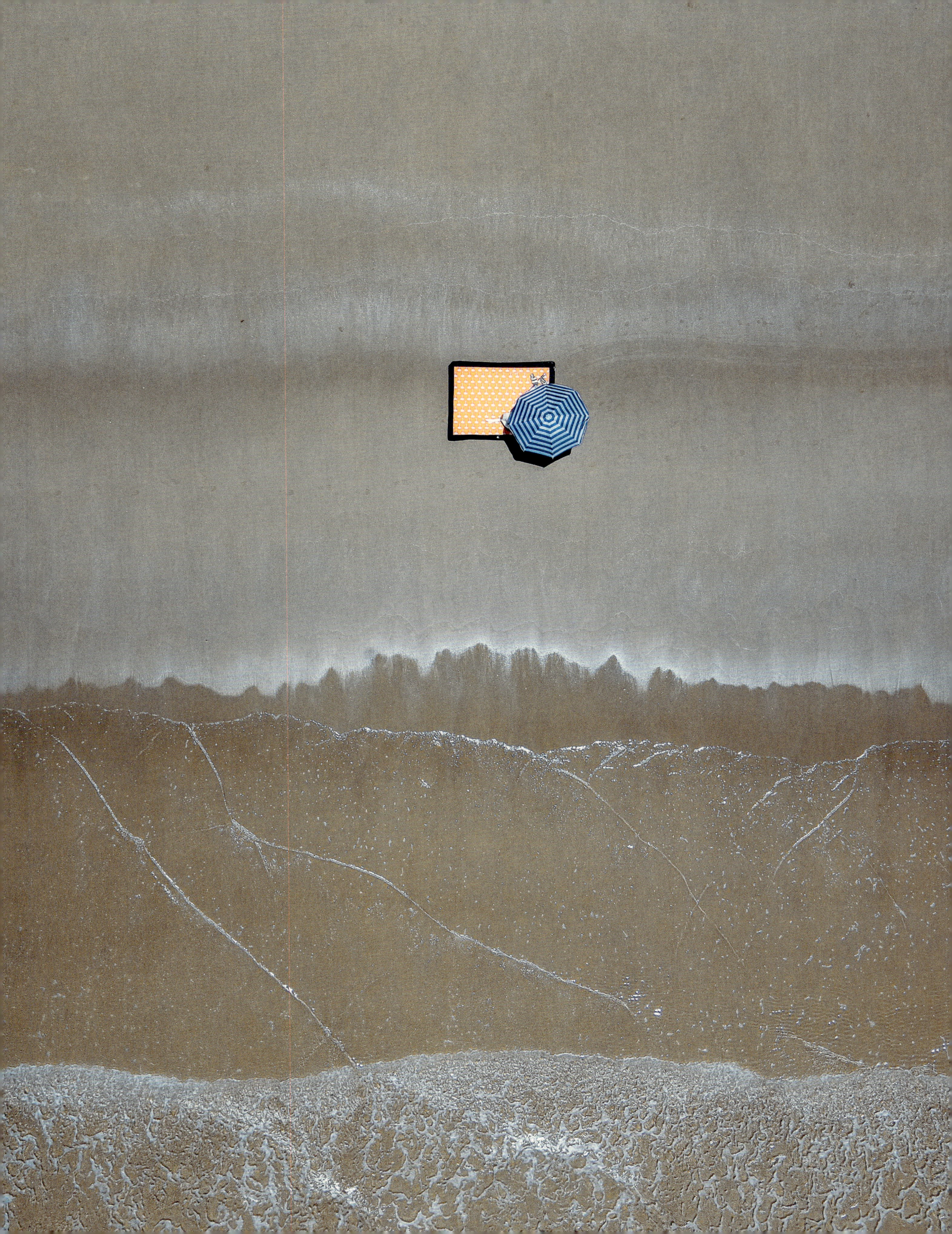

PREVIOUS PAGE

Colorful huts at the beach of Muizenberg, Cape Town, South Africa.

Farbenfrohe Hütten am Strand von Muizenberg, Kapstadt, Südafrika.

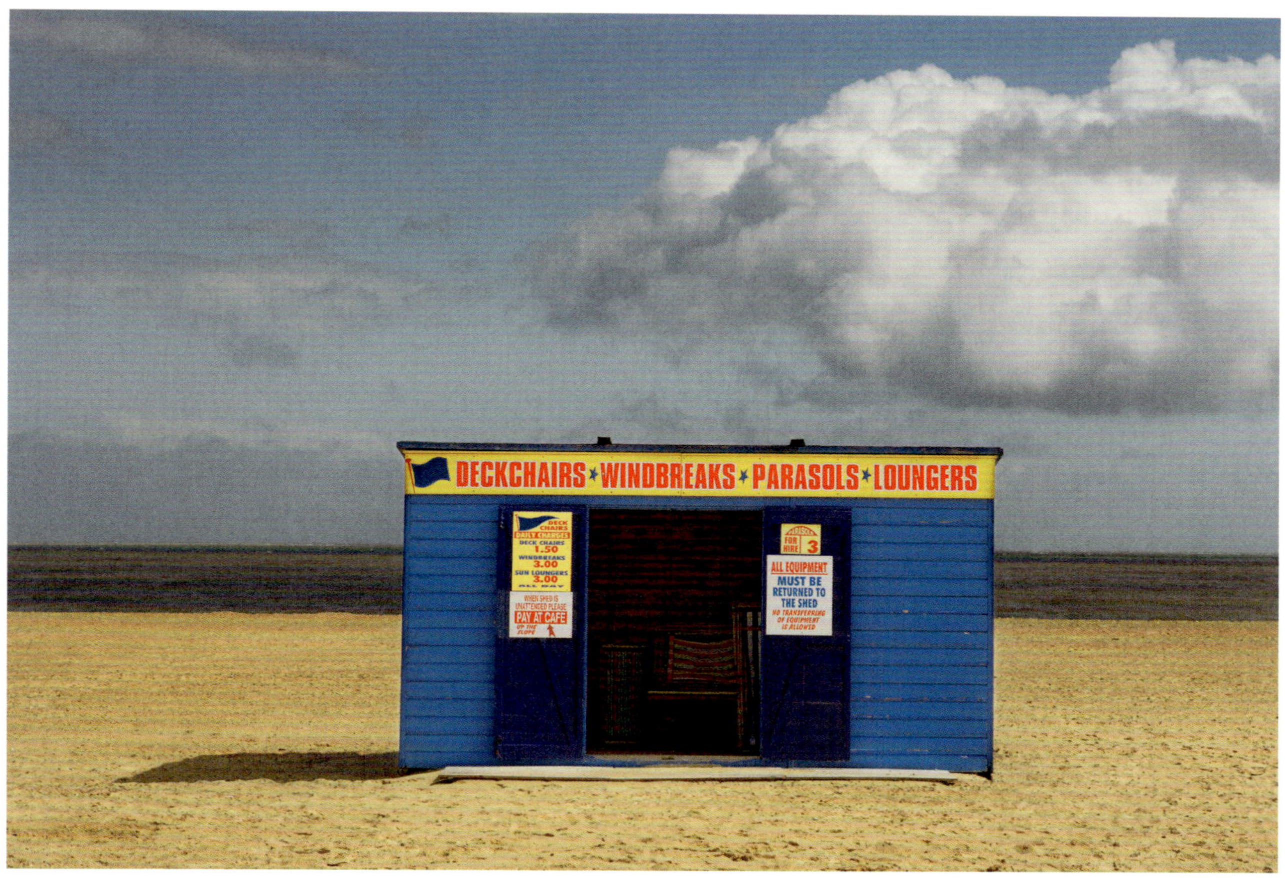

Deckchair shack at Whitby Beach, Yorkshire, UK.

Liegestuhlverleih am Strand von Whitby, Yorkshire, UK.

Beachy Keen: Cape Tribulation, Far North Queensland, Australia. This amazing beach is located in the tropics of northern Australia. With rainforest on one side and sparkling water on the other, you might find you're the only one on the beach.

Beachy Keen: Cape Tribulation, Far North Queensland, Australien. Dieser fantastische Strand liegt in den Tropen Nordaustraliens. Sie sind vielleicht der einzige Besucher am Strand, Regenwald auf der einen und glitzerndes Wasser auf der anderen Seite.

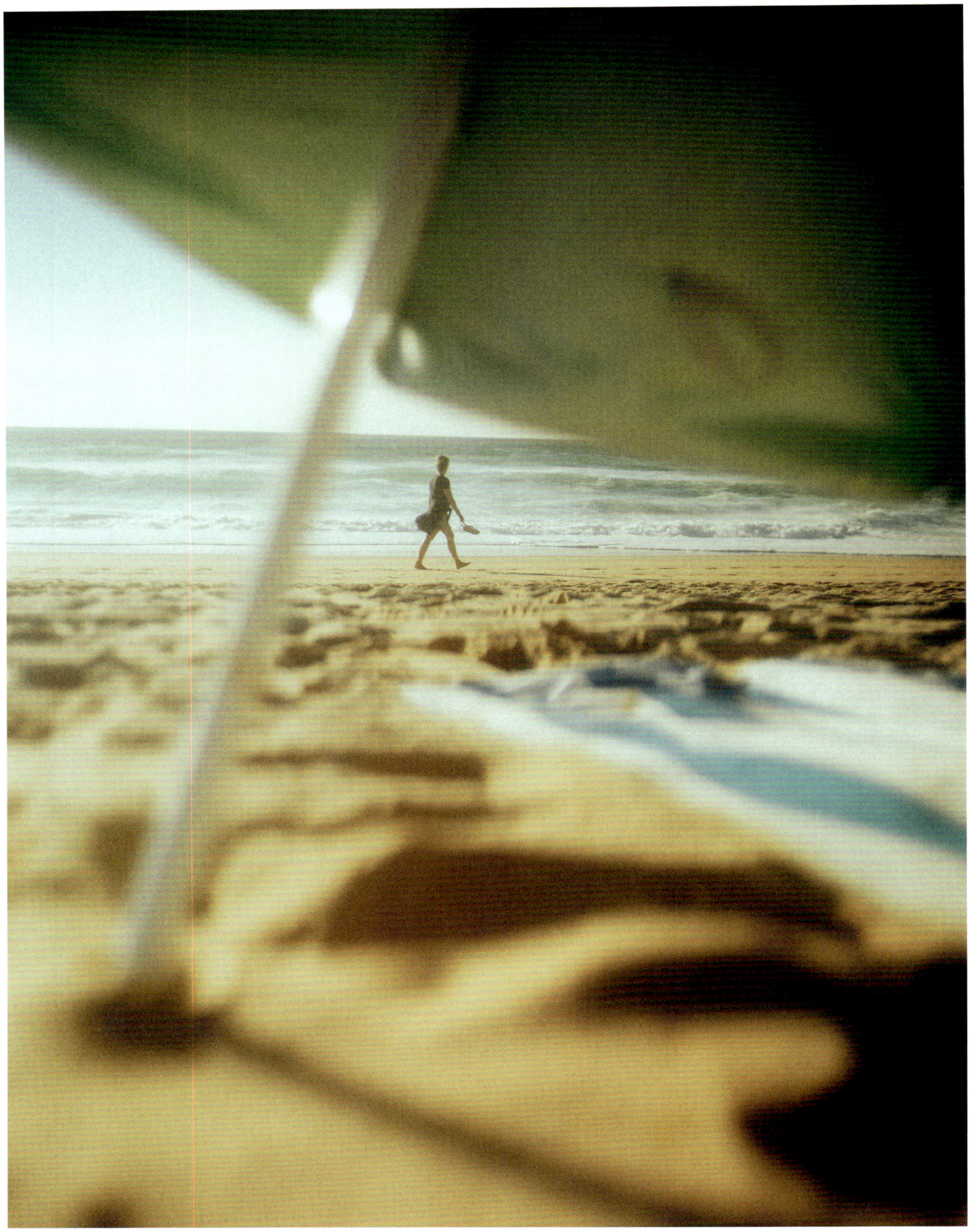

Solitary stroller at the Praia do Guincho, Portugal.

Einsame Spaziergängerin am Praia do Guincho, Portugal.

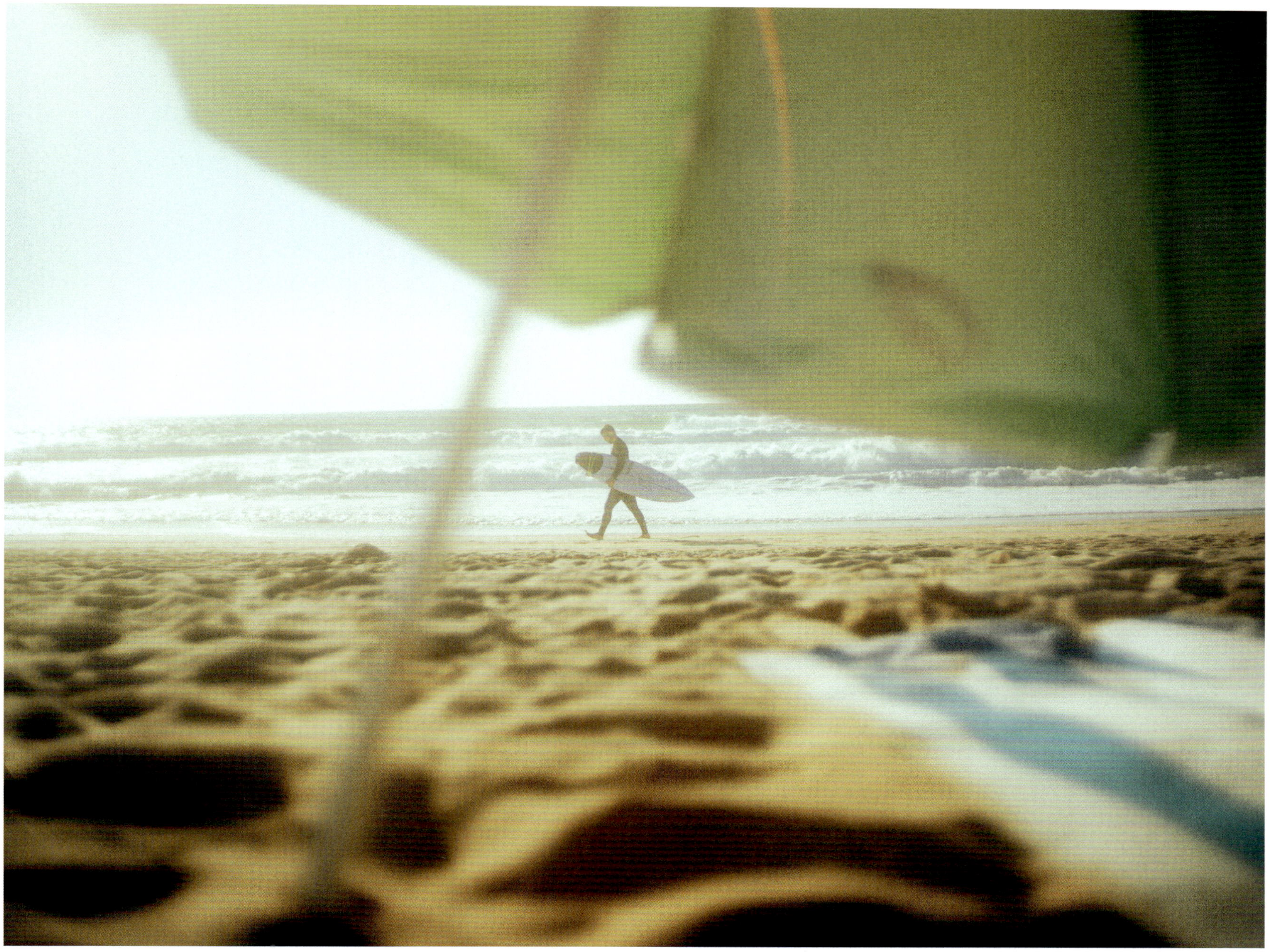

Solitary surfer at the Praia do Guincho, Portugal.

Einsamer Surfer am Praia do Guincho, Portugal.

NEXT PAGE

A warm glow covers the sandy shore while the waves glisten in the fading light of the sun. Praia Ursa, Portugal.

Ufersand und Wellen schimmern im schwindenden Licht der Sonne. Praia Ursa, Portugal.

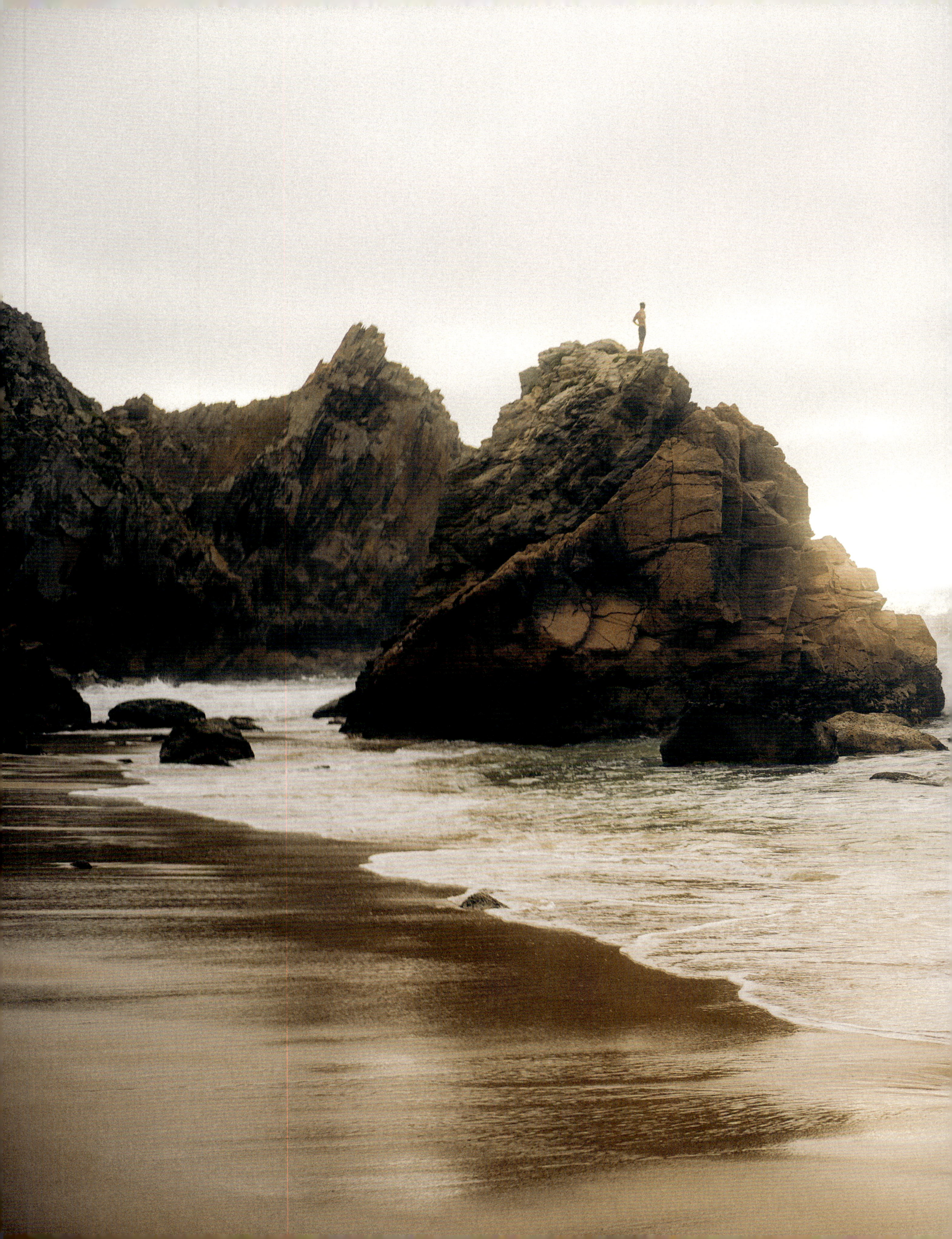

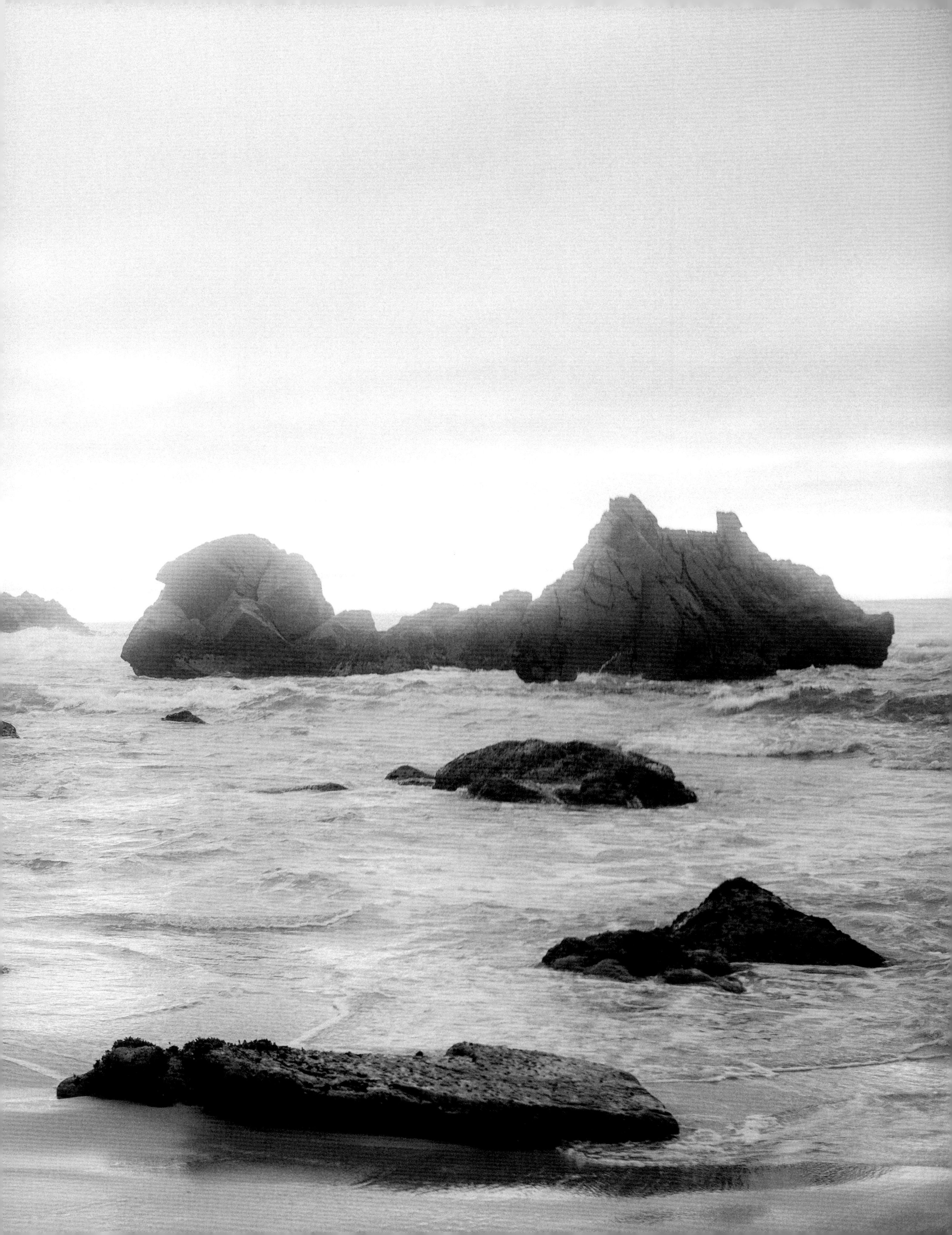

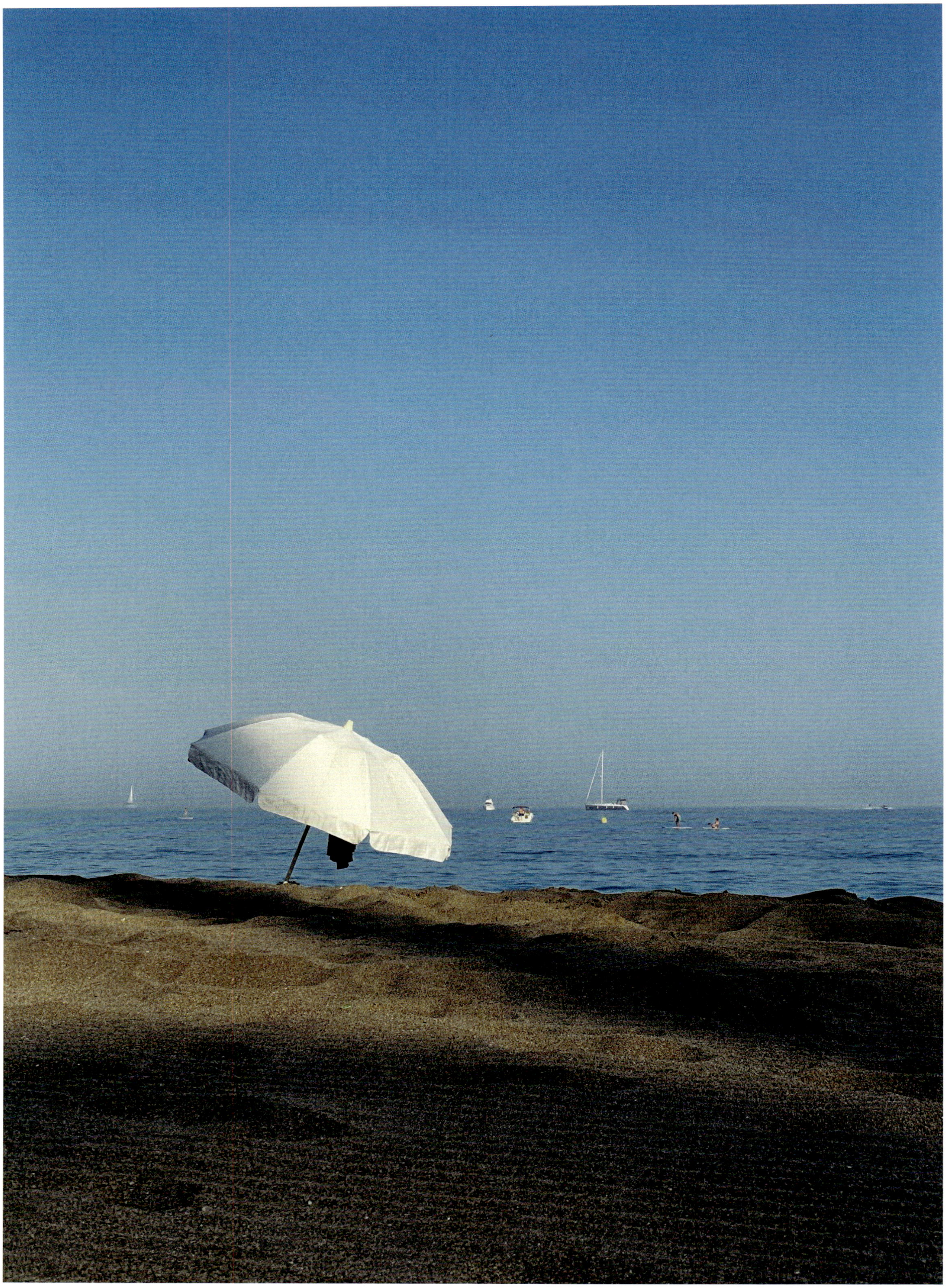

Most sandy beaches form courtesy of rivers. This is because flowing waters transport sedimentary rock from inland to the ocean. There it is carried on by the waves and left on the shallow shore, layer by layer, over hundreds of thousands of years. If wind direction, ocean currents, and the geographic shape of the shore play along, a sandy beach is very gradually formed. The raising and lowering of the sea level also plays an important role, as sand often forms where there used to be seabed and the sediment has sunk to the bottom. Nothing can be deposited on steep cliffs; here, beachgoers would have to wait a few million years for the rocks to be eroded by wind, weather, and earthquakes, to become flatter and thus more accommodating. And pebble beaches also have to be given time until the stones are crumbled by the waves and the weather–a very, very long process. But the fact that the pebbles on almost all beaches are already nicely rounded is due to the persistence of the surf. Put simply, the larger and coarser the rock, the younger the pebble beaches are in geological terms.
The snow-white beach in parts of the South Seas, Asia, Australia, and the Caribbean is largely the result of crushed shells and deposits of corals and crustaceans off the coast. Their main component is a whitish limestone, while most other beaches consist of light brown quartz sand, one of the most common minerals on Earth. Some photogenic exceptions are the black beaches on volcanic islands such as Santorini, Fuerteventura, and Bali, where the sand has formed from hardened lava and ash. And the reddish shade of sand in many deserts is because of the high iron content of the sand.
The sand we associate with vacation, happiness, and unwinding has been defined quite precisely in international science: The "stone particles" must have a diameter between 0.0625 and 2 millimeters to be called "sand." Everything smaller than that is silt, and everything larger than that is considered uncomfortable pebbles.

Die meisten Sandstrände bilden sich mit Hilfe von Flüssen. Denn die fließenden Gewässer transportieren Sedimentgestein aus dem Inland bis ins Meer. Dort wird es von den Wellen weitergetragen und über hunderttausende von Jahren, Schicht für Schicht, am flachen Ufer zurückgelassen. Wenn Windrichtung, Meeresströmung und die geografische Form des Ufers mitspielen, entsteht ganz allmählich ein Sandstrand. Auch das Heben und Senken des Meeresspiegels spielt eine wichtige Rolle: Sand bildet sich oft dort, wo früher Meeresgrund war und die Sedimente zu Boden gesunken sind. An Steilklippen kann sich nichts ablagern; hier müssten Badende ein paar Millionen Jahre warten, bis die Felsen durch Wind, Wetter und Beben abgetragen, flacher und damit aufnahmefähiger werden. Und Kieselstränden müssen wir ebenfalls noch Zeit geben, bis die Steine von den Wellen und der Witterung zerbröselt werden – ein sehr, sehr langer Prozess. Aber dass die Kiesel an fast allen Stränden immerhin schon recht rundgeschliffen sind, liegt an der Hartnäckigkeit der Dünung. Vereinfacht gesagt, sind Kieselstrände erdgeschichtlich umso jünger, je größer und gröber ihr Gestein ist.
Der schneeweiße Strand an Teilen der Südsee, in Asien, Australien oder der Karibik stammt zu großen Teilen aus zerkleinerten Muschelschalen und Ablagerungen von Korallen und Krebstieren vor der Küste. Deren Hauptbestandteil ist weißlicher Kalkstein, während die meisten anderen Stände aus hellbraunem Quarzsand bestehen, einem der häufigsten Mineralien der Erde. Fotogene Ausnahmen sind die schwarzen Strände auf Vulkaninseln wie Santorin, Fuerteventura oder Bali, wo sich der Sand aus verhärteter Lava und Asche gebildet hat. Und der rötliche Sandton in vielen Wüsten liegt am hohen Eisengehalt des Sandes.
Der Sand, den wir mit Urlaub, Glück und Entschleunigung verbinden, ist in der internationalen Wissenschaft ganz genau definiert: Die „Steinpartikel" müssen einen Durchmesser zwischen 0,0625 und 2 Millimetern haben, um „Sand" heißen zu dürfen. Alles darunter heißt Schluff, alles darüber sind schon die unbequemen Kiesel.

To ground, a mourning cloud.
From the series *Marbelland.*
Marbella, Spain.

Trauerwolke zu Boden.
Aus der Serie *Marbelland.*
Marbella, Spanien.

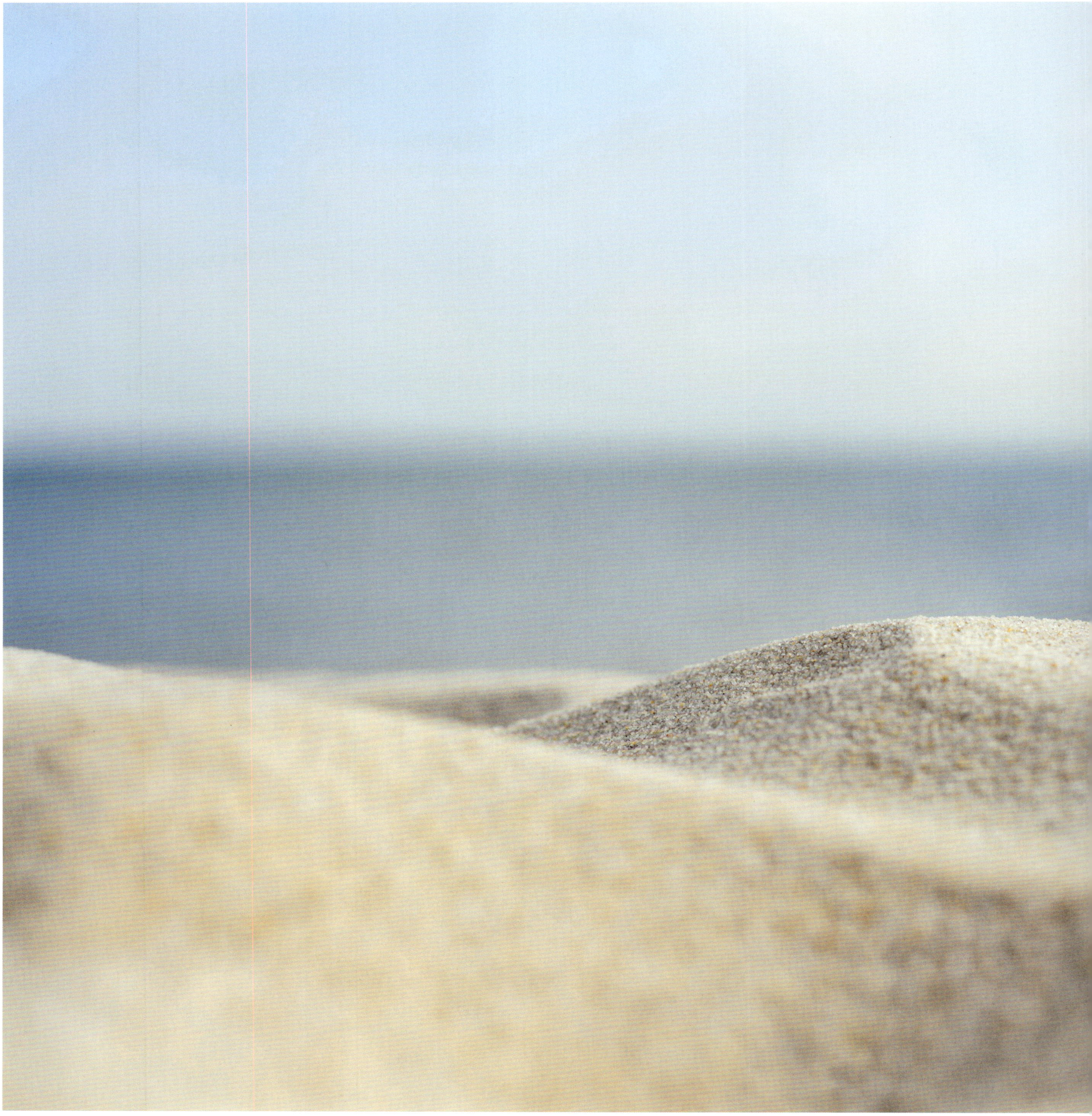

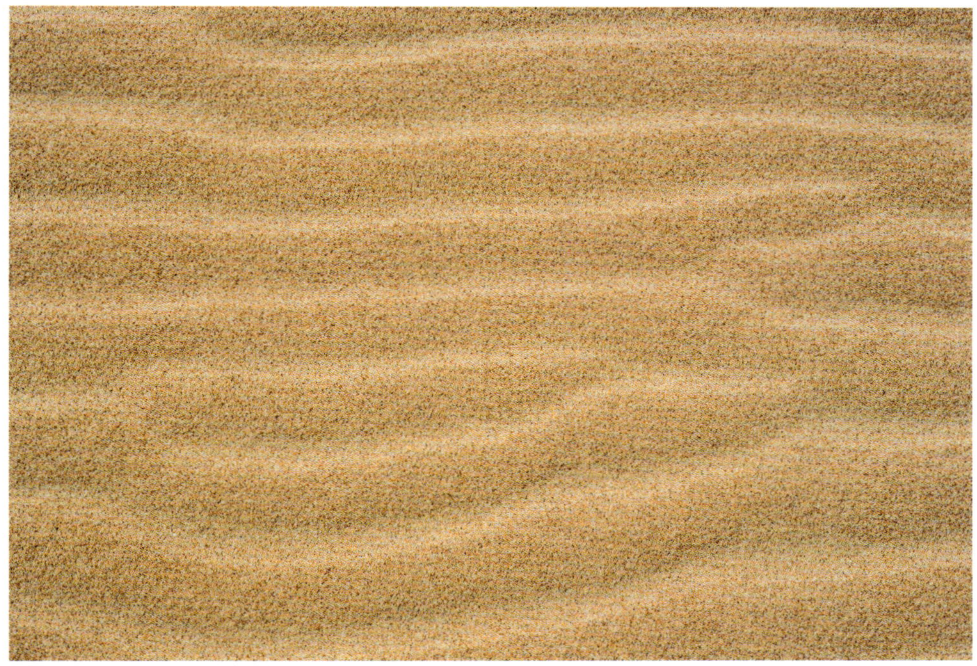

Sandy texture on the beach of Tarifa, Cadiz, Spain.

Sandtextur am Strand von Tarifa, Cadiz, Spanien.

LEFT PAGE

Sandy beach on the isle of Sylt, Germany.

Feiner Sandstrand auf Sylt, Deutschland.

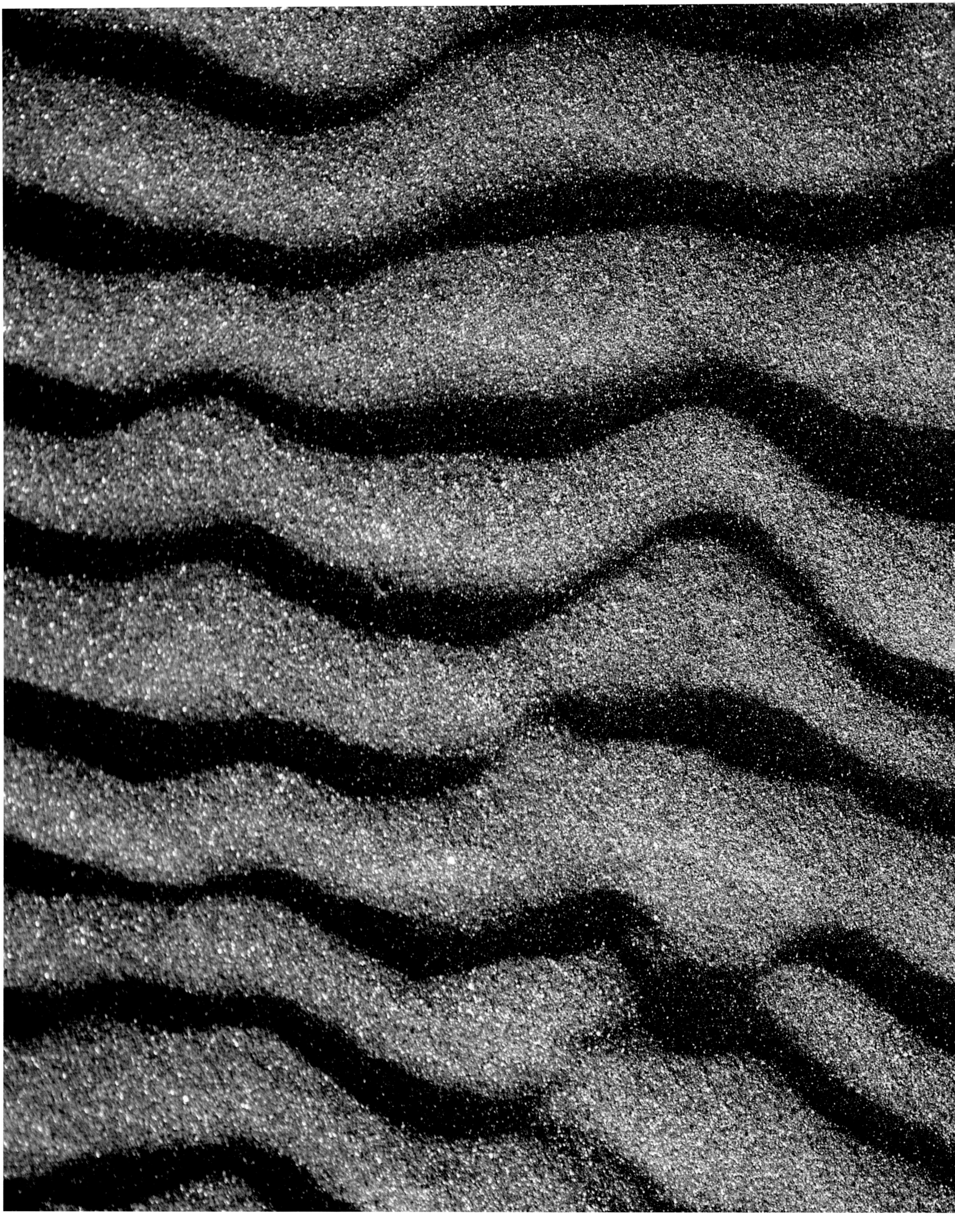

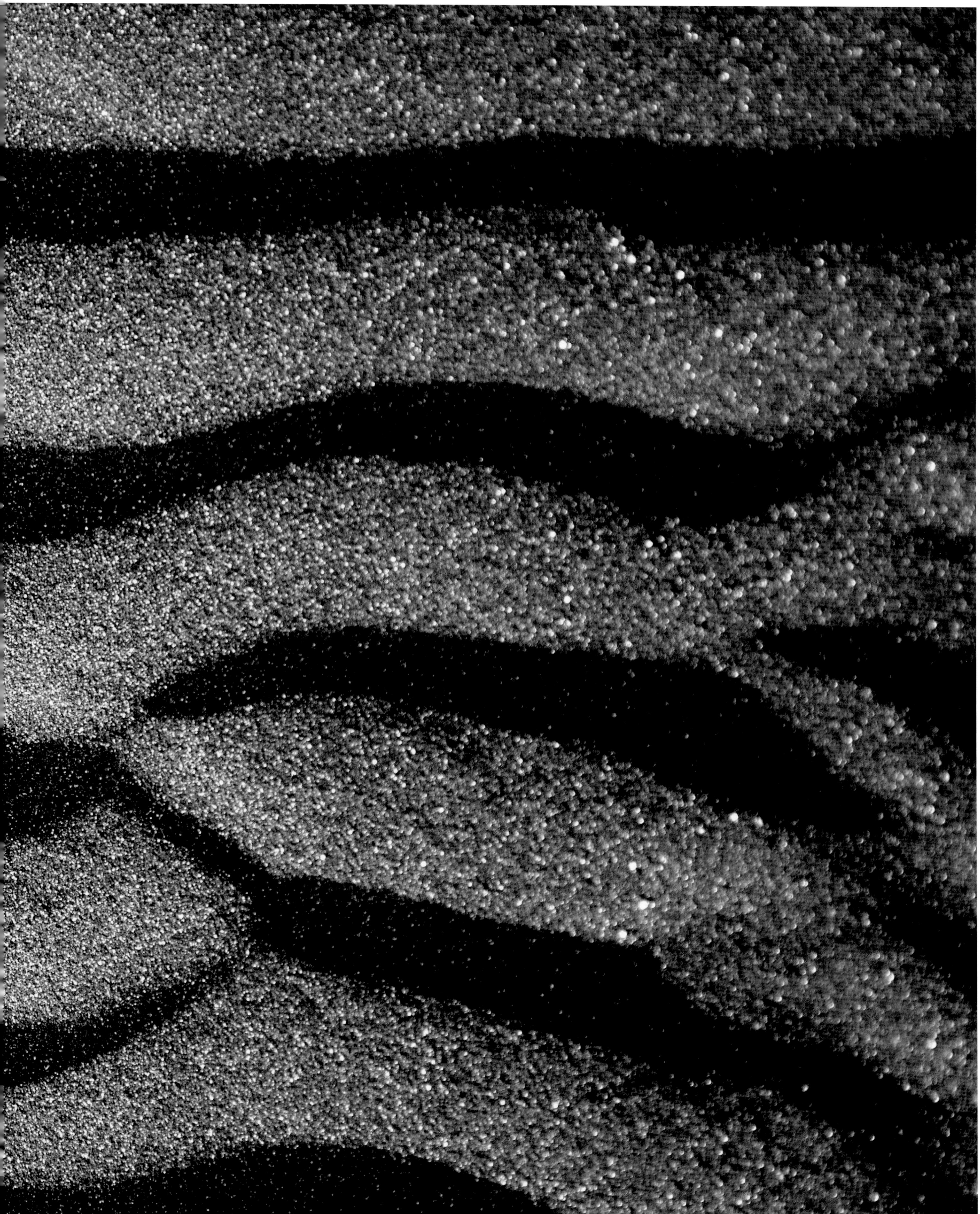

PREVIOUS PAGE

Volcanic black sand beach of Tarajalejo, Fuerteventura, Spain.

Schwarzer Vulkanstrand in Tarajalejo, Fuerteventura, Spanien.

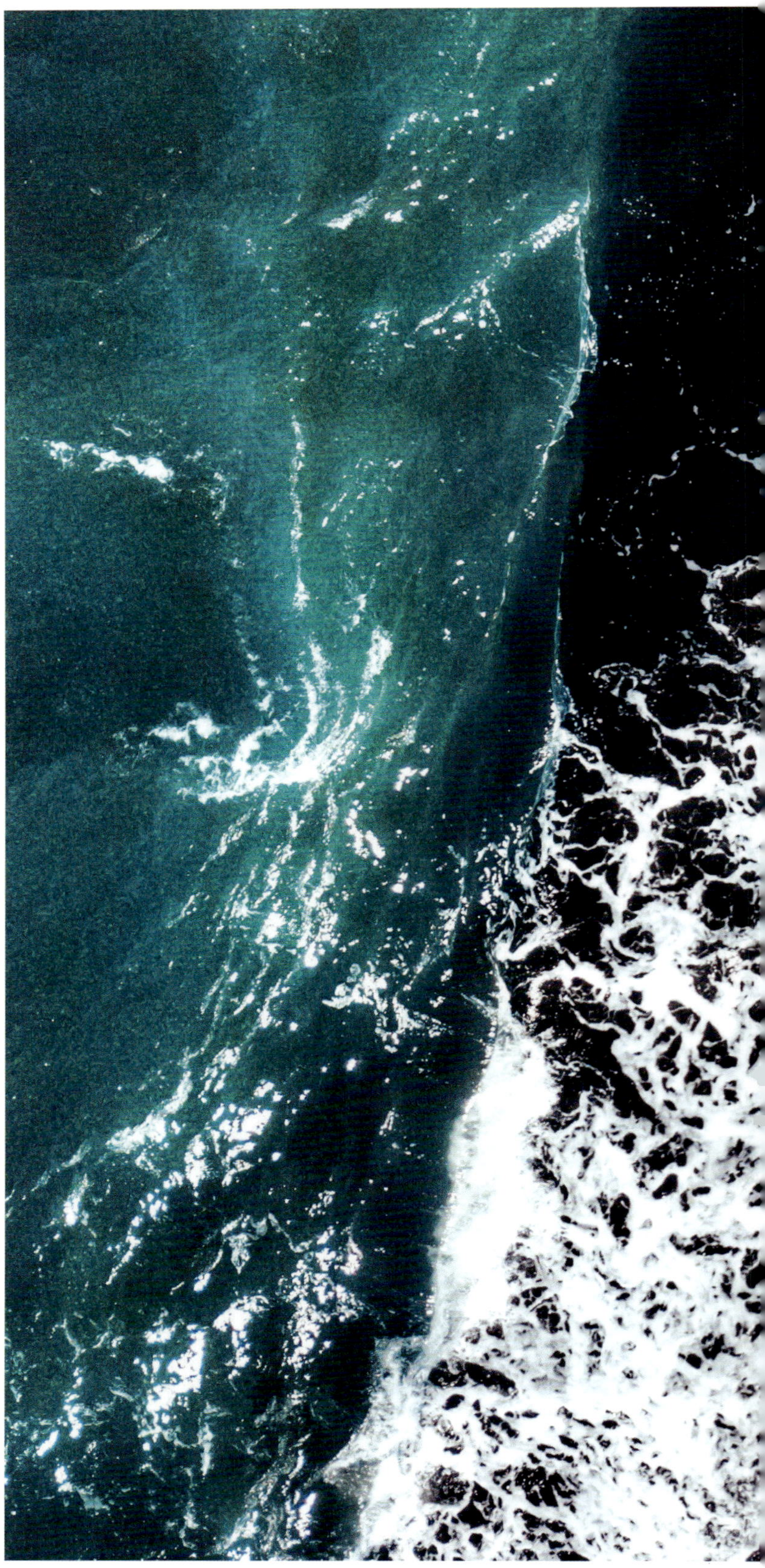

Black beach and white sea foam in Bali, Indonesia.

Schwarzer Strand und weißer Meeresschaum auf Bali, Indonesien.

LAVA, QUARTZ AND PEBBLES: HOW ARE SAND BEACHES FORMED?

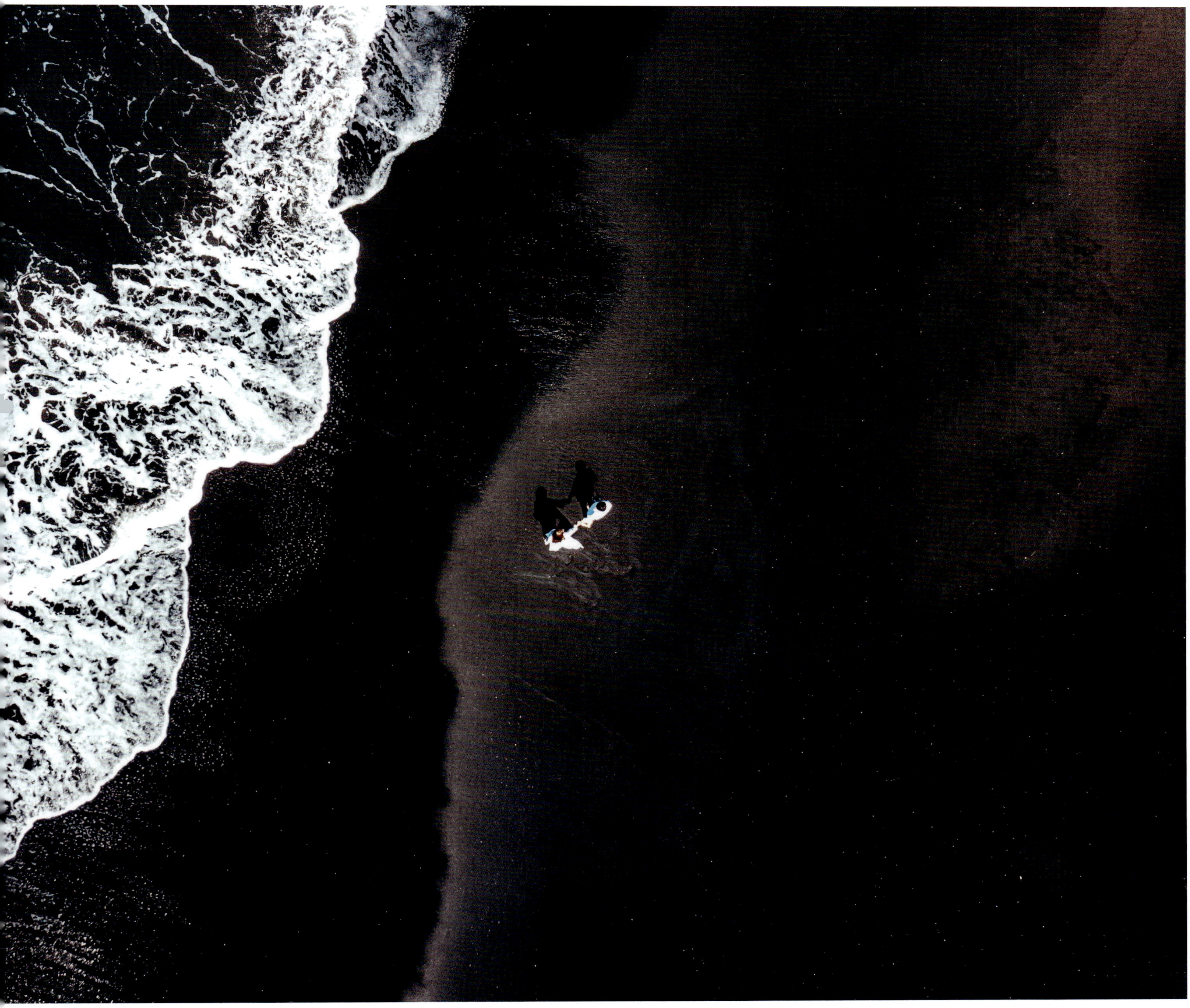

LAVA, QUARZ UND KIESEL: WIE ENTSTEHT EIN SANDSTRAND?

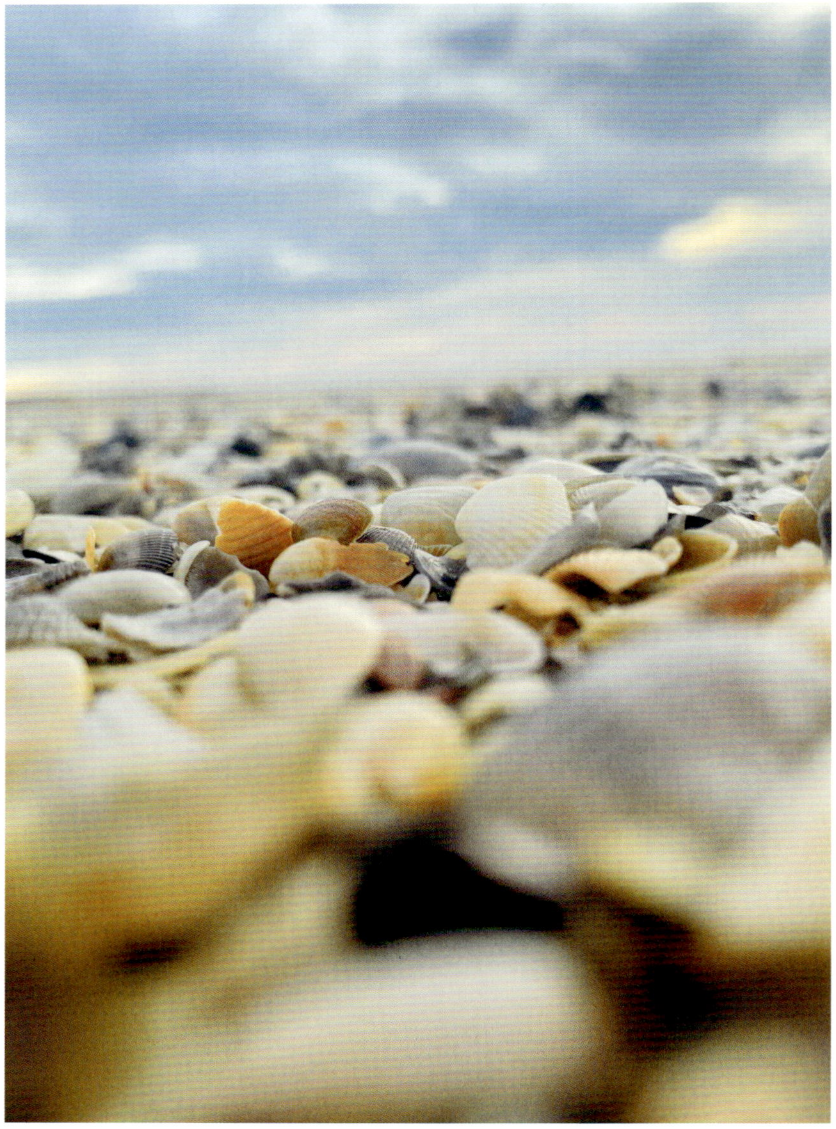

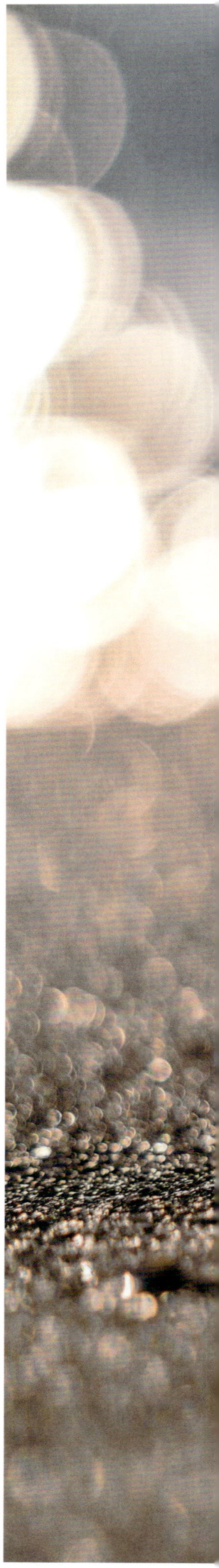

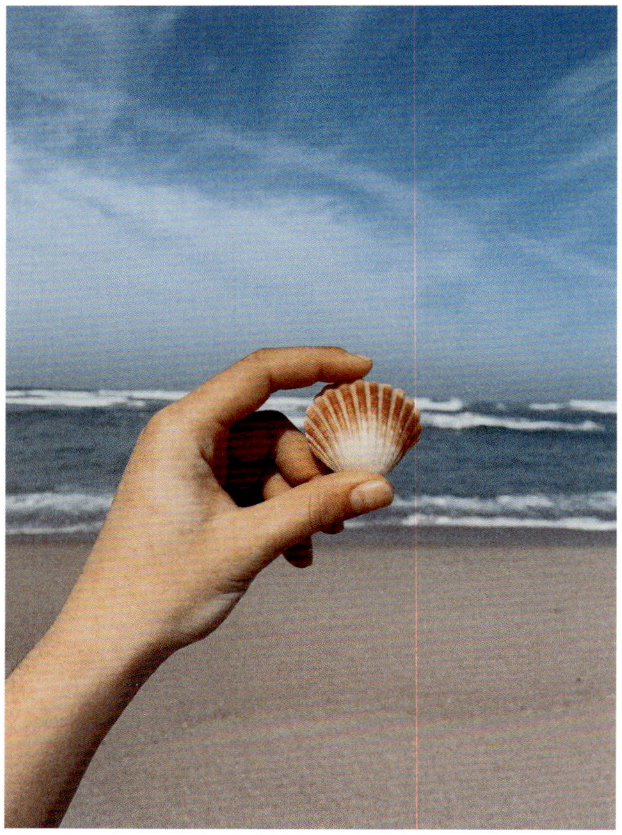

Source material: top: Shells at Cocoa Beach, Florida, United States (top) and on sand of the Island Rügen, Germany (right page)

Ausgangsmaterial: Muscheln am Strand von Cocoa Beach, Florida, Vereinigte Staaten von Amerika (oben) und am Strand auf Rügen, Deutschland

Seashell against waves and sky.
Foz do Arelho, Portugal.

Muschel vor Wellen und Himmel.
Foz do Arelho, Portugal.

LAVA, QUARTZ AND PEBBLES: HOW ARE SAND BEACHES FORMED?

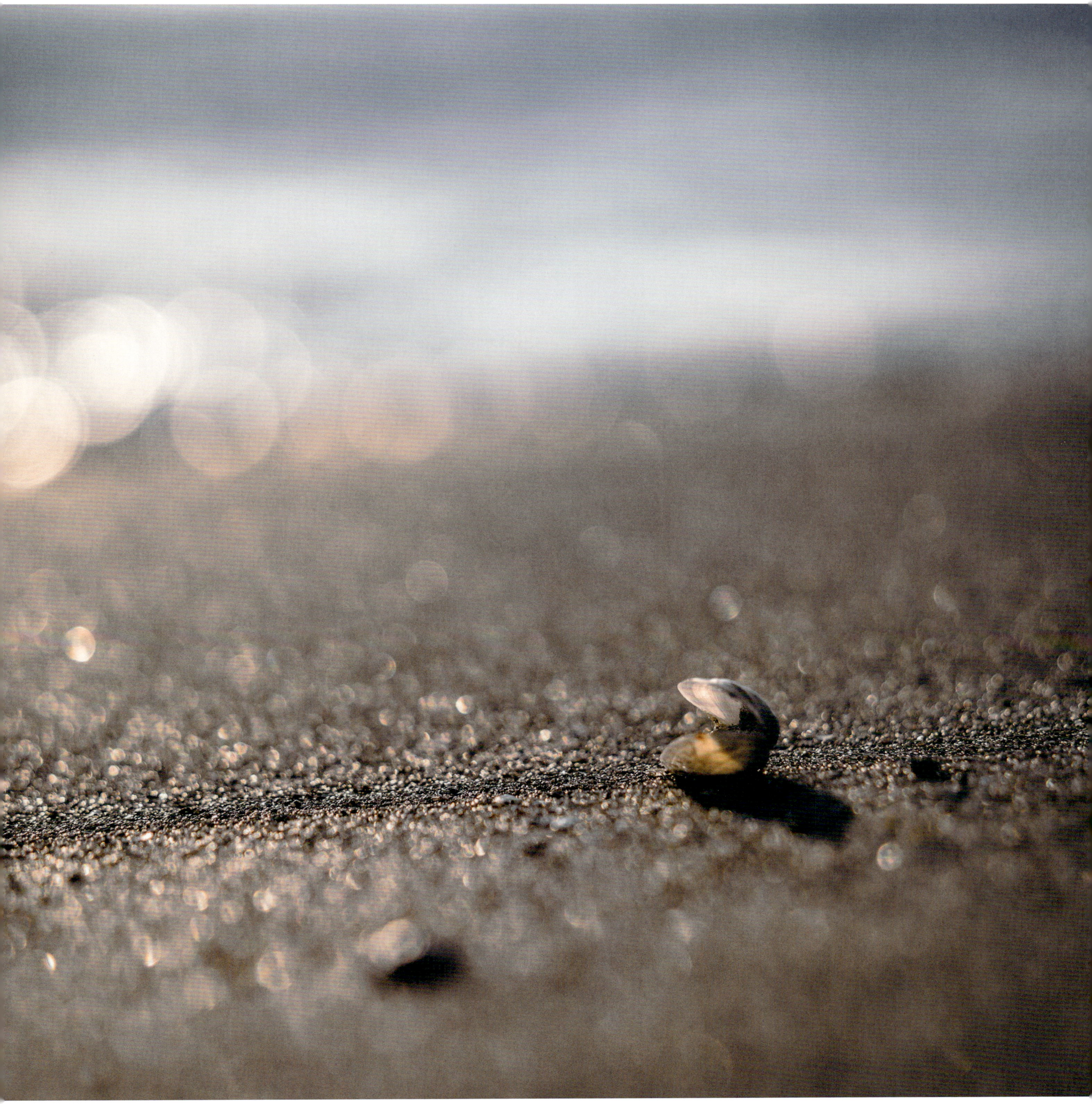

LAVA, QUARZ UND KIESEL: WIE ENTSTEHT EIN SANDSTRAND?

First, waves need a good start, and a long way for it. Preferably the entire Atlantic and Pacific. That's why the waves on the French Atlantic coast, for example, are so powerful. In the Mediterranean (or on small inland lakes), the waves are rather moderate, even during heavy storms, because the water simply lacks the many kilometers needed to build up properly.
Even if some particularly fatal waves, such as the mighty tsunamis, build up after undersea earthquakes–virtually all other waves originate from the wind sweeping over the ocean surface and stirring up the water.
The stronger and longer the wind blows, the more energy is transferred to the waves. The waves merge with one another, becoming larger and larger. The international Douglas scale, which ranges from 0 (calm (glassy)) to 9 (phenomenal), indicates the average wave height: The degree of 9 means waves in excess of 14 meters high–not uncommon on the high seas.
Surprisingly, the water molecules themselves hardly move from their spot. This means the wave crest that arrives in southern France or Portugal has definitely not traveled thousands of kilometers across the Atlantic. Instead, we have to imagine the waves like countless rotating turbines one after the other, which simply pass on their energy. And that's also the reason for the dangerous undertow after breaking–the water returns, dragging the swimmer along with it.
The wave breaks when it runs onto the beach because the seabed rises rapidly. The bottom part of the wave slows down, the kinetic energy rises steeply until the crest becomes too high and is pulled down by gravity–the height of emotion especially for children in the water, but also for surfers who are able to accurately read the breakers and glide along in their tunnels.
But what is the surge that prevails when there is no wind? Here the experts speak of swell. These are choppy seas from the last storm–the sea needs a few days to calm down.

Als erstes brauchen Wellen Anlauf. Viel Anlauf. Am besten den gesamten Atlantik und Pazifik. Deswegen sind die Wellen etwa an der französischen Atlantikküste auch so mächtig. Im Mittelmeer (oder gar auf kleinen Binnenseen) ist der Wellengang auch bei heftigem Sturm eher mäßig., denn dem Wasser fehlen einfach die vielen Kilometer, um sich ordentlich aufzuschaukeln.
Auch wenn einige besonders fatale Wellen, etwa die mächtigen Tsunamis, sich nach unterseeischen Beben aufbäumen – so gut wie alle übrigen Wogen entstehen aus dem Wind, der über die Meeresoberfläche streicht und das Wasser aufwühlt. Je stärker und länger der Wind bläst, desto mehr Energie überträgt sich auf die Wellen. Die Wellen verschmelzen miteinander und werden immer größer. Die internationale Douglas-Skala, die von 0 („spiegelglatte See“) bis 9 („schwerer Seegang“) reicht, gibt die durchschnittliche Wellenhöhe an: Stärke 9 sind Wogen von mehr als 14 Metern Höhe – keine Seltenheit auf hoher See.
Erstaunlich: Die Wassermoleküle selbst bewegen sich dabei kaum von der Stelle. Der Wellenkamm, der in Südfrankreich oder Portugal ankommt, ist also keineswegs tausende Kilometer durch den Atlantik gereist. Wir müssen uns den Wellengang stattdessen wie unzählige rotierende Turbinen hintereinander vorstellen, die ihre Energie einfach weitergeben. Und deswegen gibt es auch den gefährlichen Sog nach der Brechung – das Wasser kehrt zurück und reißt den Badenden mit sich.
Die Welle bricht, wenn sie auf den Strand läuft, weil der Meeresboden rasch ansteigt. Der untere Teil der Welle wird langsamer, die Bewegungsenergie steigt steil nach oben, bis der Kamm zu hoch wird und von der Schwerkraft nach unten gezogen wird – das Höchste der Gefühle vor allem für badende Kinder, aber auch für Surfer, die die Wellenbrecher genau lesen können, um in ihren Tunneln dahinzugleiten.
Aber was ist das für ein Wellengang, der bei Windstille herrscht? Hier sprechen die Experten von Dünung. Es handelt sich um unruhige See vom letzten Sturm – das Meer braucht ein paar Tage, bis es sich wieder beruhigt.

The sea at Mondragone, Italy.

Das Meer bei Mondragone, Italien.

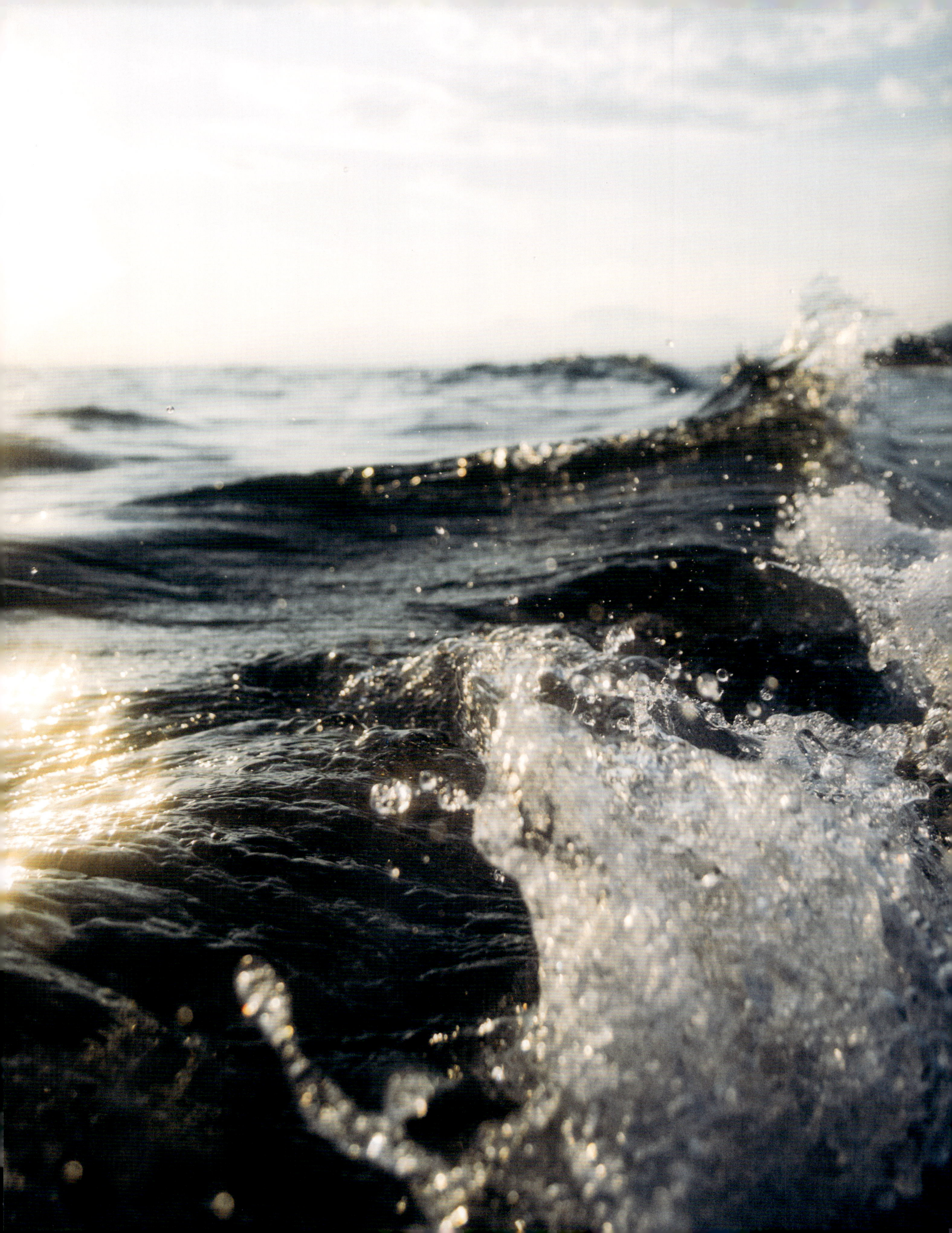

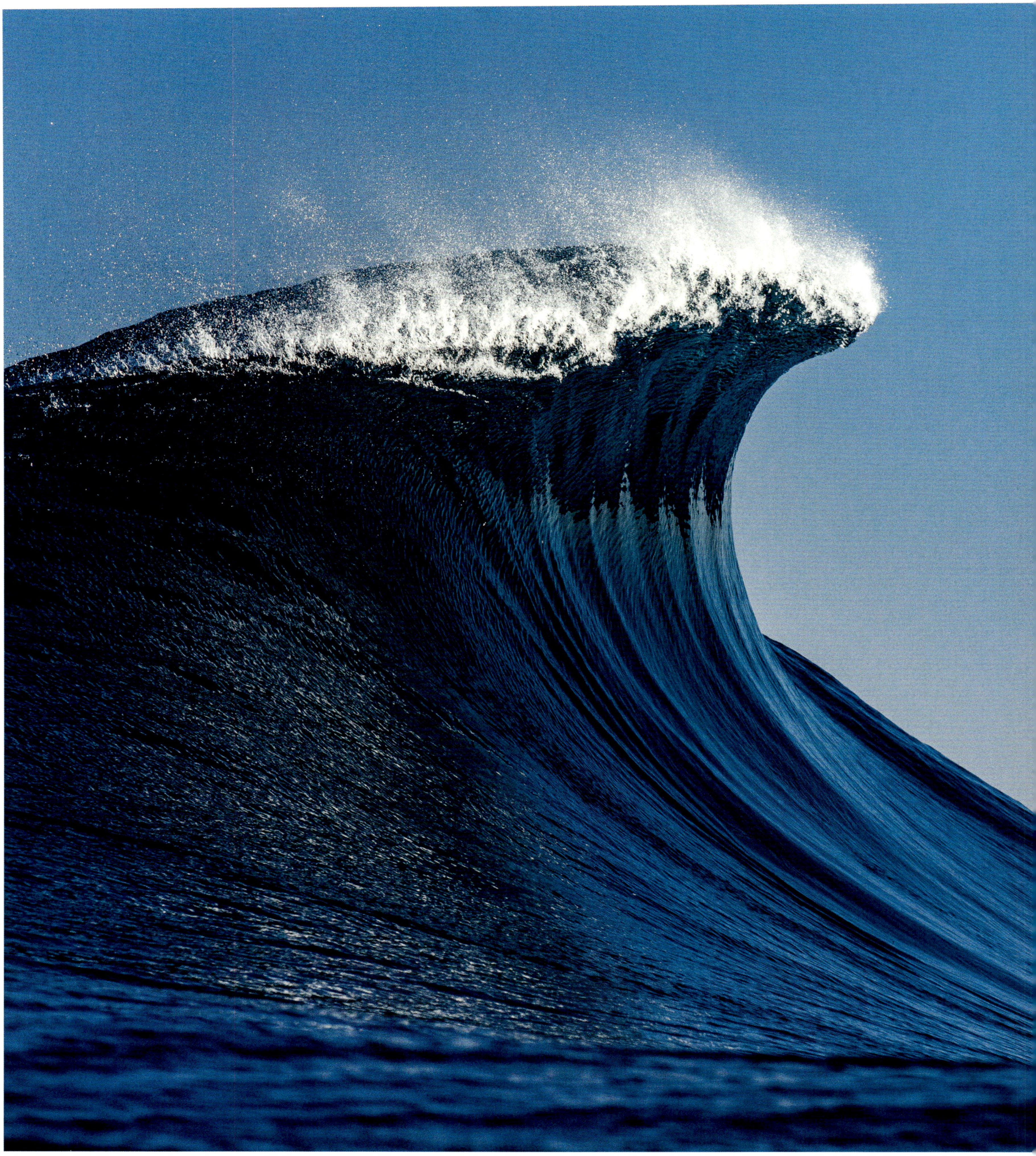

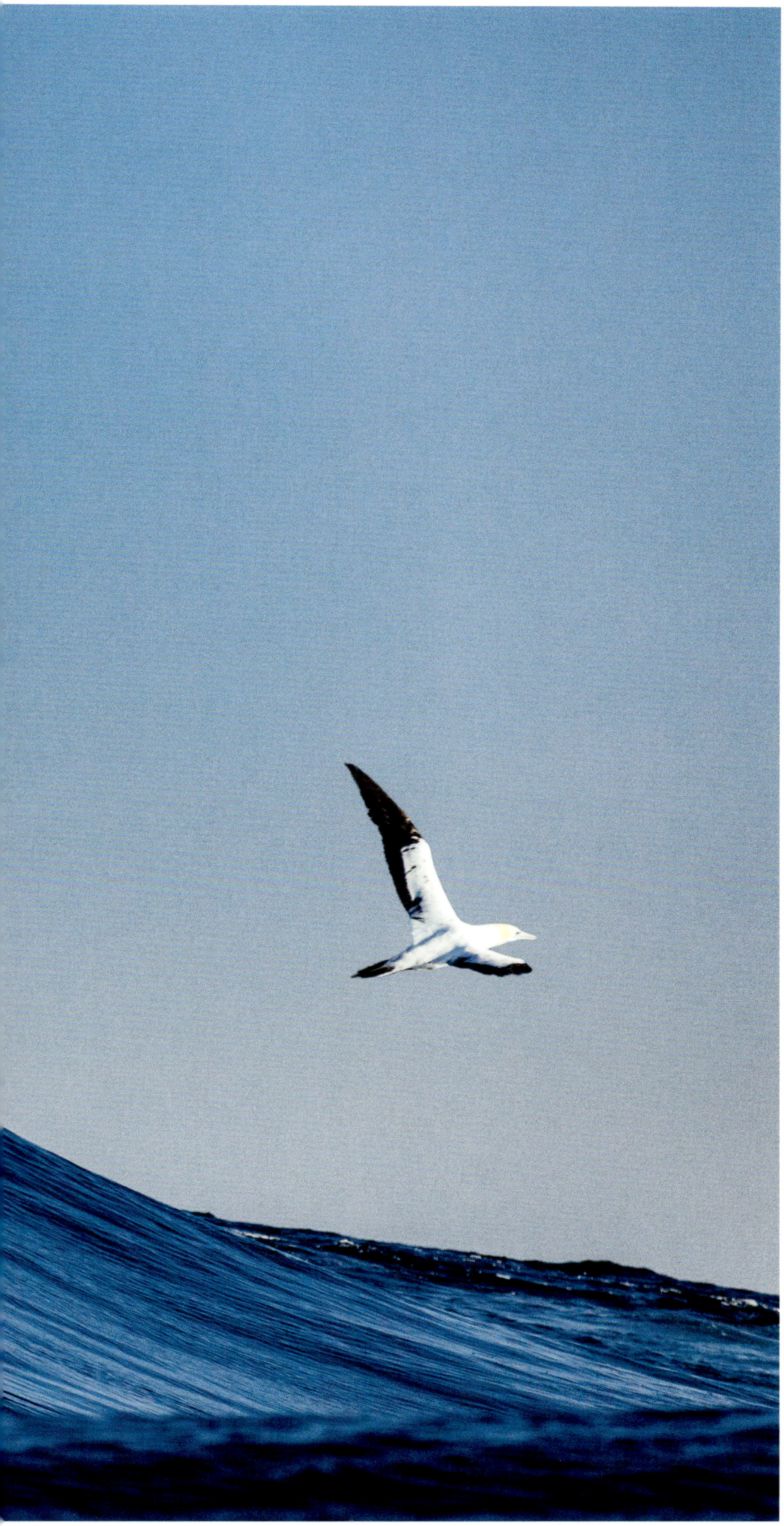

Albatros being chased by a giant cresting wave in the open ocean of Australia.

Ein Albatros wird von einer riesigen Kammwelle auf dem offenen Ozean Australiens gejagt.

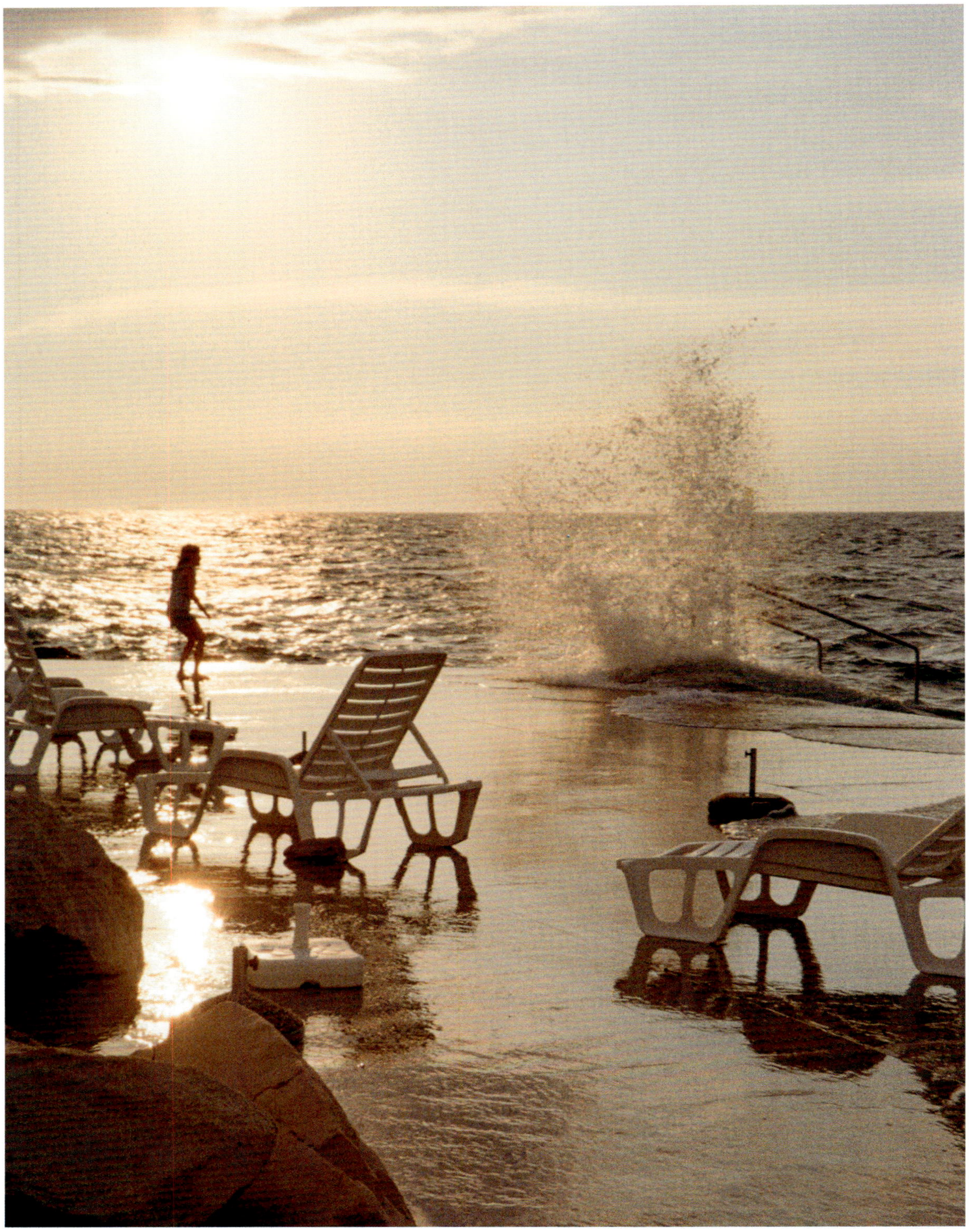

Youth. Piran, Slovenia.

Jugend. Piran, Slowenien.

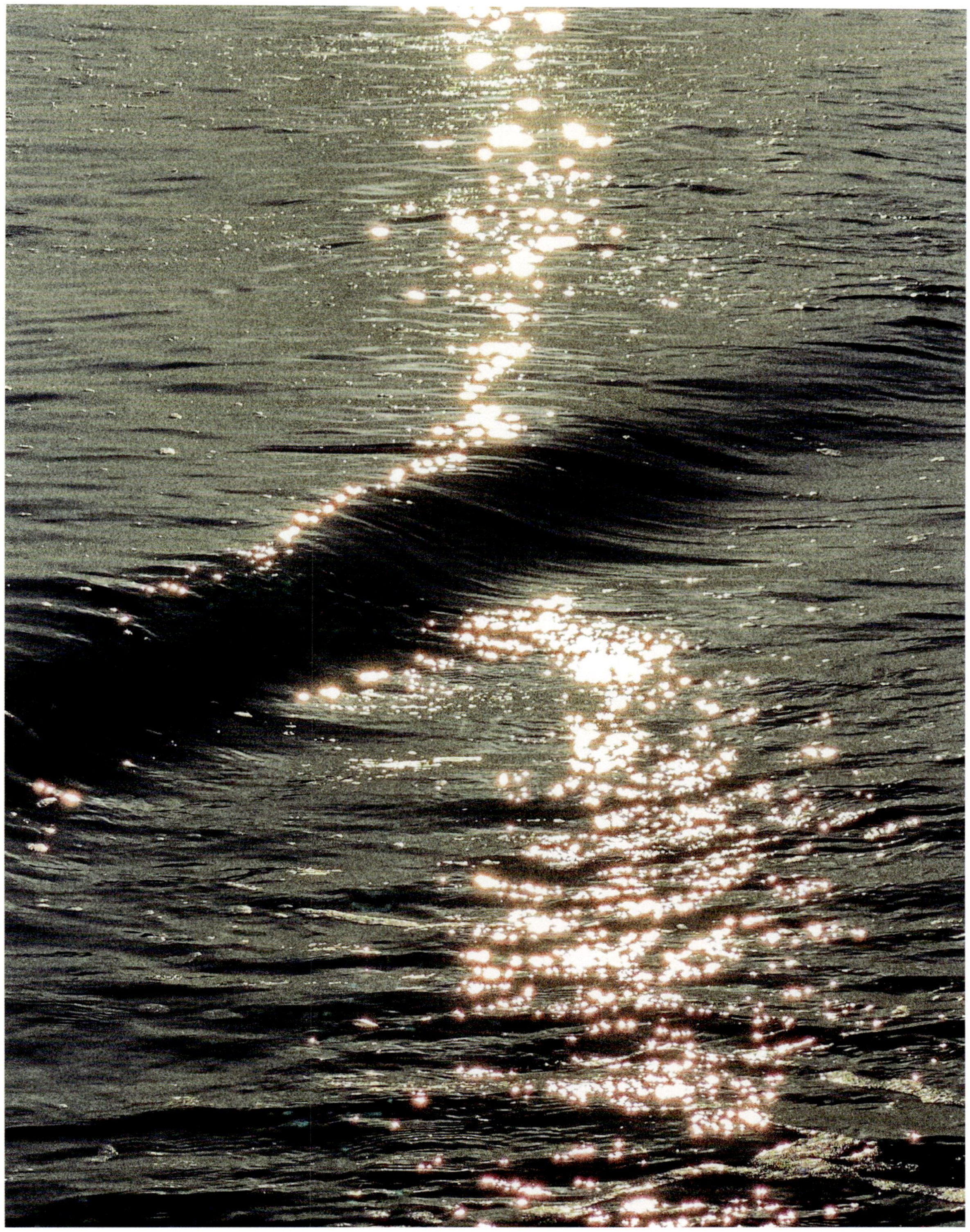

You see, all I my electricity will make it across your sea. From the series *The Gathering.* Marbella, Spain.

Siehst du, all meine Energie wird dich über dein Meer erreichen. Aus der Serie *The Gathering.* Marbella, Spanien.

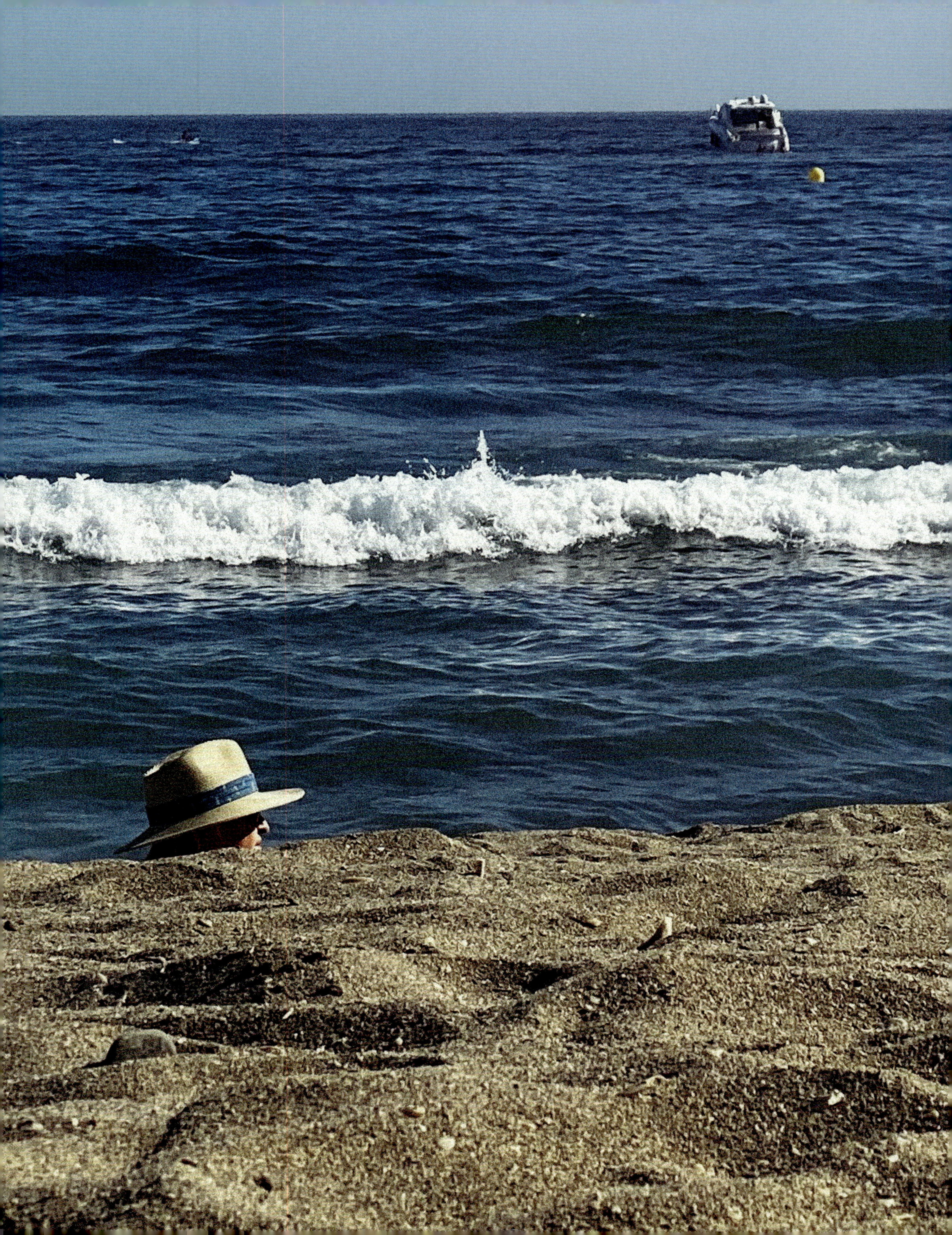

Mad.A.M. show: Marbelland is full of solar wonder, mystery and danger. Some say, to survive it, you need to be a hatter as madam. From the series *Marbelland*. Marbella, Spain.

Mad.A.M. show. Marbelland steckt voller Wunder, Mysterien und Gefahr. Manche sagen, man muss eine Hutmacherin sein, um zu überleben. Aus der Serie *Marbelland*. Marbella, Spanien.

STYLE

STILVOLL

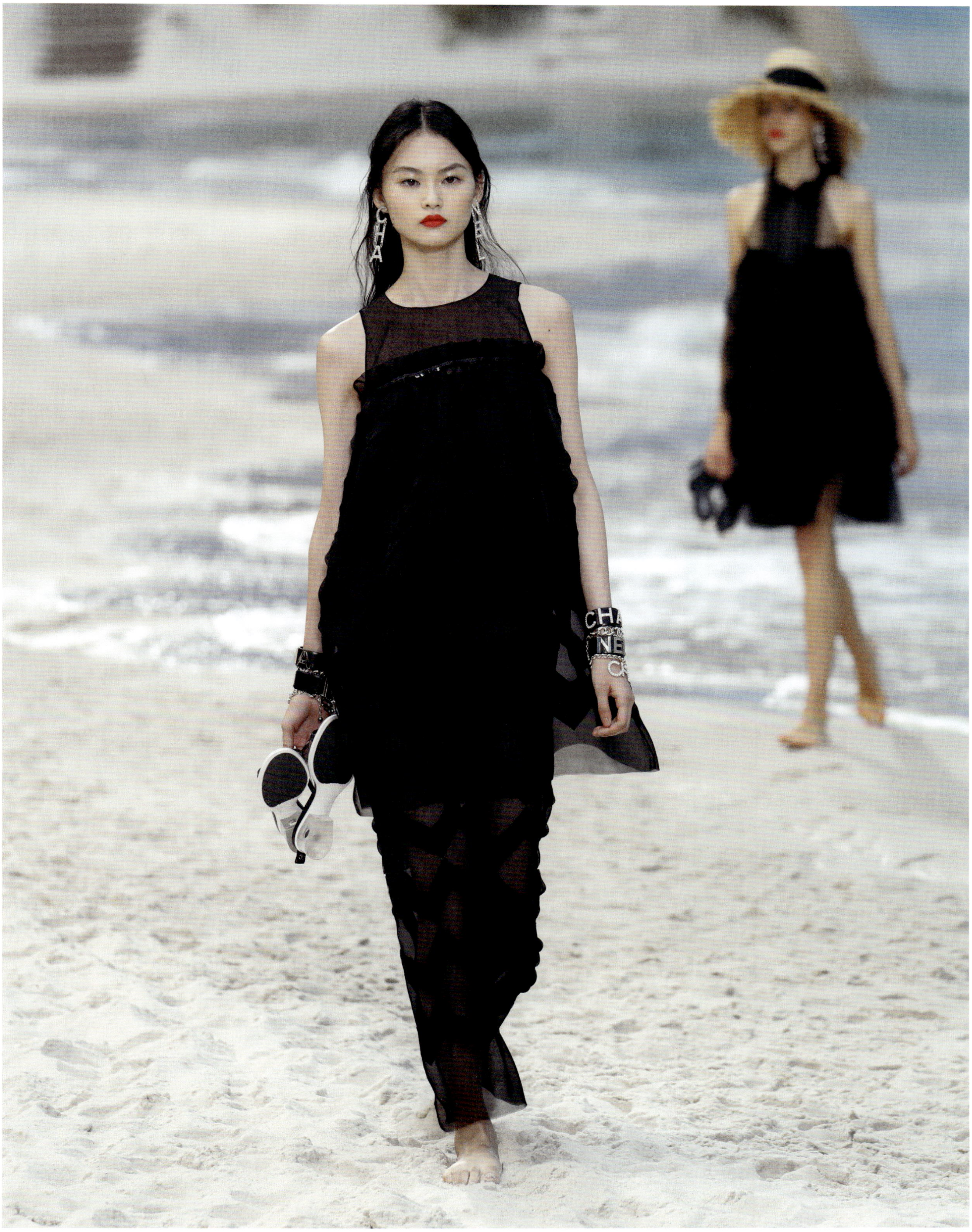
CHA
CHA
NE

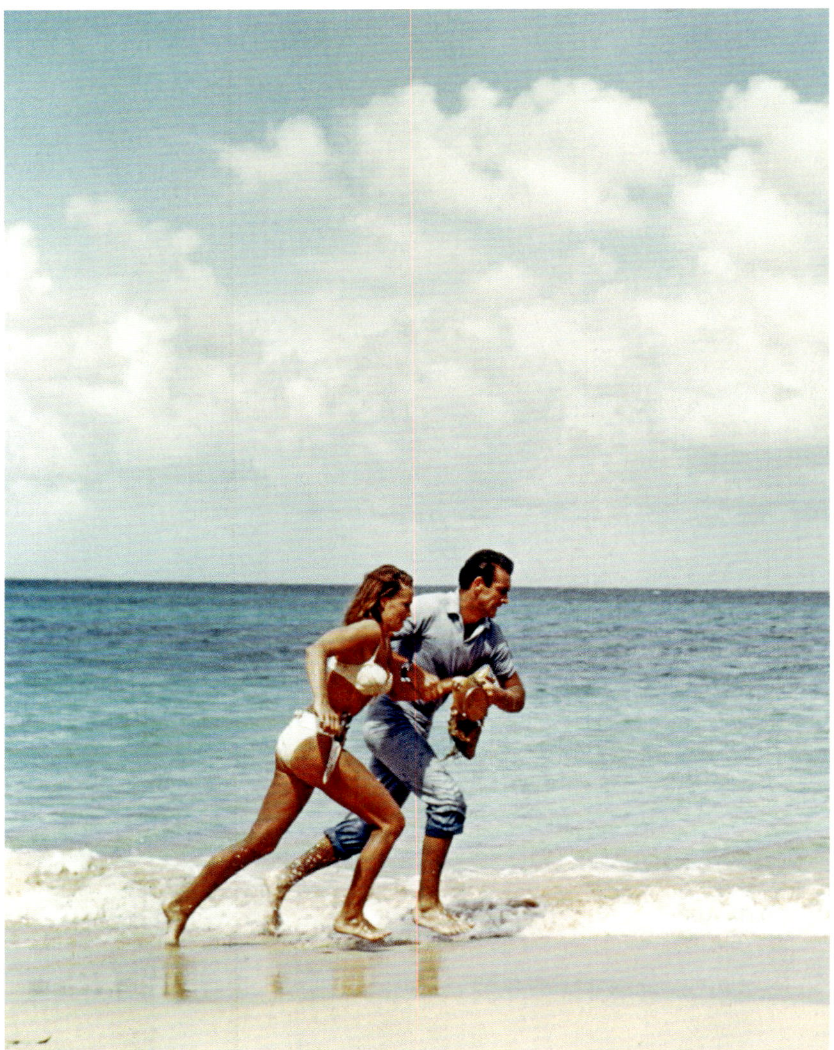

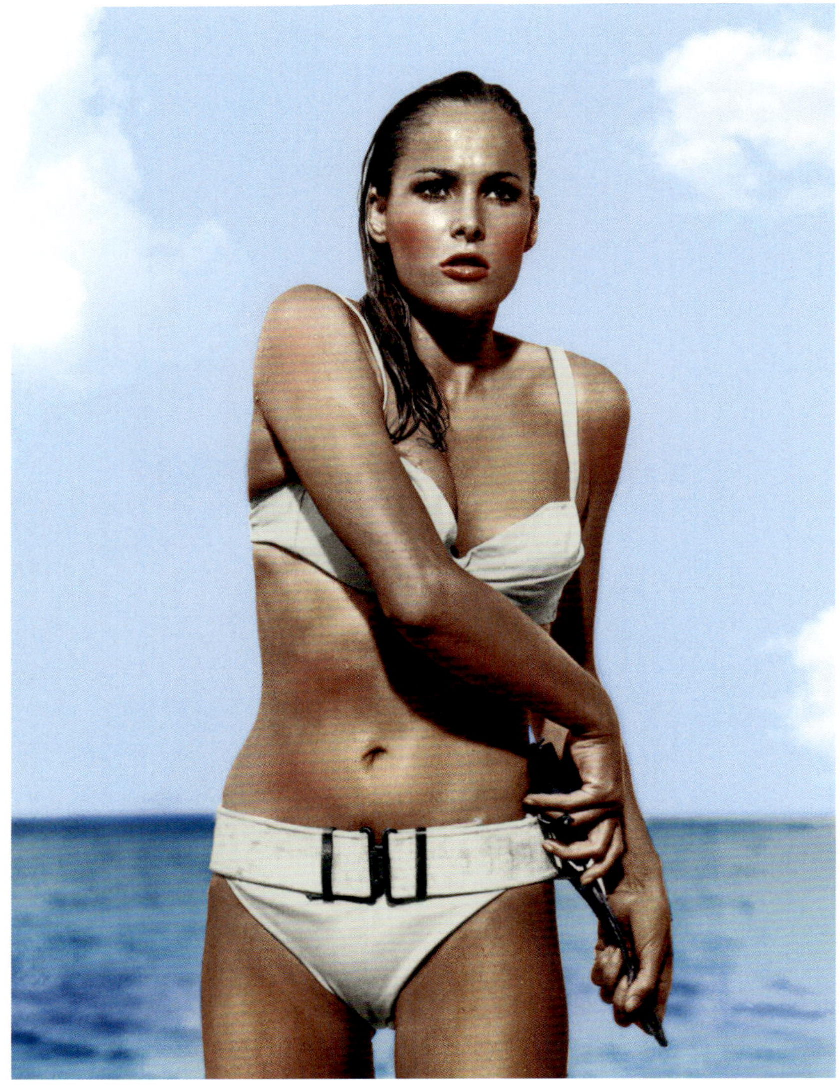

Ursula Andress and Sean Connery in *Dr. No.* Oracabessa, Jamaica 1962.

Ursula Andress und Sean Connery in *James Bond – 007 jagt Dr. No.* Oracabessa, Jamaica 1962.

PREVIOUS PAGE

A model walks the beach runway for Karl Lagerfelds *Chanel à 'a plage.*

Model auf dem Strandlaufsteg für Karl Lagerfelds *Chanel à 'a plage.*

It was a strange time, when atomic bomb tests symbolized progress, momentum, and a carefree future. That's why the new, skimpy two-piece for the beach, that revolutionary piece of fabric, was named after Bikini Atoll, where a U.S. Air Force B29 plane had just dropped an atomic bomb. Bikini–the name not only sounded exotic, the "bi" (two) also nicely expressed the difference between it and the one-piece swimsuit.
One surprise, though: The bikini is much older than nuclear fission. Roman mosaics show women playing sports in tight two-piece outfits. And even in the nudist movement around 1900, full nudity was renounced and the ladies covered up with a two-piece.
But the bikini truly entered the public consciousness on July 5, 1946, when designer Louis Réard introduced it to the public, presented by nude dancer Michelin Bernardini, a few days after the abovementioned bombing. It caused heated debates and even protests on beaches around the world. Lawmakers were completely overwhelmed–so the skimpy two-piece was permitted on the French Mediterranean coast, but at the same time banned on the French Atlantic coast. Surprisingly, the bikini went out of fashion again a few years later, because one-piece suits were a better solution to tackling alleged body problems.
But then Ursula Andress emerged from the waters in Jamaica's Oracabessa in 1962 in the first James Bond film, *Dr. No*, and everything changed. (The beach was later renamed James Bond Beach, by the way, and the bikini worn by Andress was auctioned off for $150,000 in 2001.)

Shortly thereafter, young people unleashed a wave of sexual liberation, while the advertising industry discovered how good it was to advertise products even beyond the beach, be it whiskey, stereos or car tires, using as much bare skin as possible.
At present, two opposing trends can be observed. In the U.S., Europe and South America, self-confident women like to appear in the extra skimpy microkini, while in the Islamic world the burkini dominates, covering the body almost completely.

Das waren merkwürdige Zeiten, als Atombombentests Fortschritt, Dynamik und eine sorglose Zukunft symbolisierten. Deswegen wurde der neue, knappe Zweiteiler für den Strand, jenes revolutionäre Stückchen Stoff, nach dem Bikini-Atoll benannt, auf dem just ein B29-Flugzeug der US Air Force eine Atombombe abgeworfen hatte. Bikini – der Name klang nicht nur exotisch, das „Bi" (zwei) drückte auch gut den Unterschied zum einteiligen Badeanzug aus.
Überraschung: Der Bikini ist schon viel älter als die Kernspaltung. Auf römischen Mosaiken sind sporttreibende Frauen im engen Zweiteiler zu sehen. Und auch in der FKK-Bewegung verzichtete man um das Jahr 1900 auf völlige Nacktheit und verhüllte die Damen mit einem Zweiteiler.
Doch so richtig ins öffentliche Bewusstsein kam der Bikini am 5. Juli 1946, als Designer Louis Réard ihn der Öffentlichkeit vorstellte, präsentiert von der Nackttänzerin Michelin Bernardini, ein paar Tage nach dem oben erwähnten Bombenabwurf. Und er sorgte an den Stränden weltweit für heftige Debatten und sogar Protestaktionen. Die Gesetzgeber waren völlig überfordert – so war der knappe Zweiteiler an der französischen Mittelmeerküste erlaubt, zugleich an der französischen Atlantikküste jedoch verboten. Erstaunlicherweise kam der Bikini ein paar Jahre später wieder aus der Mode, denn mit Einteilern ließen sich vermeintliche figürliche Probleme besser beheben.
Doch dann entstieg Ursula Andress im ersten James-Bond-Film *007 jagt Dr. No* 1962 im jamaikanischen Oracabessa den Fluten, und alles veränderte sich. (Der Strand wurde später übrigens in James-Bond-Beach umgetauft, und der Andress-Bikini wurde 2001 für 150.000 Dollar versteigert). Kurz danach trat die Jugend die Welle der sexuellen Befreiung los, und parallel dazu entdeckte die Werbung, wie gut sich mit möglichst nackter Haut für Produkte auch jenseits der Strandes werben ließ, ob für Whisky, Stereoanlagen oder Autoreifen.
Derzeit sind zwei gegenläufige Trends zu beobachten: In den USA, Europa und Südamerika treten selbstbewusste Frauen gern im extra knappen „Mikrokini" auf, während in der islamischen Welt der „Burkini" dominiert, der den Körper fast vollständig verhüllt.

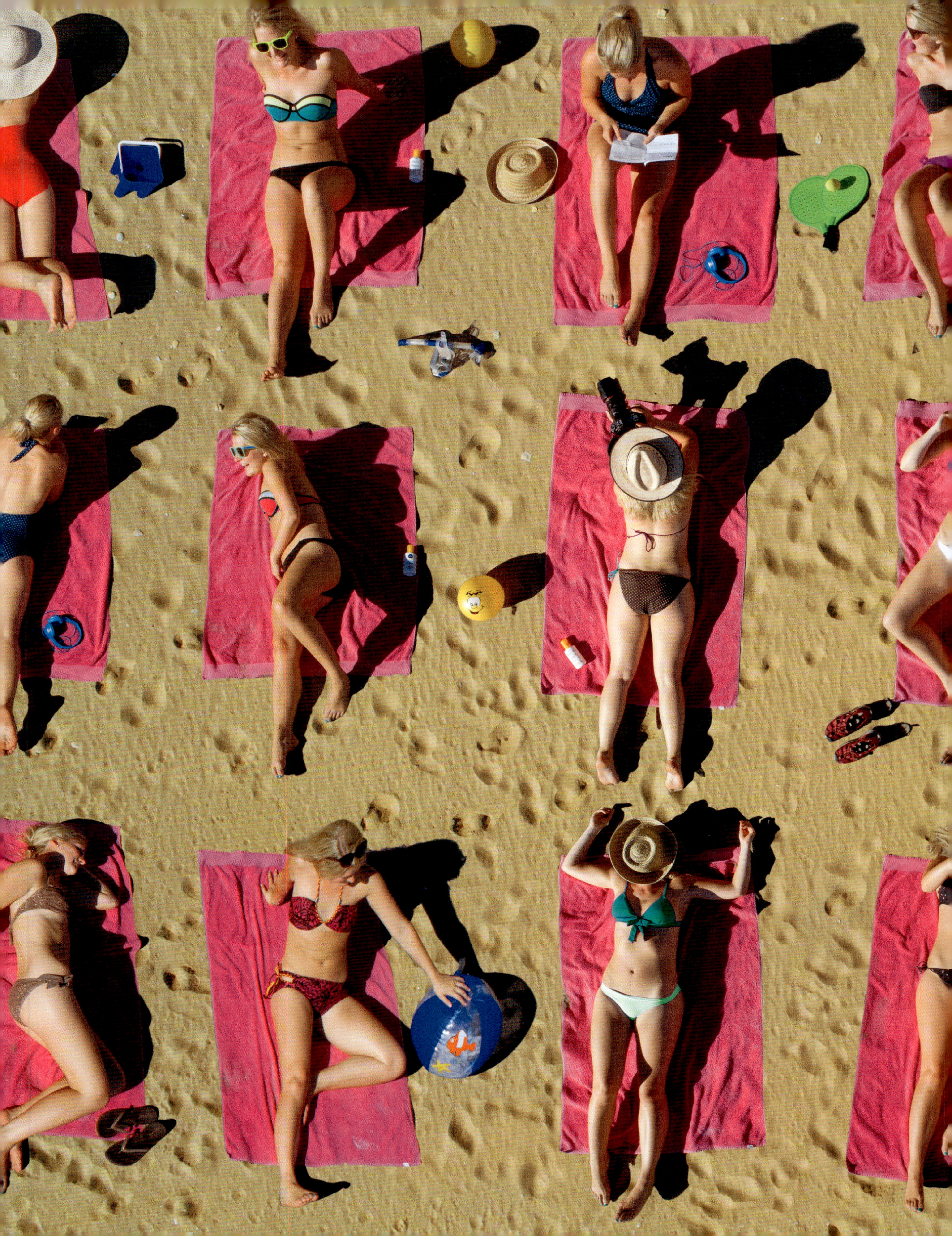

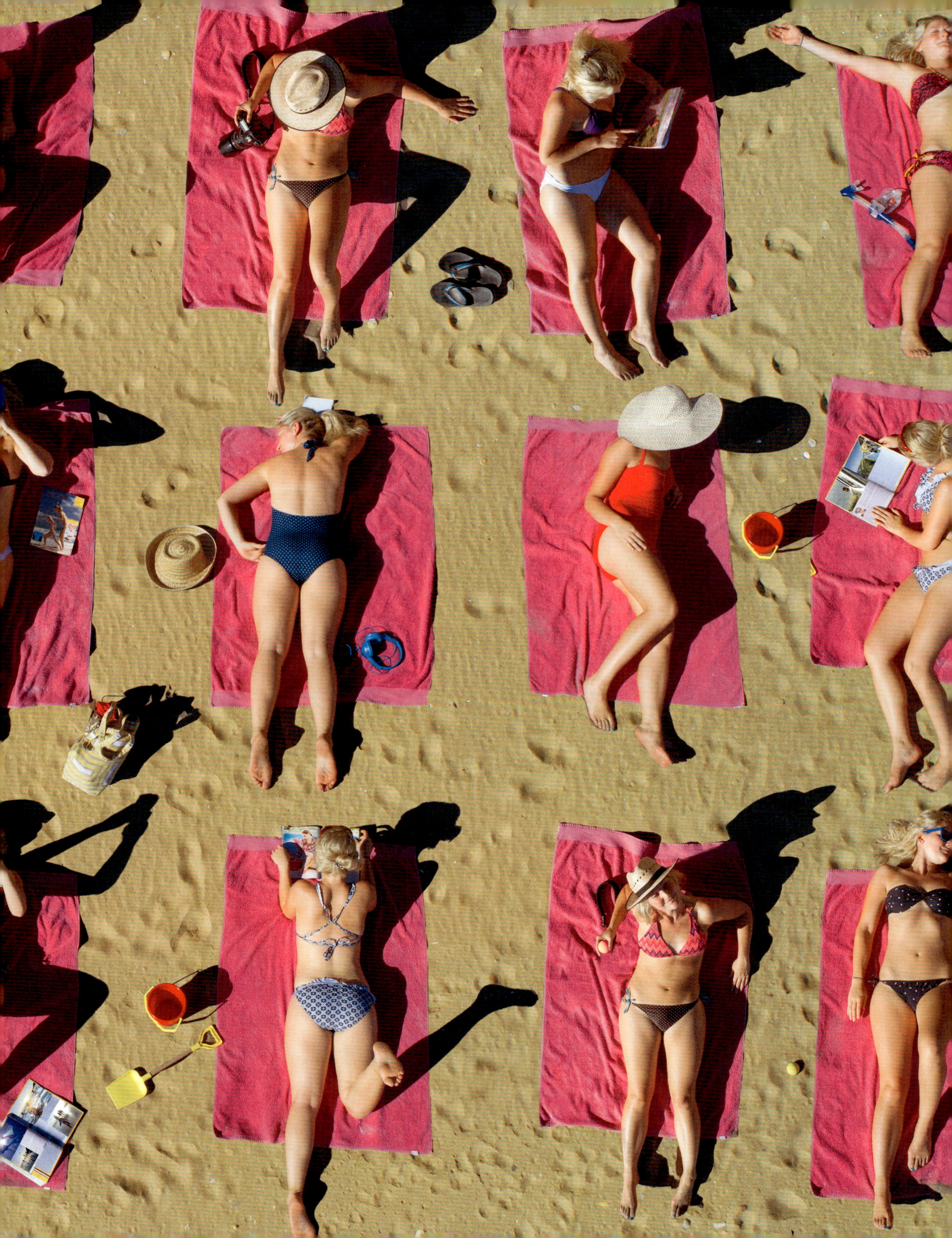

PREVIOUS PAGE

Bikini variations. Tregantle, Cornwall, UK.

Bikinivariationen. Tregantle, Cornwall, UK.

RIGHT PAGE

Marilyn Monroe at the beach near Hollywood, USA. Record date unknown.

Marilyn Monroe am Strand in der Nähe von Hollywood, USA. Aufnahmedatum nicht bekannt.

Micheline Bernardini presents the bikini by designer Louis Réard to the public. Paris, France 1946.

Micheline Bernardini präsentiert der Öffentlichkeit den Bikini von Designer Louis Réard. Paris, France 1946

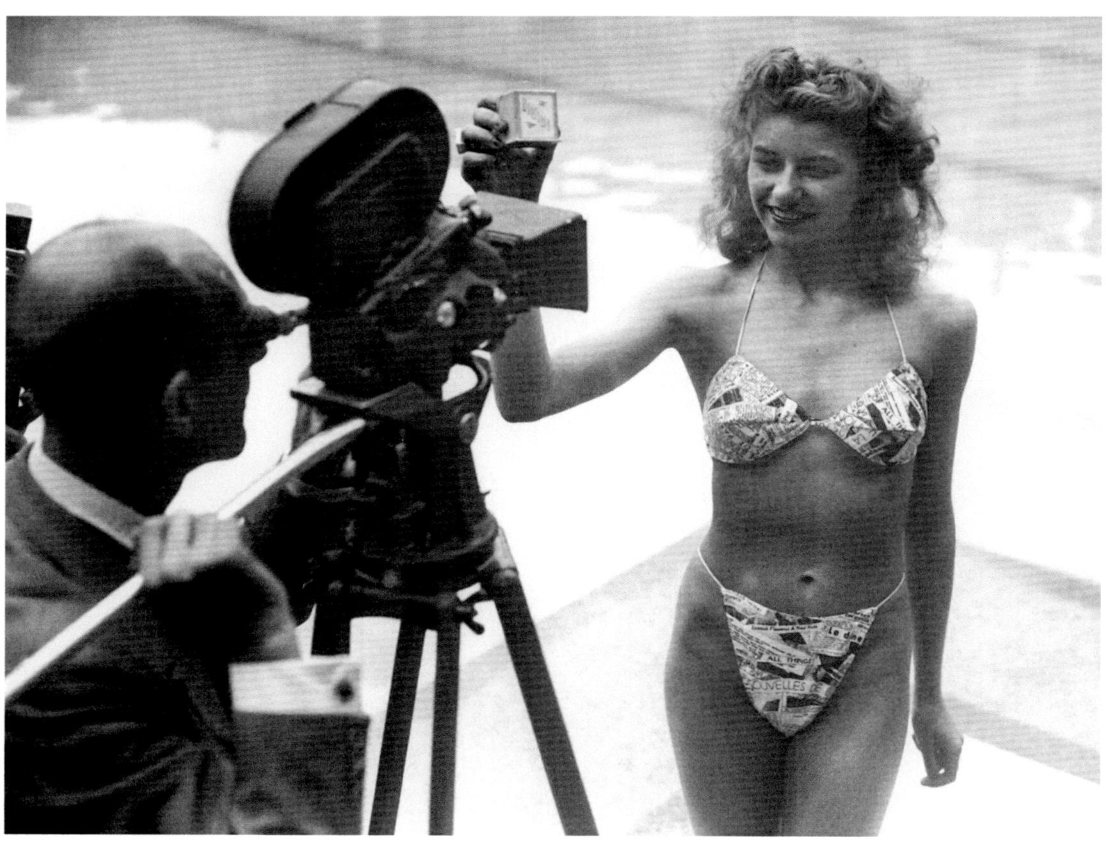

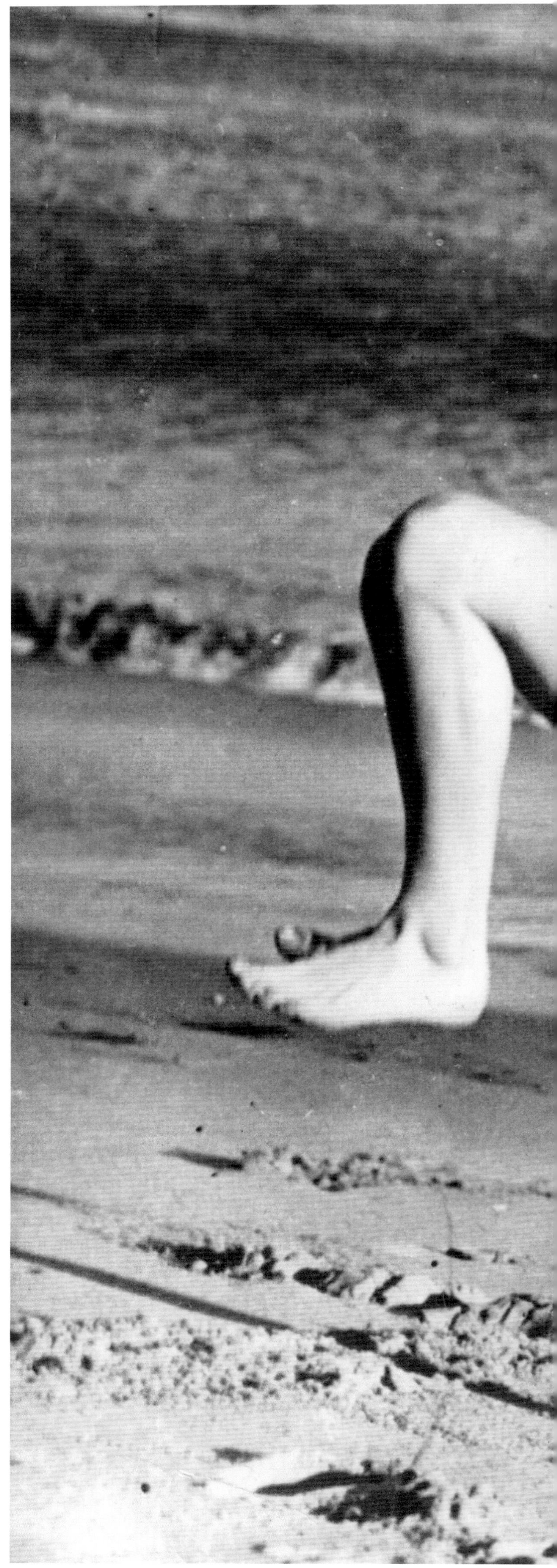

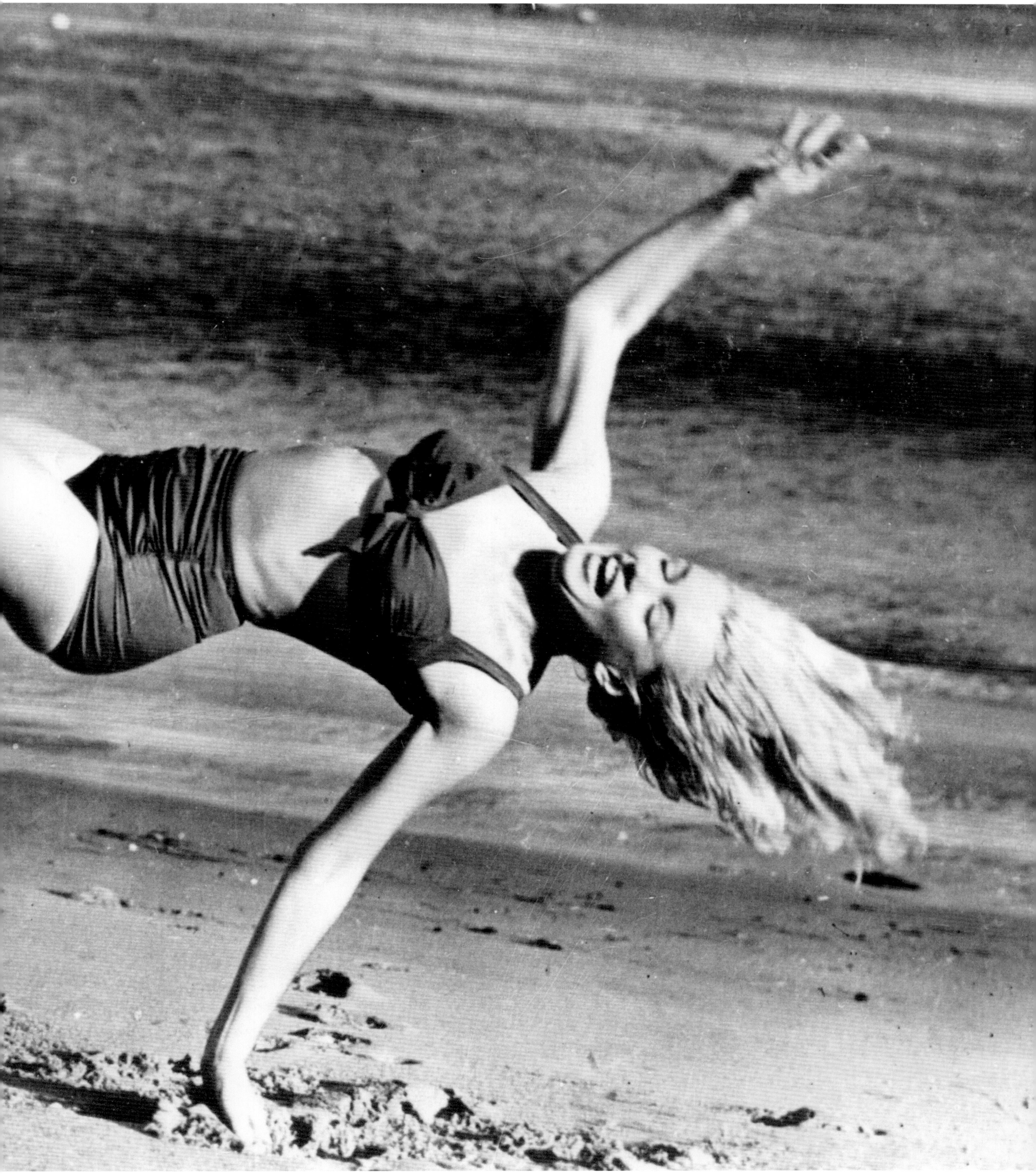

EINE ZIEMLICH PROVOKANTE GESCHICHTE: DER BIKINI

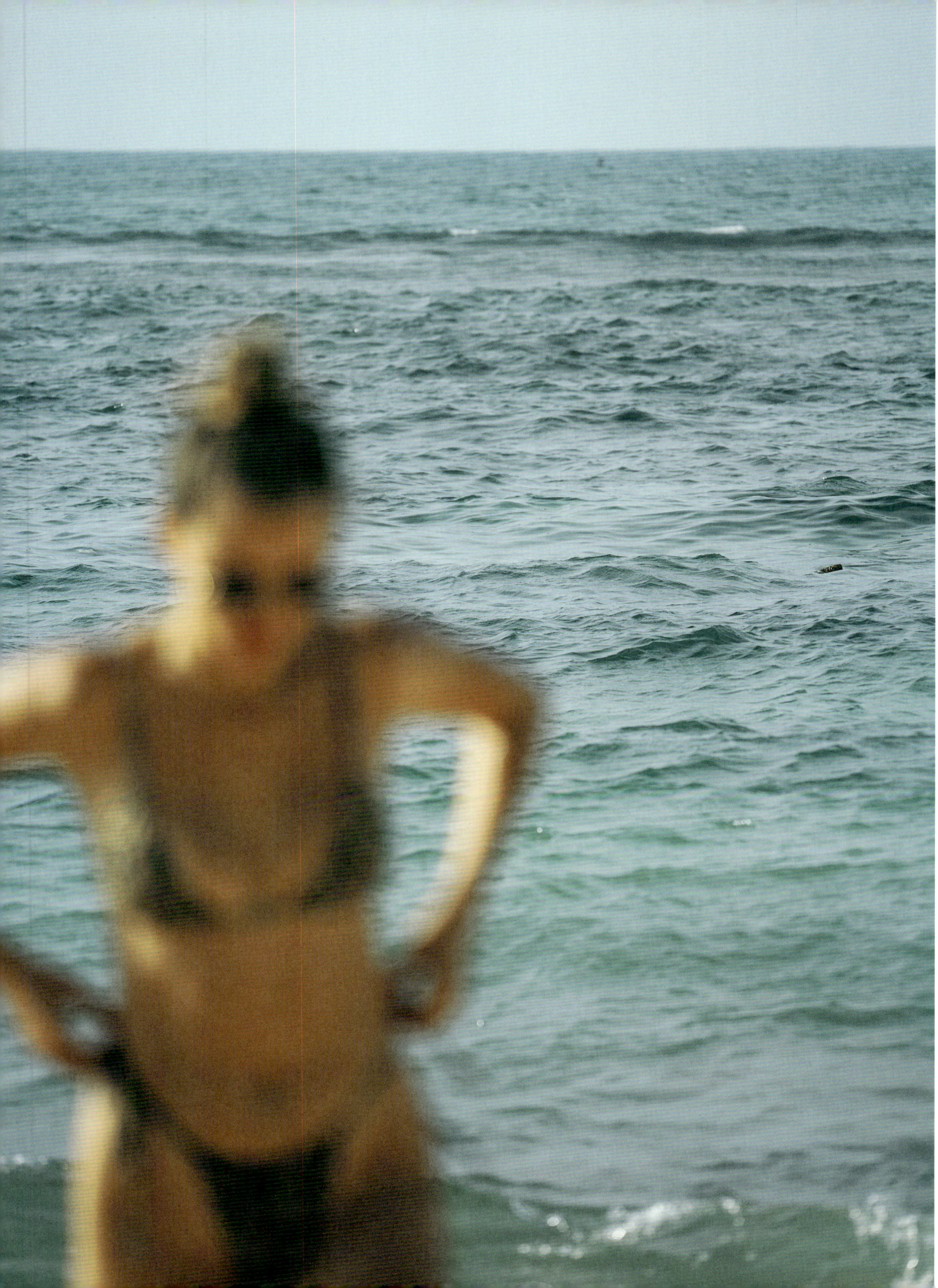

Out of the waves. Bingin Beach,
Bali, Indonesia.

Aus den Wellen. Bingin Beach,
Bali, Indonesien.

Bikini variations part II. Location unknown.

Bikinivariationen Teil II. Ort unbekannt.

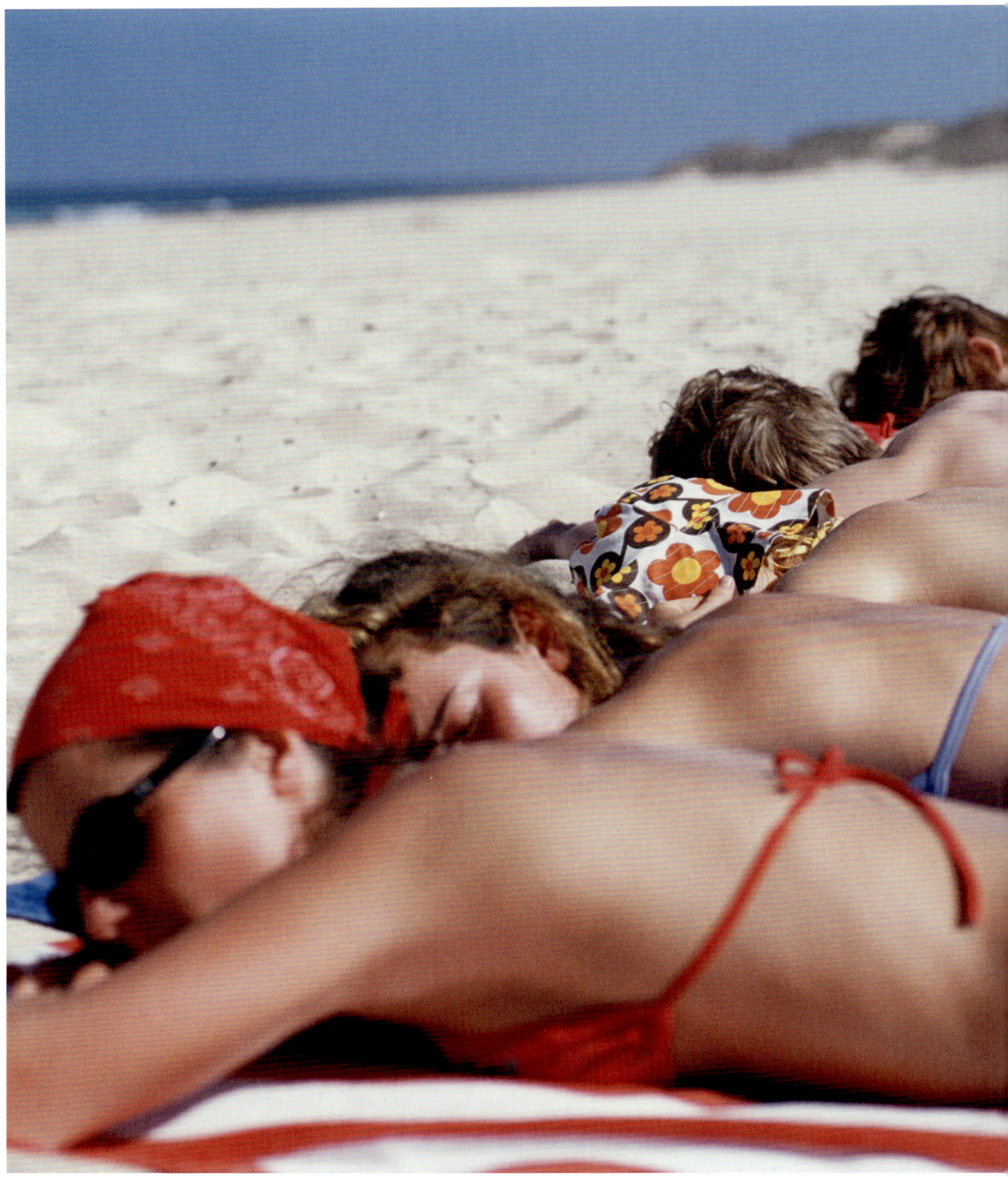

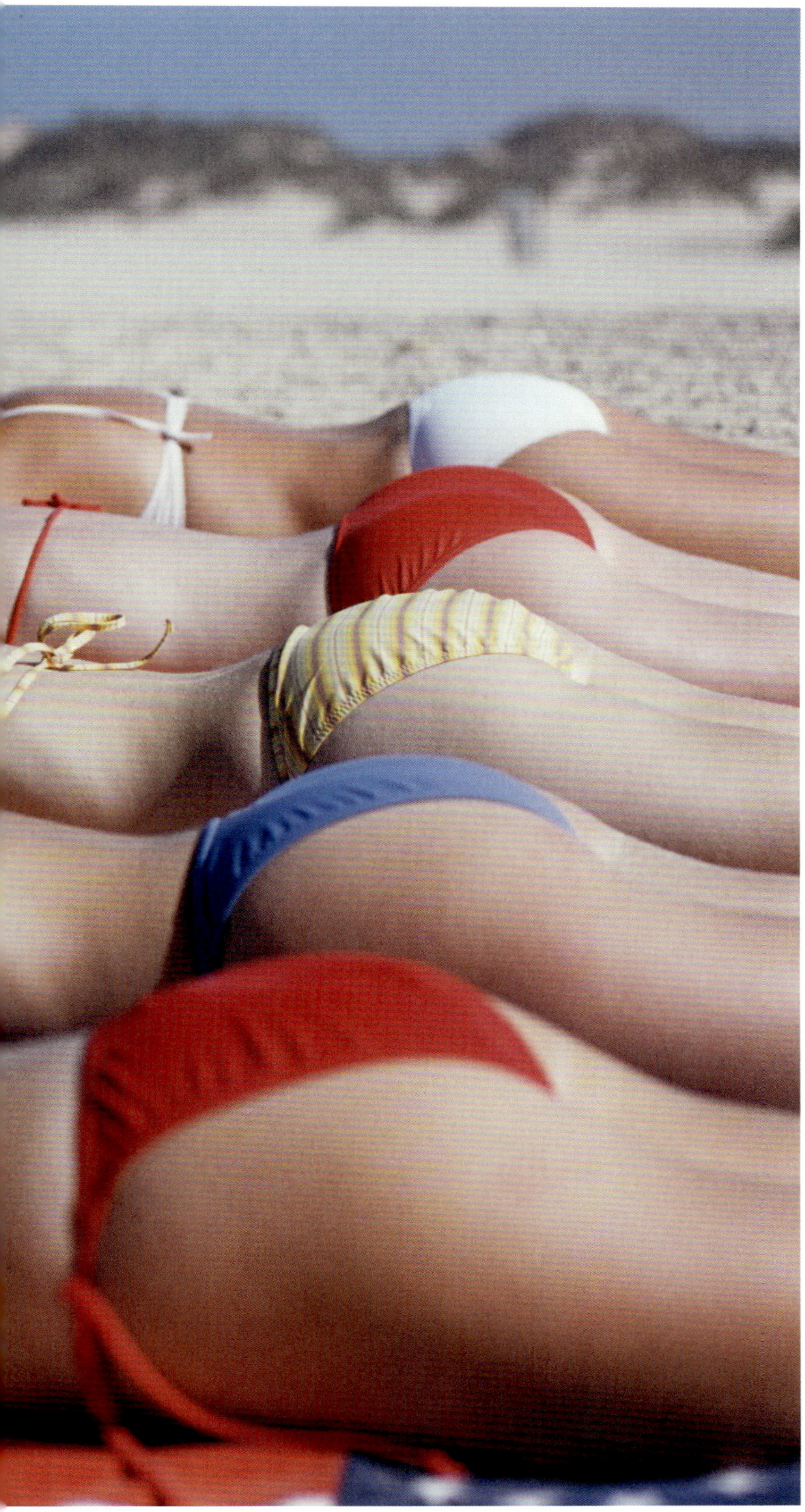

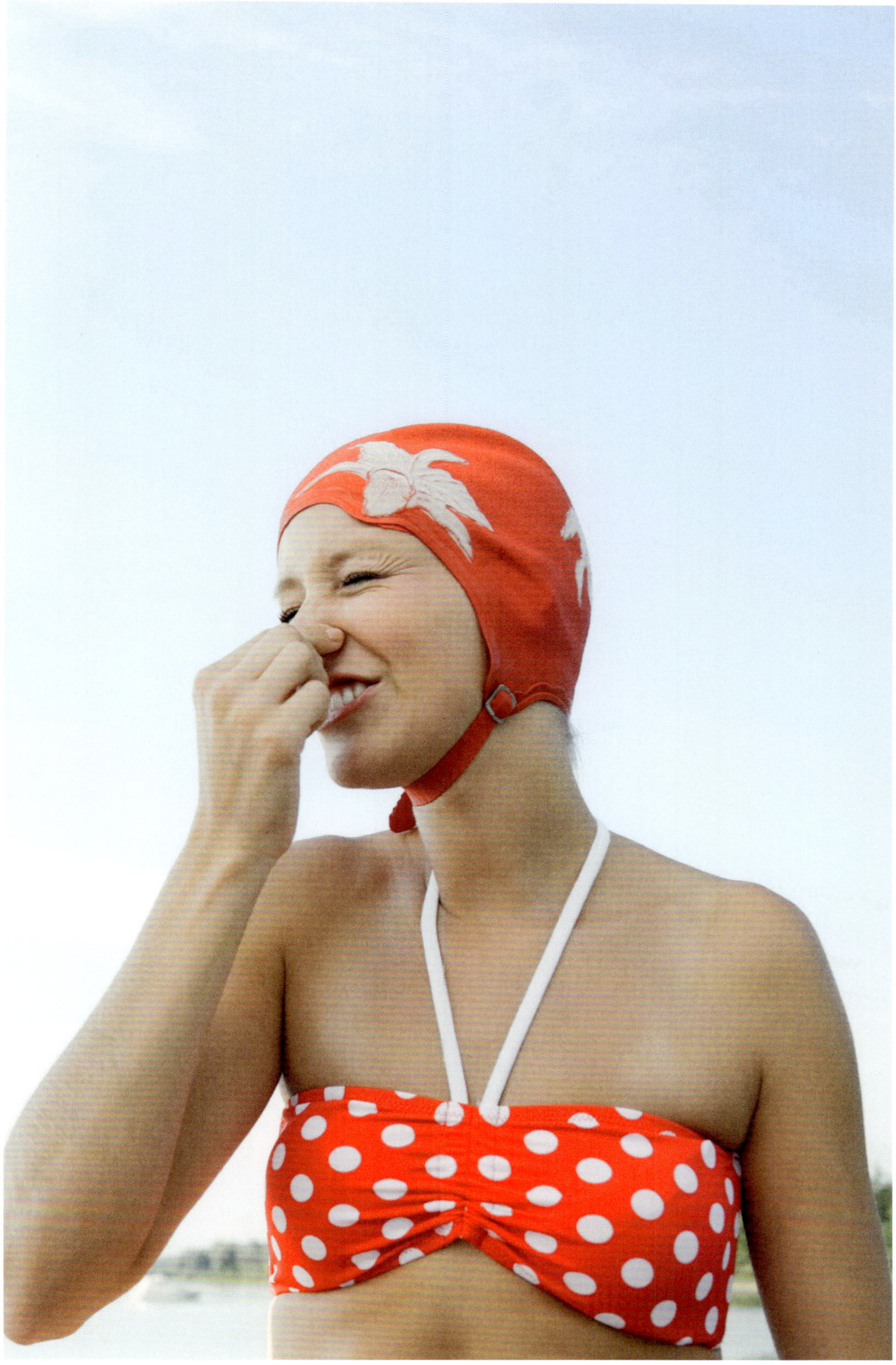

Let's not beat around the bush: Obviously, the beach is a very erotic place to see and be seen, of voyeurism and exhibitionism. When our age and physical shape cooperate–or that spring diet actually worked–we courageously slip into skimpy clothing that we really only accept on ourselves and others here in this place. Vacationers who only get to go to the beach two weeks a year make do with department store merchandise. But the more frequently life takes place at the beach, the more value is placed on the right outfit. No wonder major fashion companies like Burberry and Gucci feature bikinis. But there are also high-priced shorts for men as a well-fitting status symbol: Daniel Craig's James Bond was sponsored by upscale brand Orlebar Brown. And so the beach is a kind of runway for each of us, an exciting catwalk–even if it's just from the lounge chair to the ice cream stand. Karl Lagerfeld had a bigger idea in mind. For his last major Chanel show, *Chanel à la plage*, in the fall of 2018 at the Grand Palais in Paris, he spared no expense to recreate a beach, including a wind machine, swells, and lifeguards in high chairs, and had his models stroll barefoot through the surf in billowing summer dresses.

One striking example of the exhibitionist beach principle are the Muscle Beaches, which helped bodybuilding achieve its real breakthrough in the first place. The first outdoor barbell station was built in Santa Monica, California, in 1934, and pioneers like the legendary Jack LaLanne showed off their muscles and athletic tricks on the bars. But Venice Muscle Beach, two miles to the south, became the most famous muscle beach of all. Here Arnold Schwarzenegger showcased his mountains of muscle to the public. And unlike other sports, the equipment is timeless: weights, barbells, benches, the sun and fresh air, and plenty of spectators to encourage maximum performance.

Anyone can join in: a day pass costs $10, and the annual fee is $200.

Reden wir nicht drumherum: Natürlich ist der Strand ein hocherotischer Ort des Sehens und Gesehenwerdens, des Voyeurismus und Exhibitionismus. Wenn das Alter und die Form mitspielen – oder die Frühjahrsdiät tatsächlich gewirkt hat –, dann schlüpfen wir wagemutig in knappe Kleidung, die wir wirklich nur an diesem Ort an uns und anderen akzeptieren.

Urlauber, die nur zwei Wochen im Jahr an den Strand dürfen, nehmen mit Kaufhausware vorlieb. Aber je häufiger das Leben am Strand stattfindet, desto mehr wird Wert auf das richtige Outfit gelegt. Kein Wunder, dass große Modefirmen wie Burberry und Gucci Bikinis führen. Aber auch für die Herren gibt es hochpreisige Shorts als gutsitzendes Statussymbol: Daniel Craigs James Bond wurde von der Edelmarke Orlebar Brown gesponsort. Und so ist der Strand für jeden von uns ein kleiner Runway, ein aufregender Catwalk – und sei es nur vom Liegestuhl zum Eisstand. Karl Lagerfeld dachte gleich größer. Für seine letzte große Chanel-Show *Chanel à la plage* baute er im Herbst 2018 im Pariser Grand Palais unter großem Aufwand einen Strand nach, inklusive Windmaschine, Dünung und Bademeister im Hochsitz, und ließ seine Models barfuß in wehenden Sommerkleidern durch den Spülsaum flanieren.

Ein krasser Sonderfall des exhibitionistischen Strandprinzips sind die Muscle Beaches, die überhaupt erst dem Bodybuilding zum Durchbruch verhalfen. Die erste Hantelstation im Freien entstand 1934 in Santa Monica, Kalifornien, und Pioniere wie der legendäre Jack LaLanne zeigten ihre Muckis und athletische Tricks am Reck. Zum berühmtesten Muscle Beach wurde aber der Venice Muscle Beach zwei Kilometer weiter südlich. Hier präsentierte Arnold Schwarzenegger der Öffentlichkeit seine Muskelberge. Und im Gegensatz zu anderen Sportarten ist die Ausrüstung zeitlos: Gewichte, Hanteln, Stangen, Bänke, dazu die Sonne und die frische Luft und jede Menge Zuschauer, die zu Höchstleistungen animieren. Mittrainieren darf jeder: Eine Tageskarte kostet 10 Dollar, der Jahresbeitrag 200 Dollar.

Fashionable yoga practice on the beach. Knokke, Belgium.

Modische Yogaübung am Strand. Knokke, Belgien.

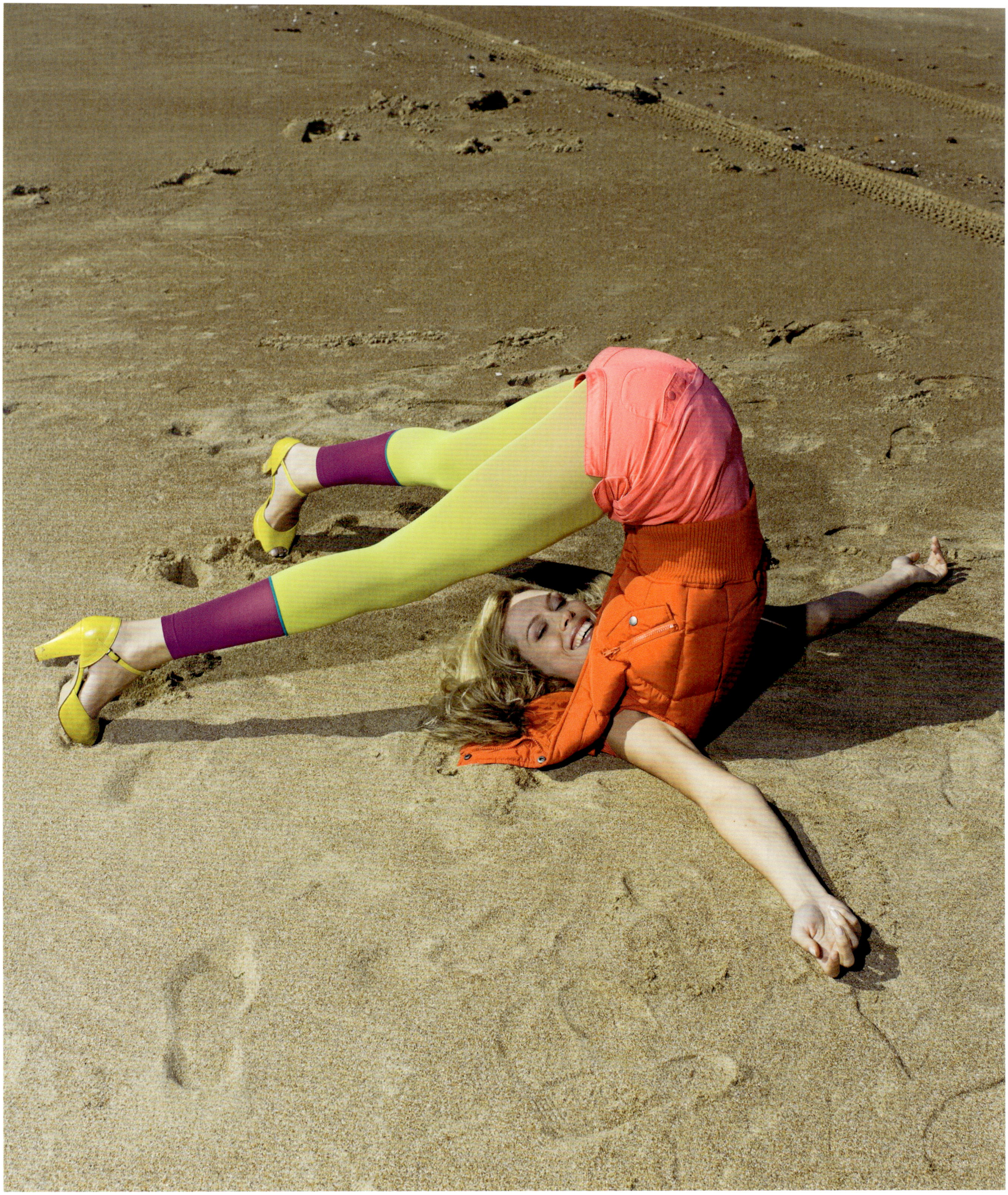

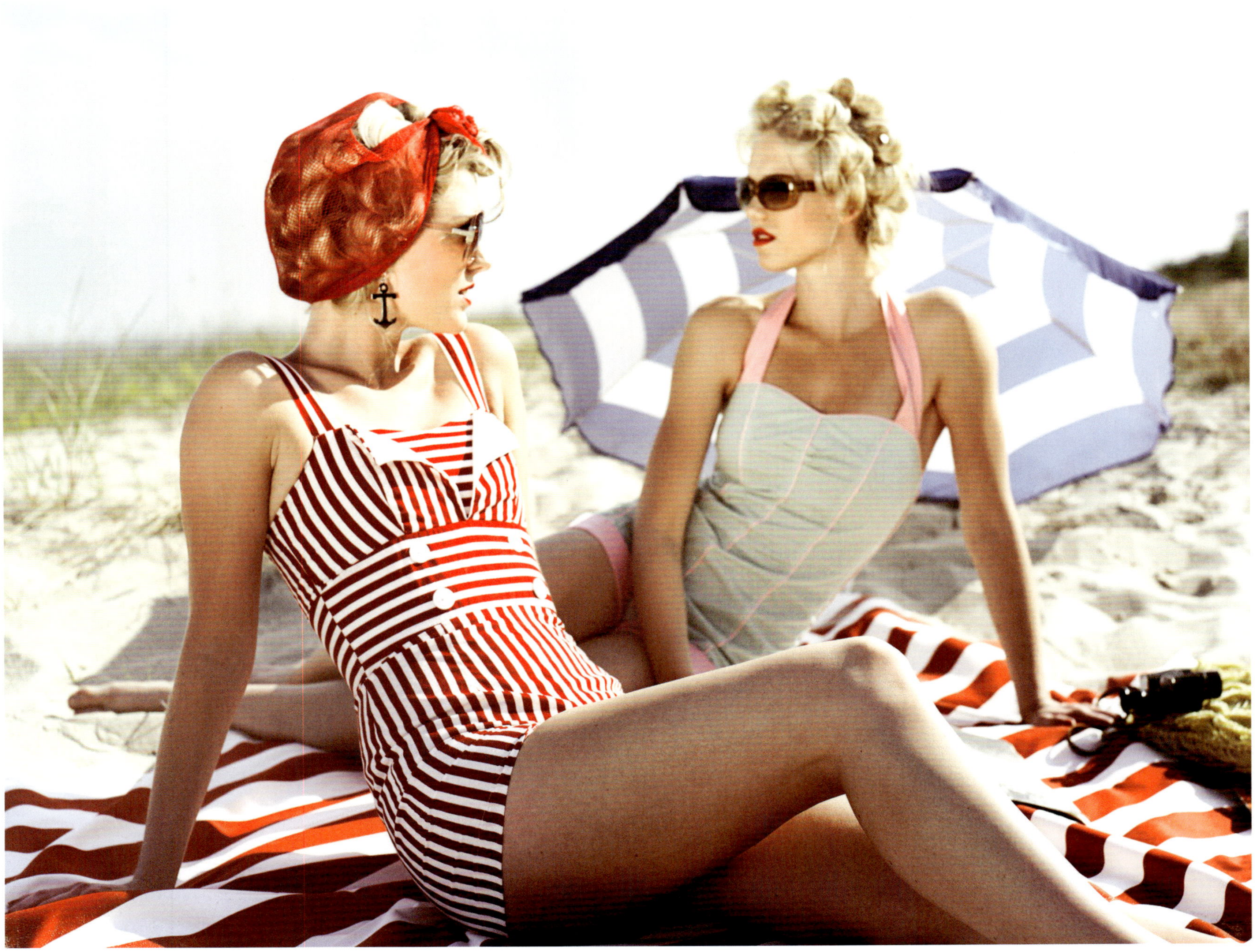

Elegance at the beach.
Location unknown.

Eleganz am Strand.
Ort unbekannt.

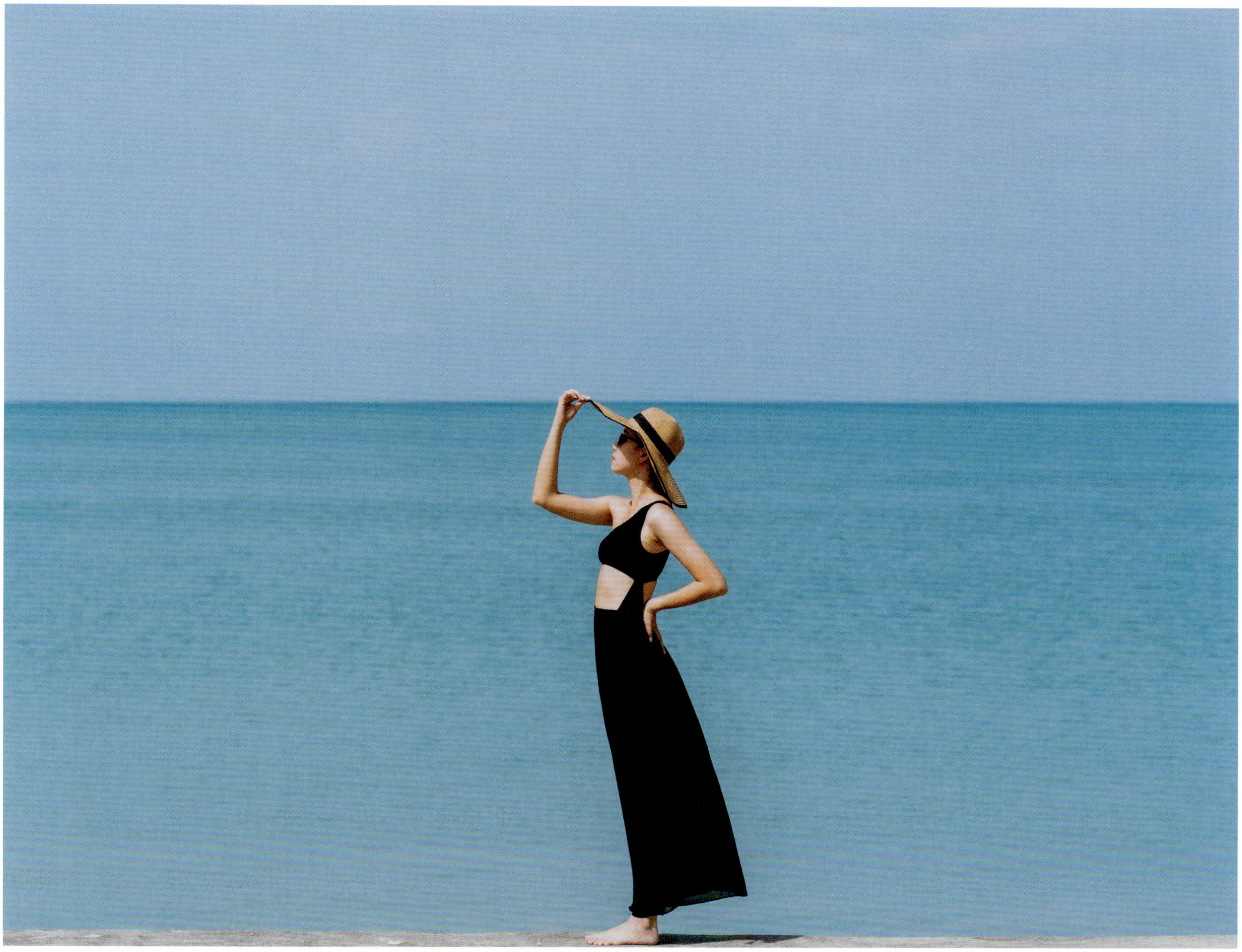

To pose or not to pose.
Muang, Phuket, Thailand.

Posieren oder Nichtposieren.
Muang, Phuket, Thailand.

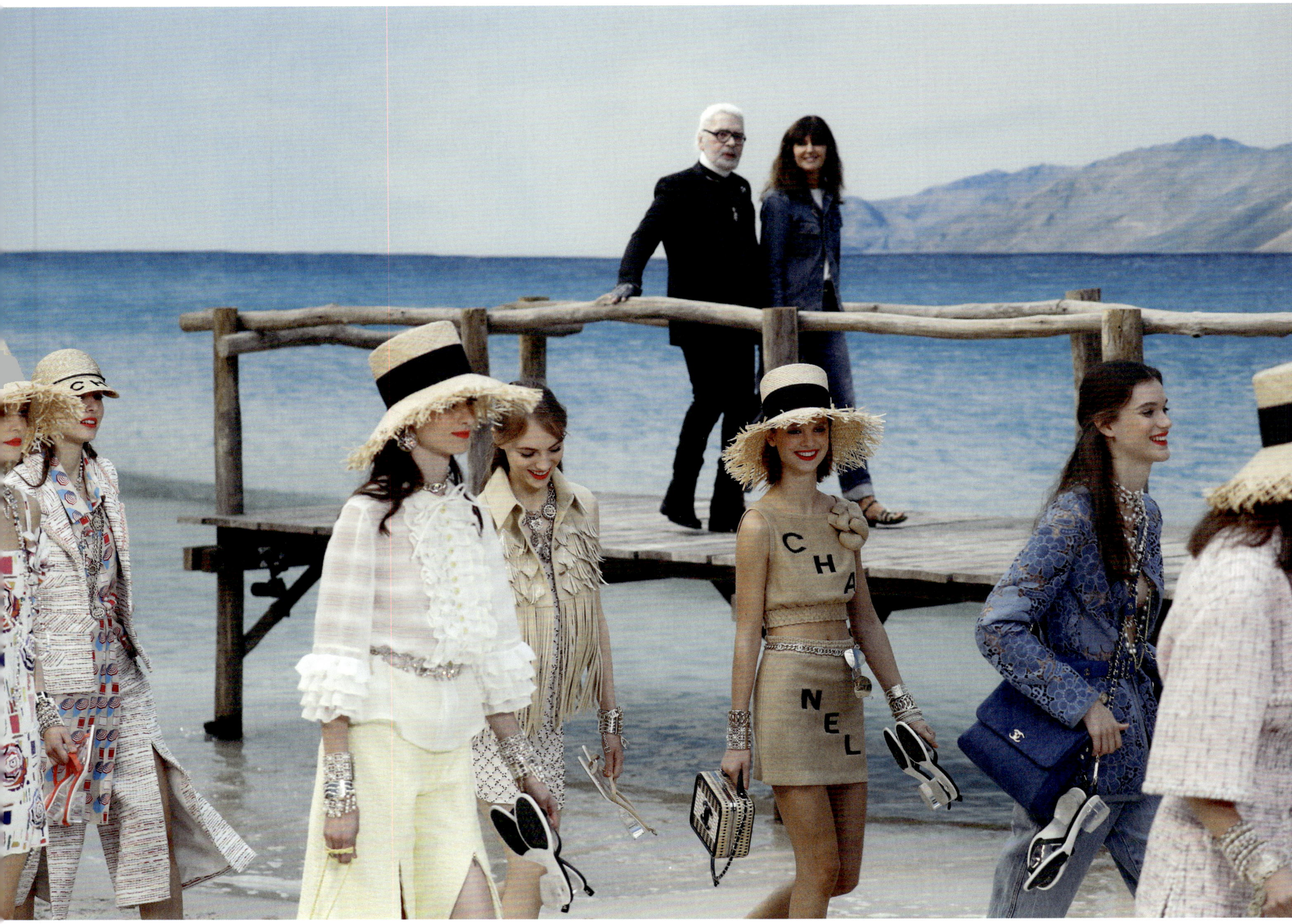

LEFT AND RIGHT PAGE

Karl Lagerfeld`s beach vision in Paris` Grand Palais during Fashion Week 2018.

Karl Lagerfelds Strandvision im Pariser Grand Palais während der Fashion Week 2018.

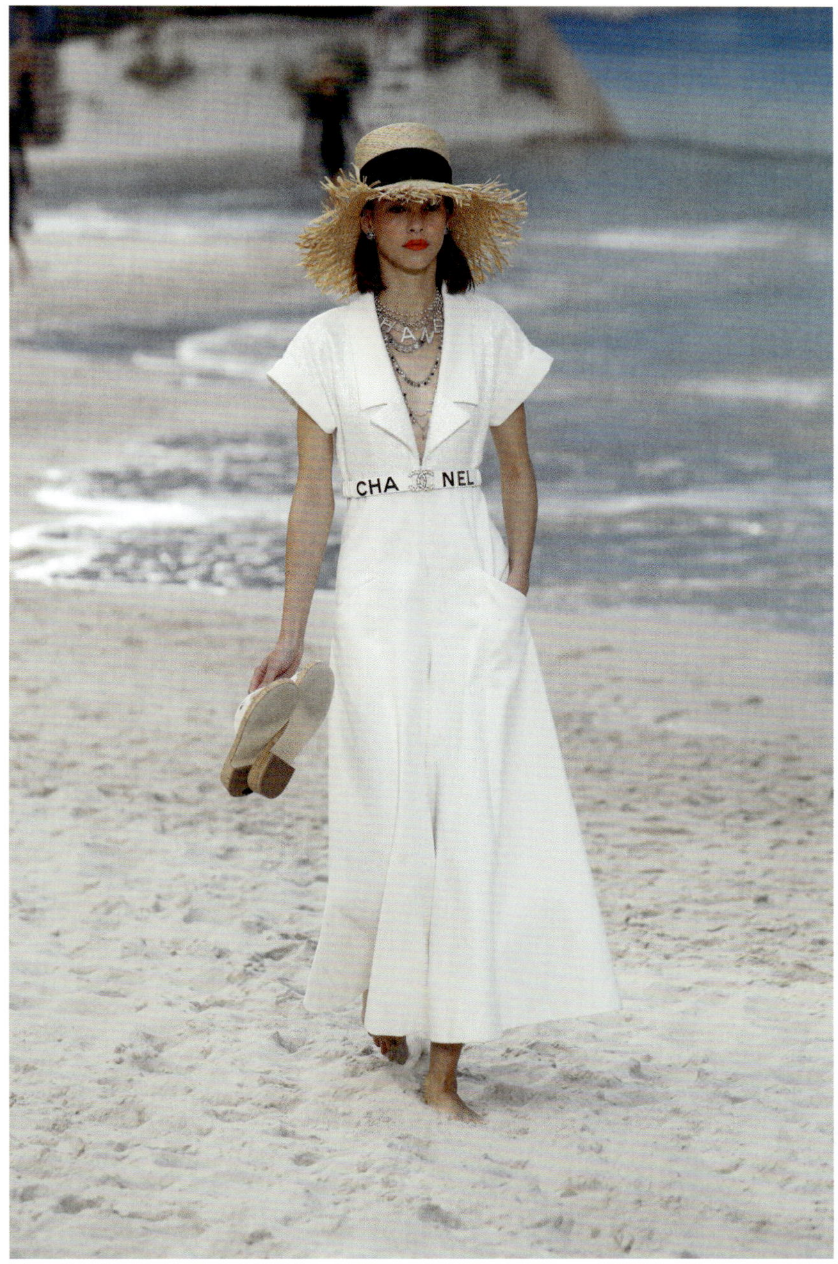
CHA NEL

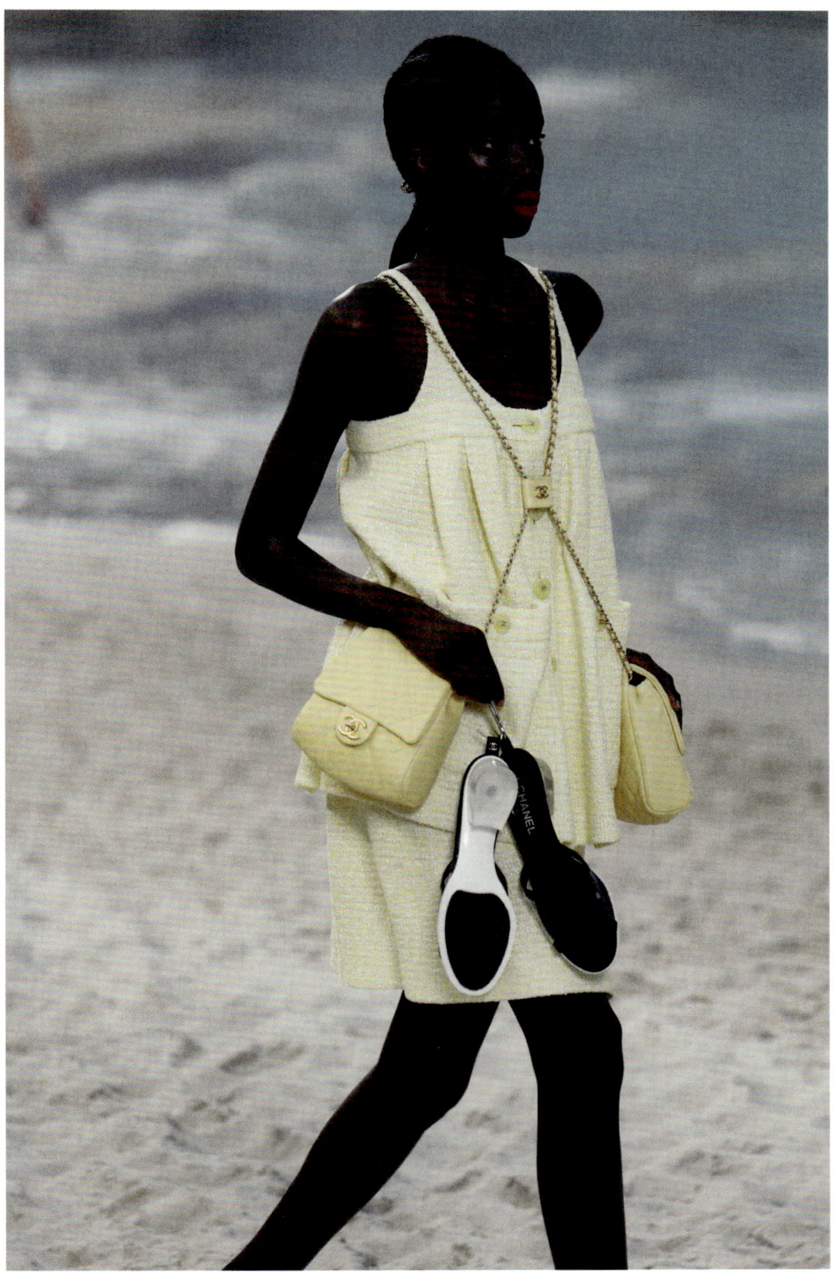

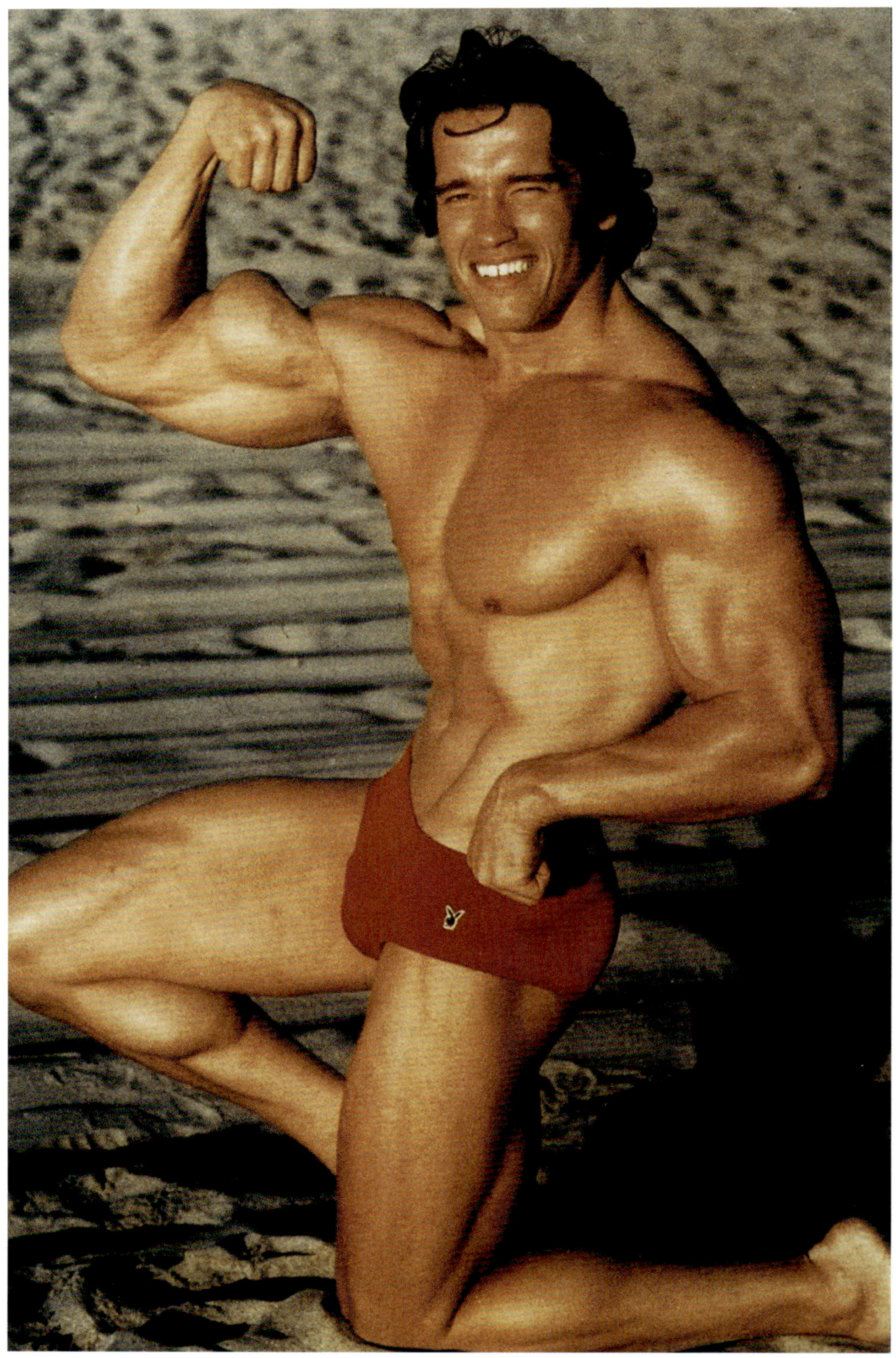

Arnold Schwarzenegger showing his muscles at Venice Beach, USA.

Arnold Schwarzenegger präsentiert seine Muskeln. Venice Beach, USA.

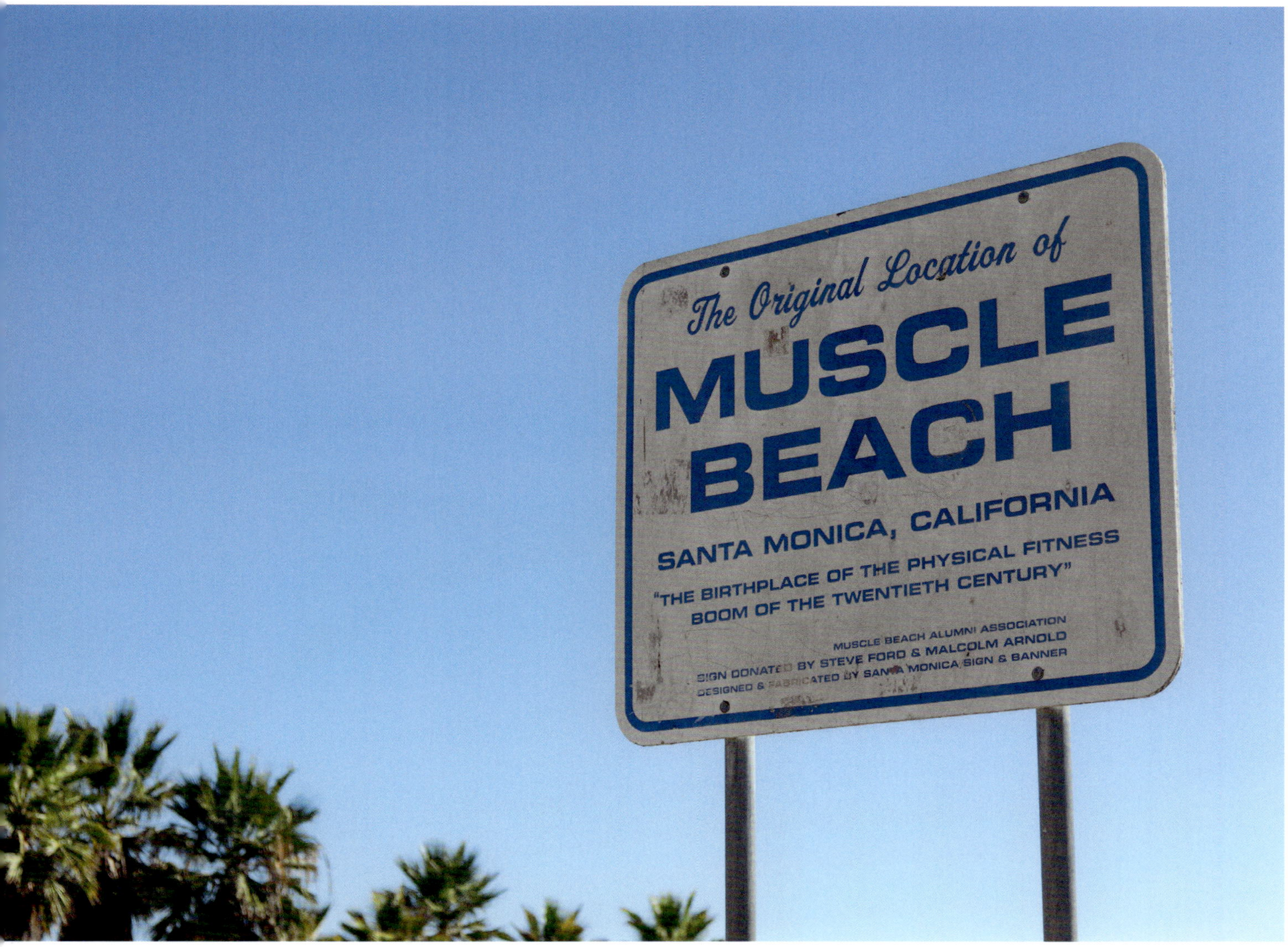

Founded in 1934, Santa Monica
Muscle Beach was the first of its kind.

Gegründet im Jahr 1934, war der
Muscle Beach von Santa Monica der
erste seiner Art.

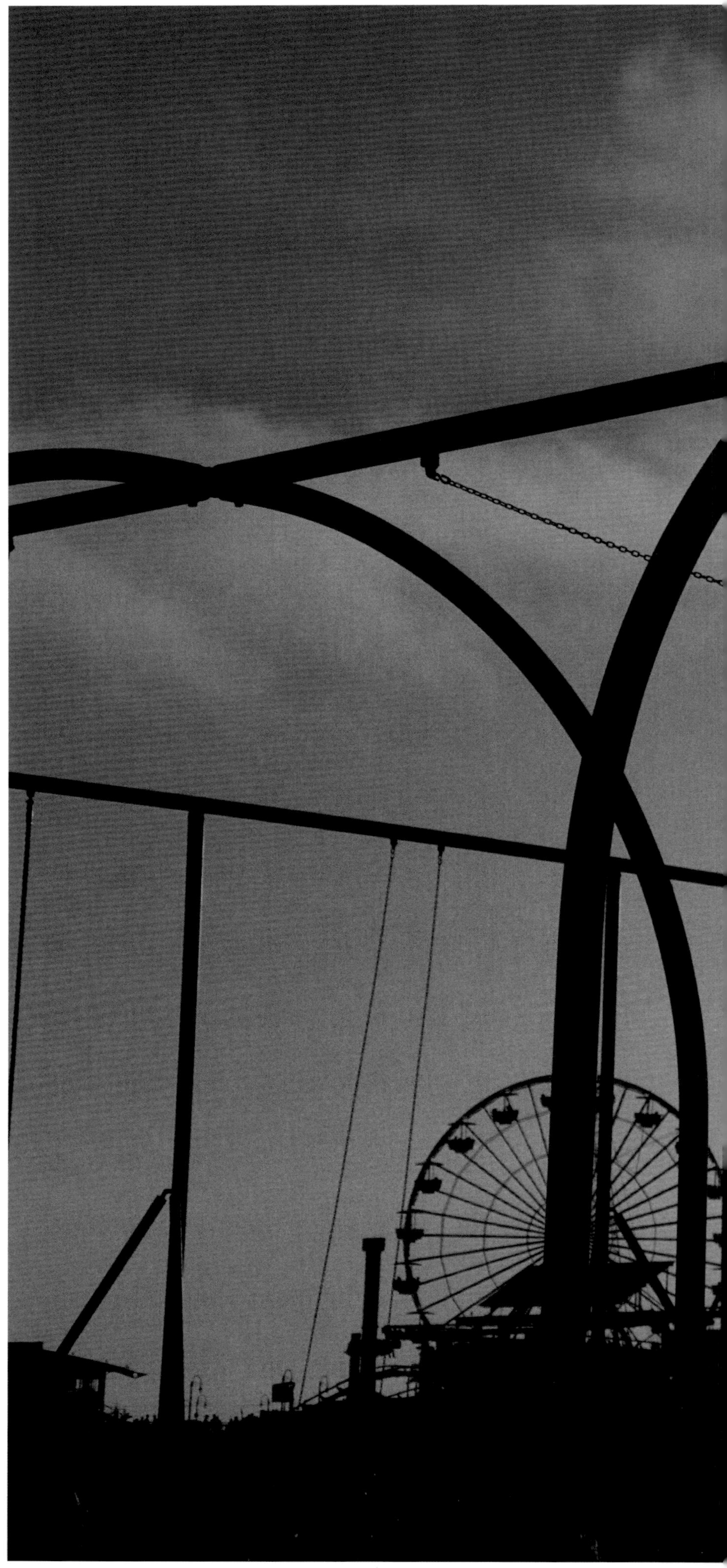

Silhouette of Santa Monica
Muscle Beach at sunset.

Silhouette des Muscle Beach
bei Sonnenuntergang.
Santa Monica, USA.

SKIMPY SHORTS AND BULGING MUSCLES: THE BEACH AS A RUNWAY

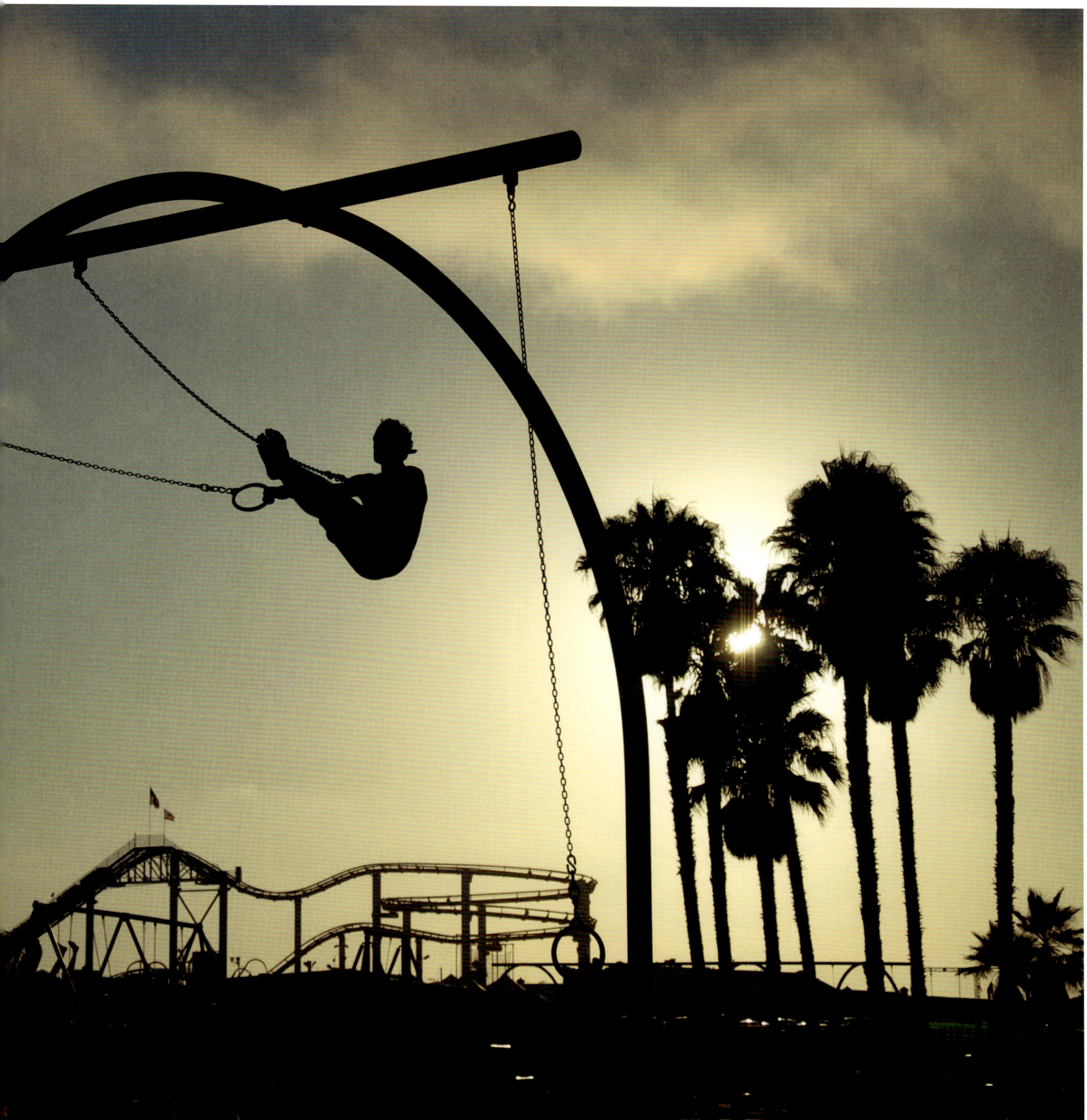

KNAPPE SHORTS UND DICKE MUSKELN: DER STRAND ALS RUNWAY

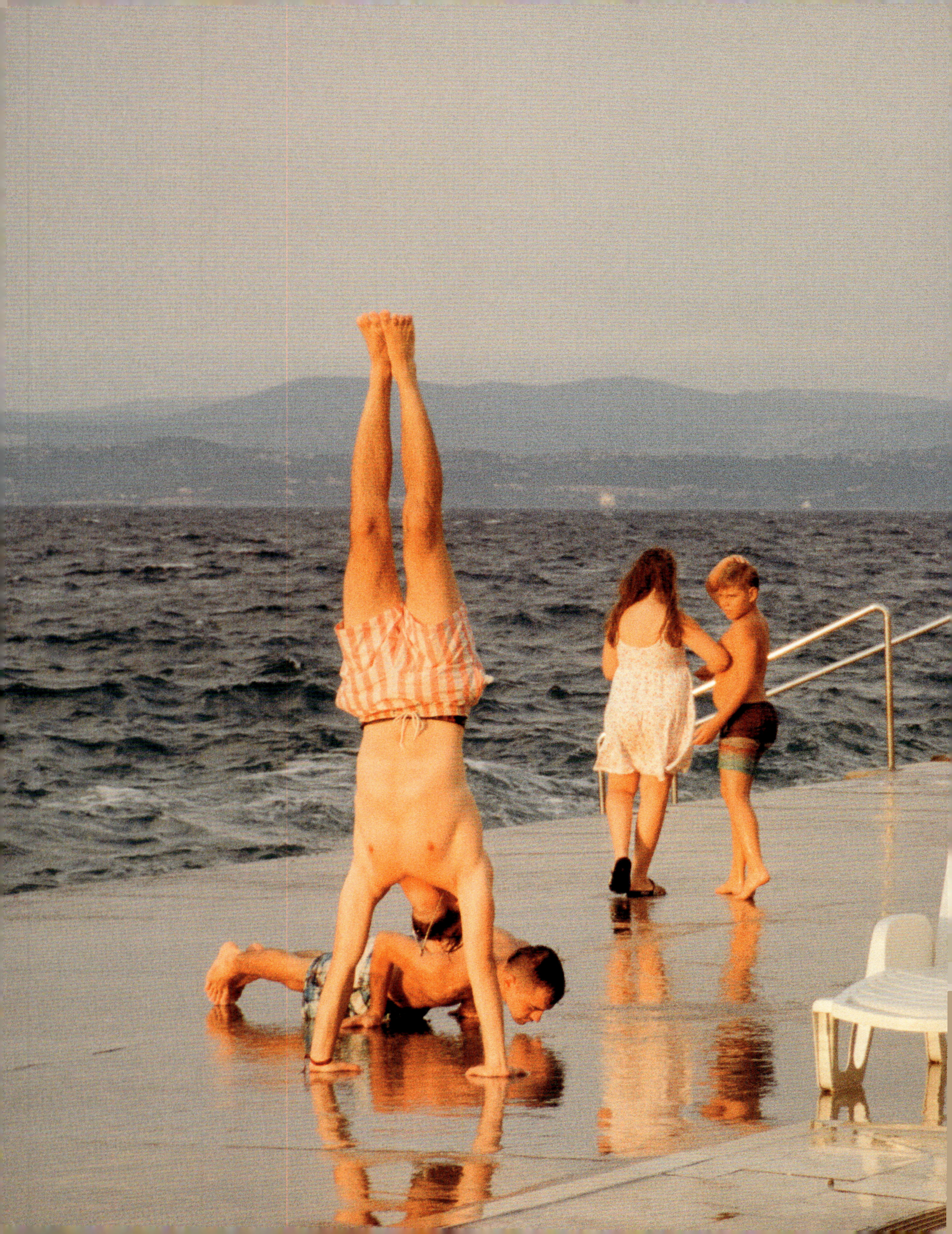

Eternal reflection. Youngsters performing their own version of the Venice Muscle Beach. Piran, Slovenia.

Unendliche Reflexion. Jugendliche und ihre eigene Version des Venice Muscle Beach. Piran, Slowenien.

Snow goggles came first: The Inuit carved narrow slits for their eyes in the bones and ivory of their prey to protect themselves from the fierce reflections of the sun in the snowy landscapes of their hunting grounds. And apparently even the ancient Romans used broken glass to avoid being blinded during gladiator fights; Nero is said to have gazed at the spectacle through beveled emerald. A Königsberg optician invented glasses made of amber in the 19th century–which brings us back to the beach. In 1908, he succeeded in proving that UV radiation was harmful.
Sunglasses not only protect against sunlight; they also prevent eye contact and with it one of the most important interpersonal non-verbal actions. In addition, they conceal the direction of the wearer's gaze. This is considered cool, and is also necessary in some professions–security personnel, for example. Sunglasses have long since become a trademark, and not just for celebrities who want to protect their sensitive eyes, like Ray Charles and Bono. Sunglasses became socially acceptable thanks to Audrey Hepburn in *Breakfast at Tiffany's*, where she wore a model by designer Oliver Goldsmith, as she did in later films, such as *Two for the Road*.
Elvis Presley and Bob Marley preferred to look at their fans through the Wayfarers made by Ray-Ban. Another important Ray-Ban model, the Aviator, was made famous by Paul McCartney and Freddie Mercury in the 1970s and has been part of pop culture since *Top Gun*.

And when it comes to style, Jackie Kennedy-Onassis must of course be mentioned: She is considered the trailblazer of the oversized glasses that are still named for her today: Jackie O sunglasses are among the most popular models, especially among women, and the other extreme, the tiny glasses known as teashades or John Lennon glasses, have made a comeback during the recent fashion revival of the 90s.
How all these celebrities manage to wear sunglasses all the time and still have a seamless tan is one of the great mysteries of humankind.

Die Schneebrille war zuerst da: Inuit schnitzten sich schmale Sehschlitze in Knochen und Elfenbein ihrer Beutetiere, um sich vor den heftigen Reflexionen der Sonne in den Schneewüsten ihrer Jagdreviere zu schützen. Und offenbar benutzten schon die alten Römer Glasscherben, um bei Gladiatorenkämpfen nicht geblendet zu werden; Nero soll durch geschliffenen Smaragd auf das Spektakel geblickt haben. Ein Königsberger Optiker erfand im 19. Jahrhundert Gläser aus Bernstein – womit wir wieder am Strand wären. Im Jahr 1908 gelang der Nachweis, dass UV-Strahlung schädlich sei.
Die Sonnenbrille schützt nicht nur vor dem Sonnenlicht, sondern verhindert den Augenkontakt und damit eine der wichtigsten zwischenmenschlichen nonverbalen Aktionen. Zudem verbirgt sie die Blickrichtung des Tragenden. Das gilt als cool und ist bei manchen Berufen – etwa bei Sicherheitspersonal – auch notwendig. Längst ist die Sonnenbrille ein Markenzeichen geworden, nicht nur bei Berühmtheiten wie Ray Charles und Bono, die ihre empfindlichen Augen schützen wollen. Die Sonnenbrille wurde dank Audrey Hepburn in *Frühstück bei Tiffany's* gesellschaftsfähig: Sie trug ein Modell des Designers Oliver Goldsmith, wie auch in späteren Filmen, etwa *Zwei auf gleichem Weg*.
Elvis Presley und Bob Marley blickten am liebsten durch die Wayfarer der Firma Ray-Ban auf ihre Fans. Das zweite wichtige Ray-Ban-Modell, die Fliegerbrille Aviator, wurde in den 1970ern durch Paul McCartney und Freddie Mercury berühmt gemacht und gehört seit *Top Gun* zur Populärkultur.
Und wenn es um Stil geht, darf Jackie Kennedy-Onassis natürlich nicht fehlen: Sie gilt als Wegebereiterin der überdimensionalen Gläser, die bis heute nach ihr benannt sind: Jackie-Os gehören besonders bei Frauen zu den beliebtesten Modellen, und das andere Extrem, die winzigen sogenannten Teashades oder John-Lennon-Brillen, feierten im Zuge des 90s-Revivals ein Comeback.
Wie es all die Berühmtheiten schaffen, stets Sonnenbrille zu tragen und dennoch über nahtlose Bräune zu verfügen, gehört zu den großen Rätseln der Menschheitsgeschichte.

Resting beach face sundial.
From the series *Marbelland.*
Marbella, Spain.

Strandlook und Sonnenuhr.
Aus der Serie *Marbelland.*
Marbella, Spanien.

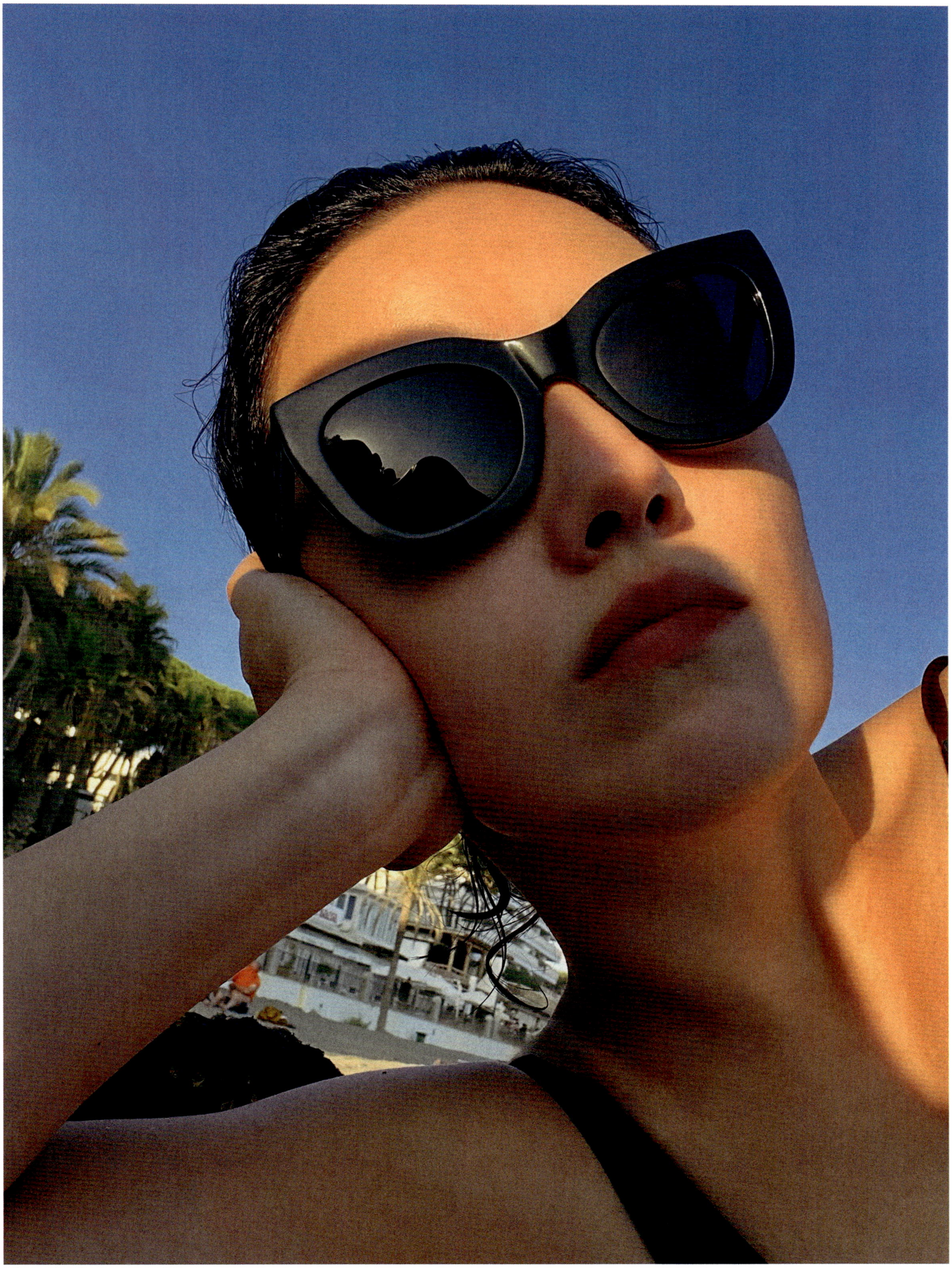

Mirror, mirror in the sun.
Palau Beach, Sardinia, Italy.

Spieglein, Spieglein in der Sonn'.
Strand von Palau, Sardinien, Italien.

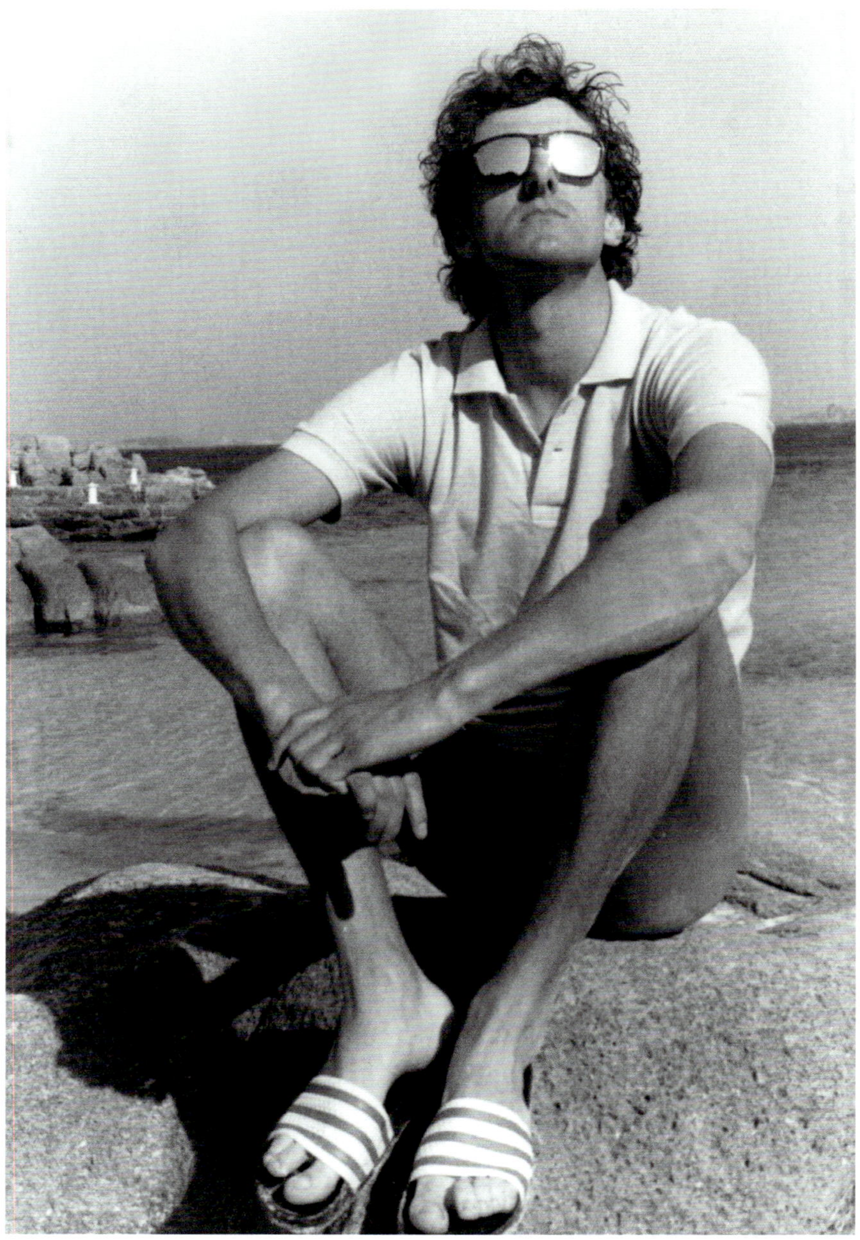

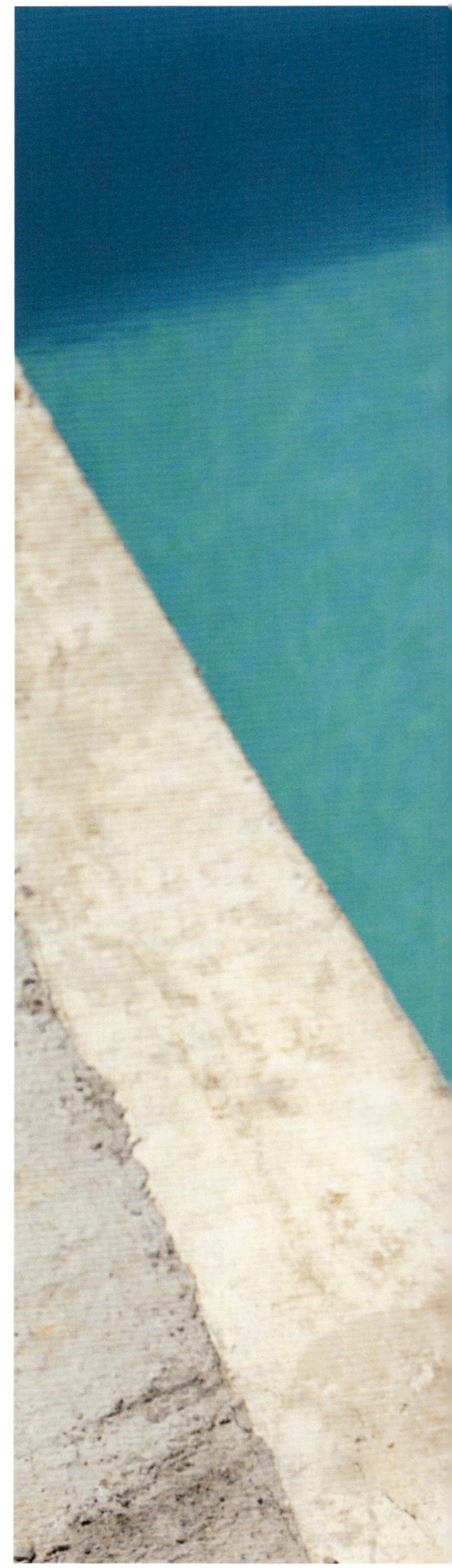

Shades in the sun. Location unknown.

Schwarz sehen im hellen Licht.
Ort unbekannt.

TINY, HUGE, COOL: MAKING A STATEMENT WITH SUNGLASSES

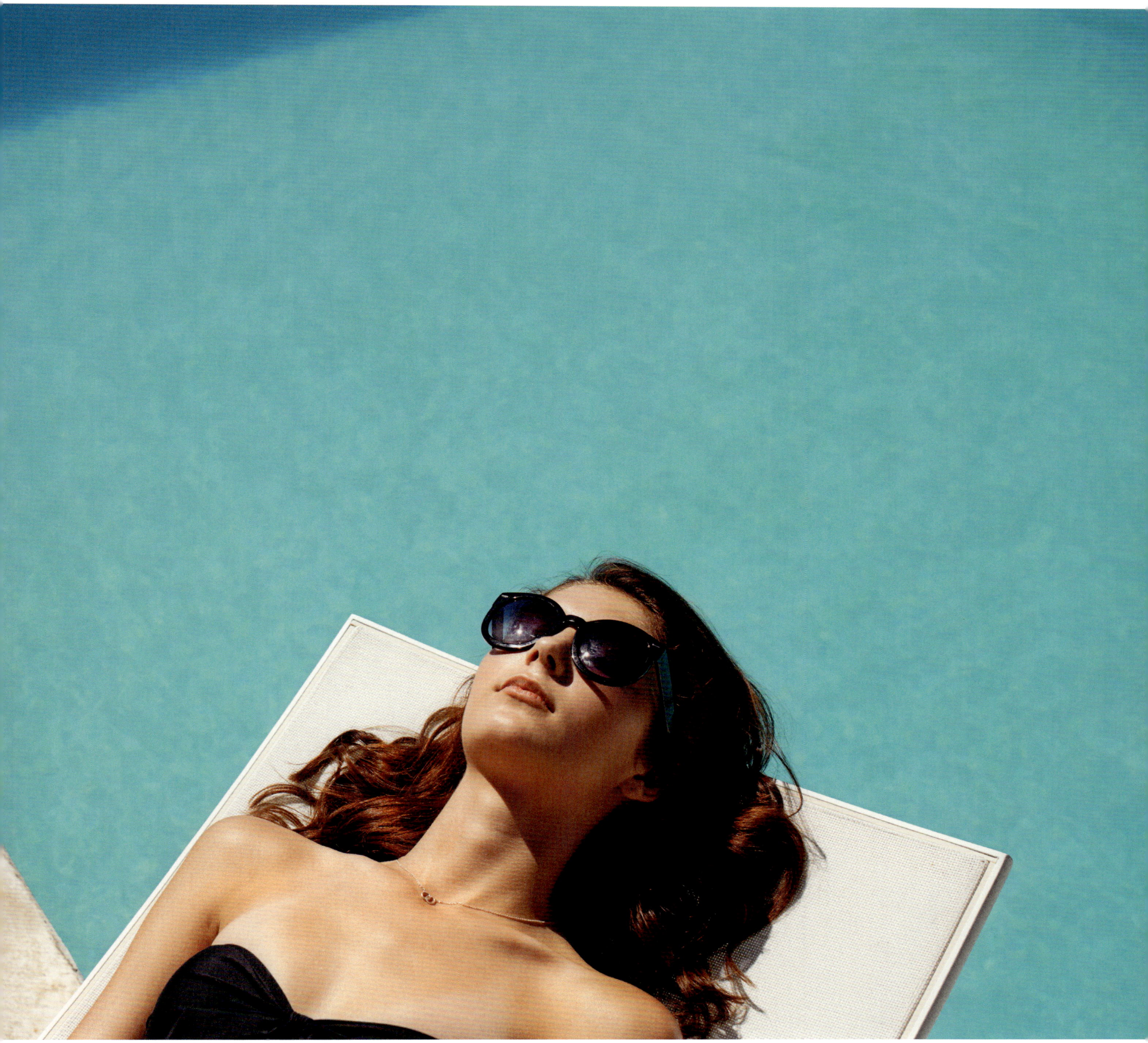

WINZIG, RIESIG, COOL: DIE SONNENBRILLE ALS STYLE-STATEMENT

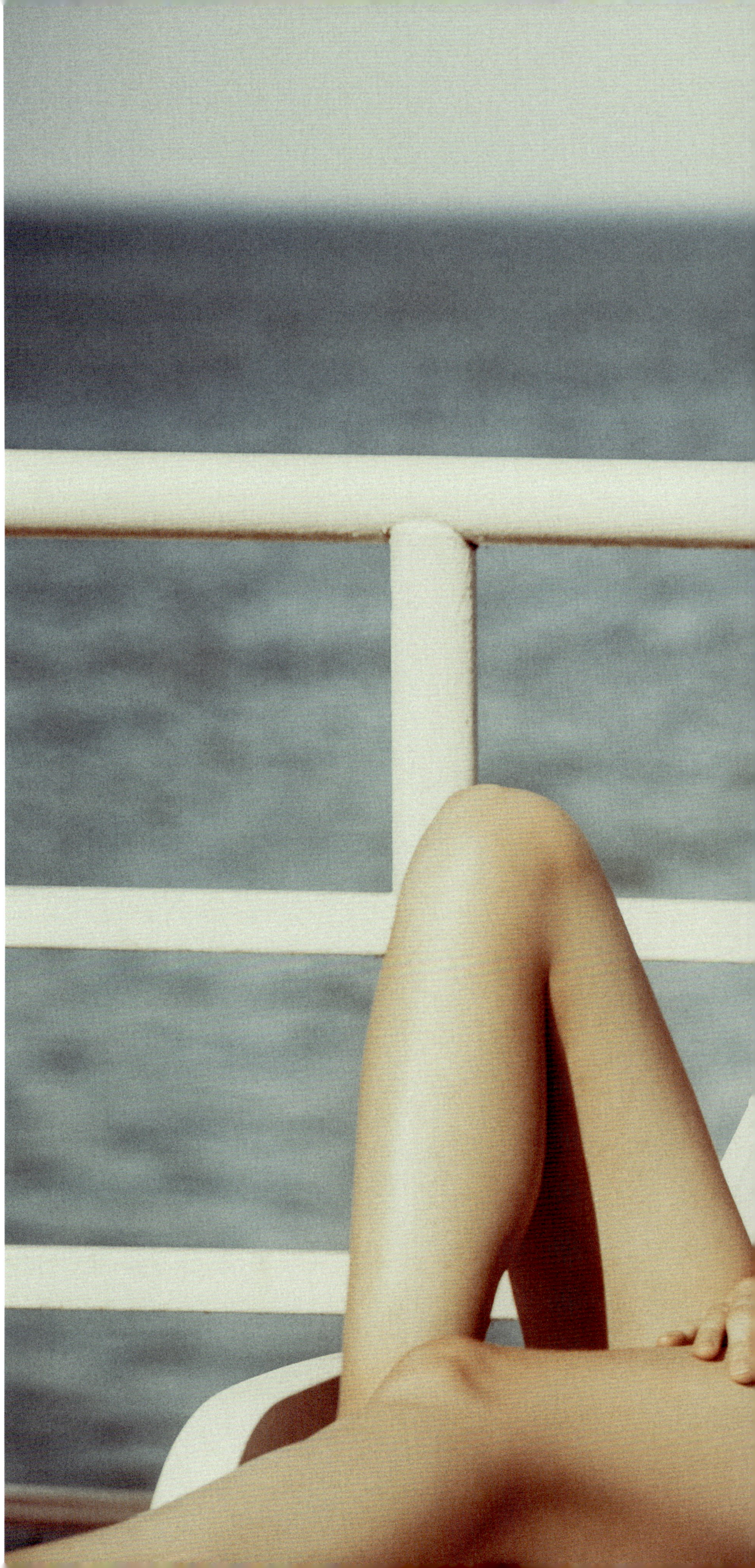

The riddle of the shielded gaze. Location unknown.

Das Rätsel des verdunkelten Blicks. Ort unbekannt.

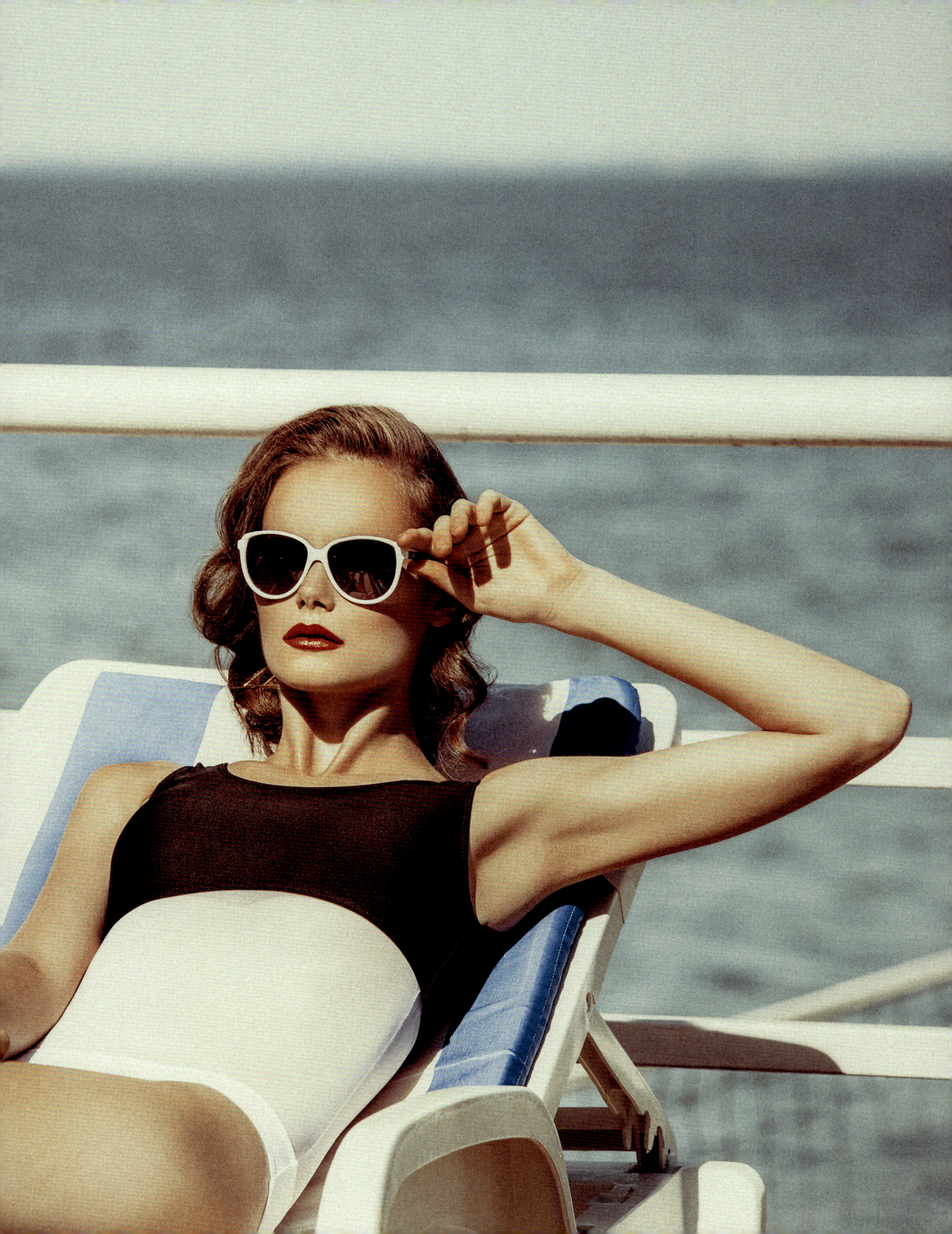

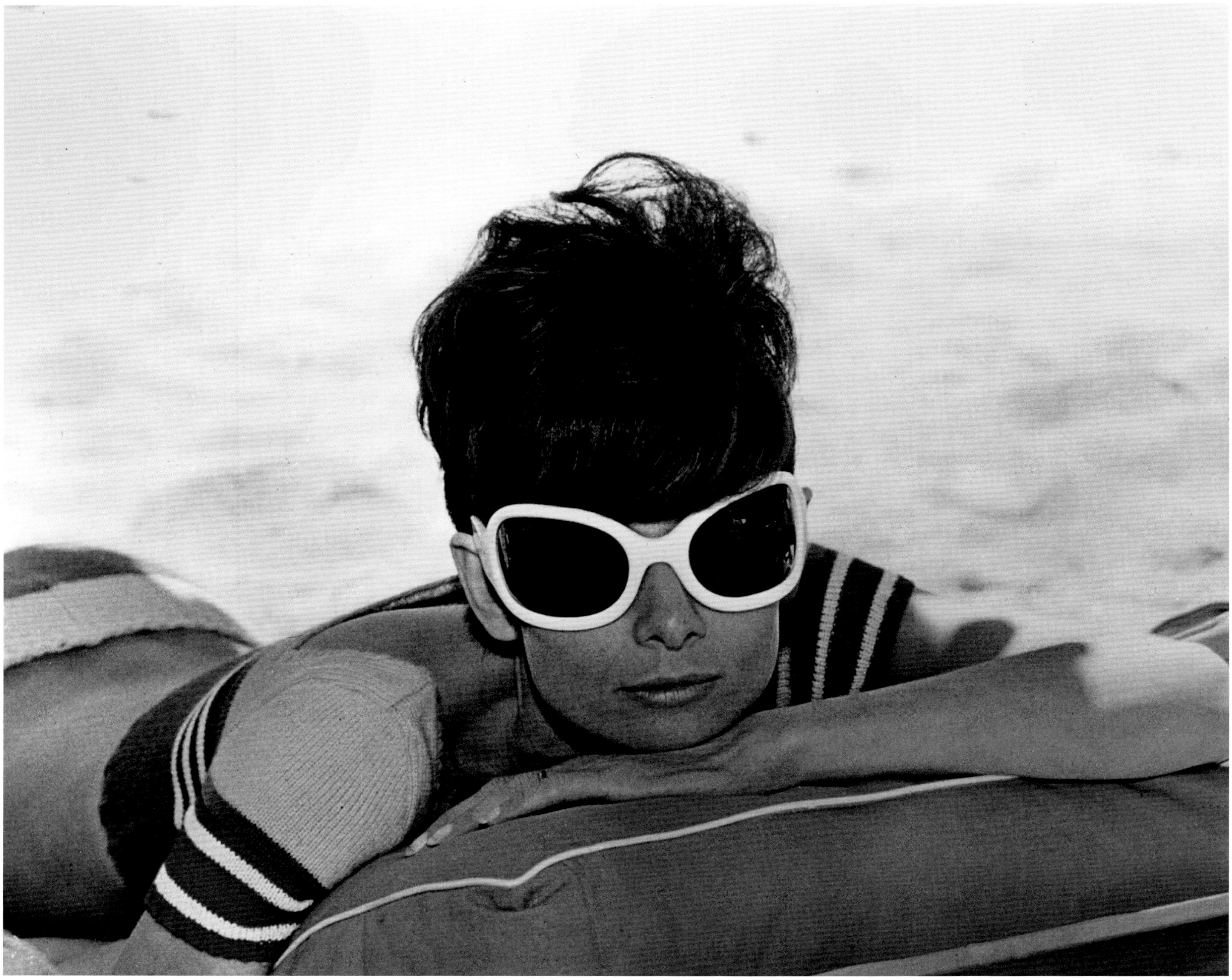

Audrey Hepburn wearing a pair of sunglasses by Oliver Goldsmith in *Two for the Road*. 1967.

Audrey Hepburn trägt in *Zwei auf gleichem Weg* eine Sonnebrille von Oliver Goldsmith. 1967.

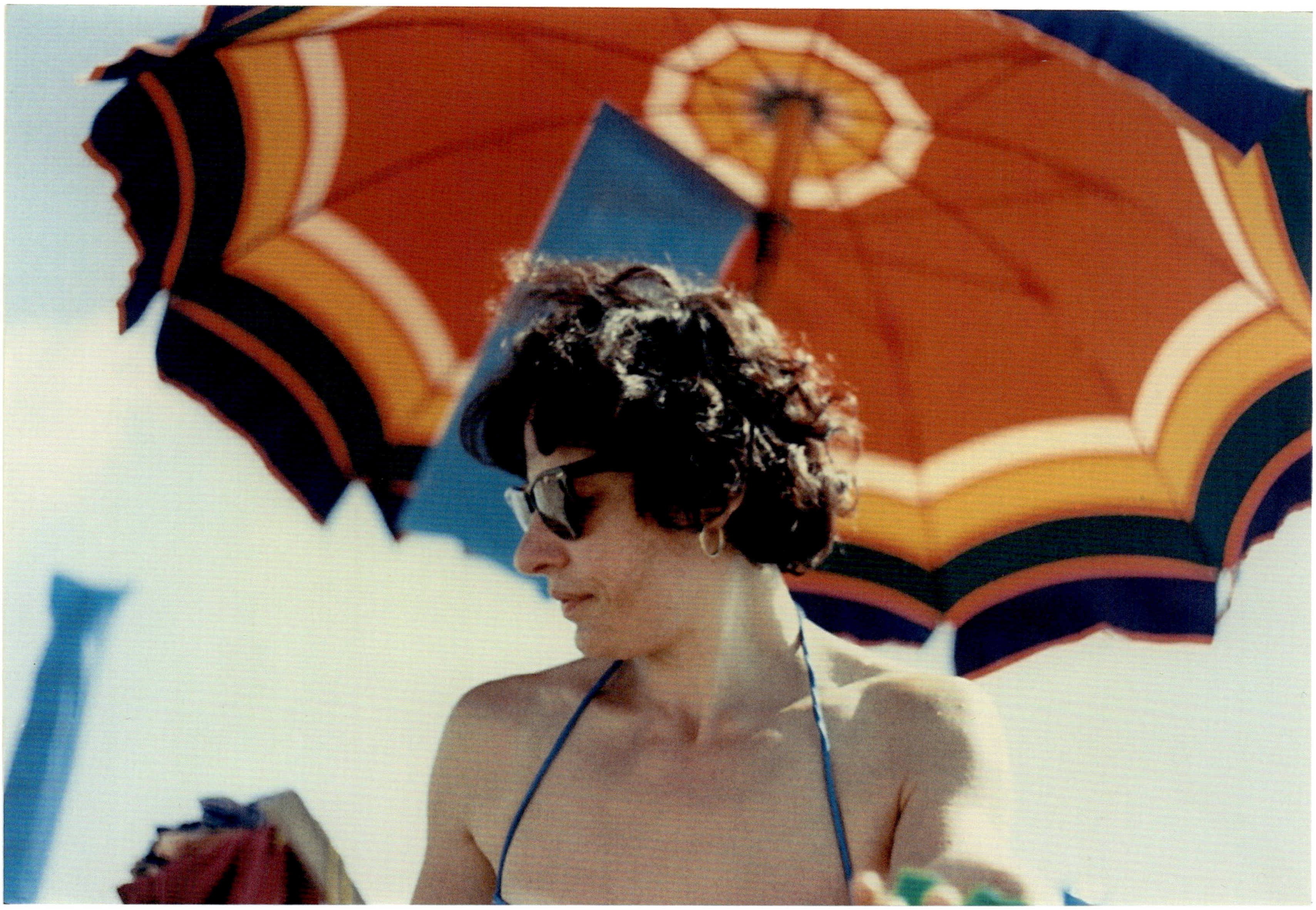

Protected. Venice, Italy.

Abgeschirmt. Venedig, Italien.

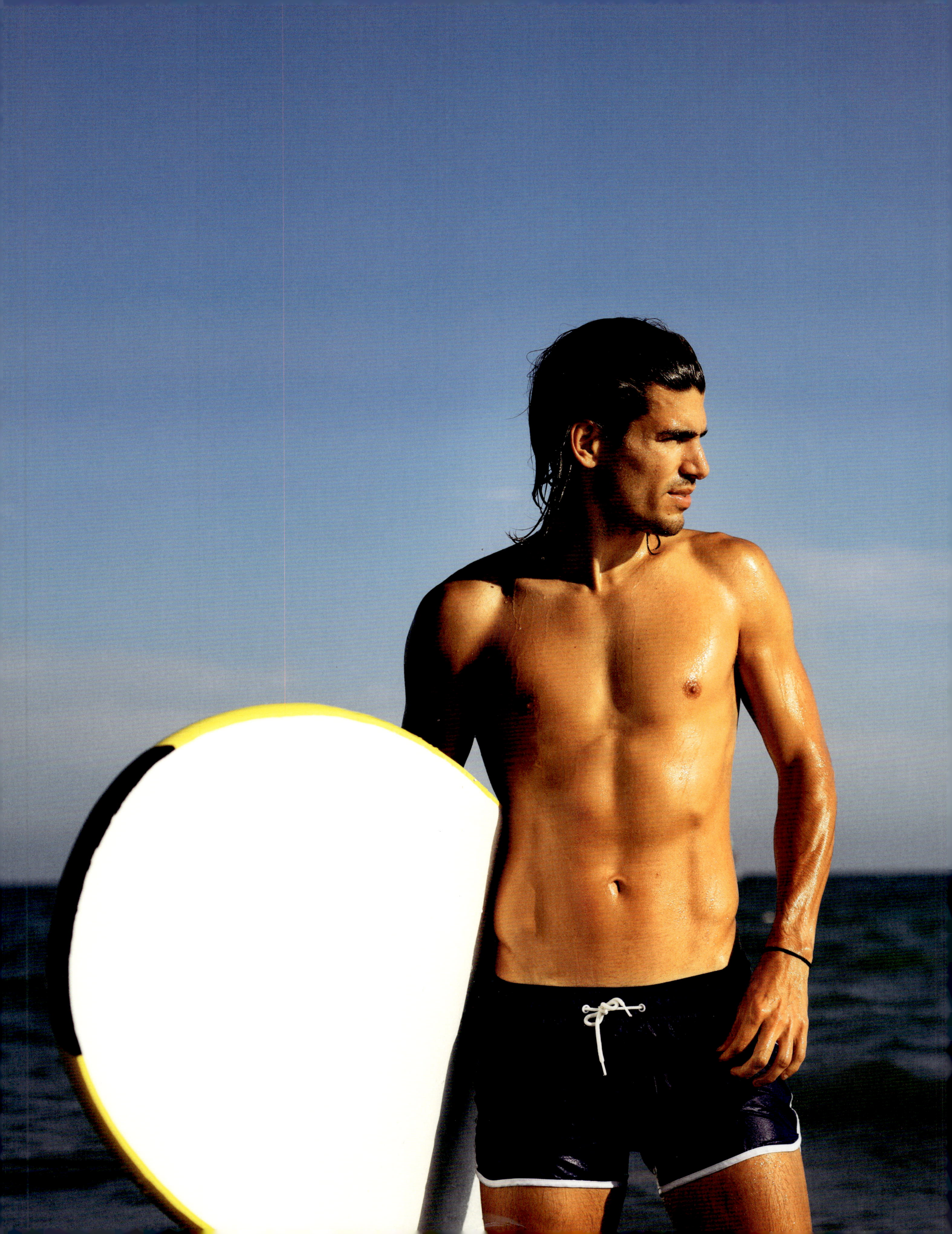

Not all revolutions need barricades. When surfing became a mass phenomenon on California beaches, more than 300,000 U.S. soldiers were stationed in Vietnam. There was something in the air–and surfing made it especially clear. Unemployed young men had taken up surfing during the Great Depression of the 1930s, an inexpensive, do-it-yourself hobby at the time. They surfed on wooden planks up to four meters long and weighing 50 kilos, made their own swimwear, and slept in their cars at the beach.
But in 1966, an incredible audience of 40,000 people turned out for the World Surfing Championships at San Diego's Ocean Beach. The media reported enthusiastically, with *Newsweek* and the *New York Times* also in attendance–and the mainstream media had a good excuse to show half-naked, well-toned men and women. Only a few days later, Brian Wilson's *Good Vibrations* would be released. Surfing magazines sprang up, as did a sophisticated, high-priced accessories industry, and advertisers were thrilled. The city leaders of San Diego and other hot spots were also delighted with the new batch of tourists who, thanks to the postwar economic boom, were bringing in a tidy stream of dollars.
But the surfers fought back. They didn't want to be marketed, and wanted to preserve their subculture. From that point on, they eschewed competitions and awards ceremonies in favor of mind-expanding drugs. As the Vietnam War escalated, surfers became California's psychedelic avant-garde.

Today, the image of the surfer still contains this contradiction: freedom, self-determination, and sensuality on the one hand, but also consumption, competition, and status symbols on the other. As an example, surfers like to drive VW buses, only older models, of course–which, well-preserved, are now as expensive and sought-after as luxurious sports cars.

Nicht alle Revolutionen brauchen Barrikaden. Als das Surfen an den kalifornischen Stränden zu einem Massenphänomen wurde, waren mehr als 300.000 US-Soldaten in Vietnam stationiert. Es lag etwas in der Luft – und beim Surfen wurde es besonders deutlich. Arbeitslose junge Männer hatten während der Großen Depression der 1930er-Jahre mit dem Surfen begonnen, einem damals preiswerten, selbstgenügsamen Hobby. Sie surften auf bis zu vier Meter langen Holzplanken, die 50 Kilo schwer waren, schneiderten sich ihre Badekleidung selbst und schliefen am Strand in ihren Autos.
Doch 1966 kamen zur Weltmeisterschaft im Wellenreiten am Ocean Beach von San Diego unfassbare 40.000 Zuschauer. Die Medien berichteten begeistert, auch *Newsweek* und die *New York Times* waren zugegen – und die Medienmacher hatten eine gute Entschuldigung, halbnackte, durchtrainierte Männer und Frauen zu zeigen. Nur wenige Tage später sollte Brian Wilsons „Good Vibrations" erscheinen. Surfzeitschriften entstanden ebenso wie eine ausgeklügelte, hochpreisige Zubehör-Industrie, und die Werbekunden waren begeistert. Auch die Stadtoberen von San Diego und anderer Hot Spots freuten sich über die neuen Touristen, die dank des ökonomischen Booms der Nachkriegszeit ordentlich Dollars mitbrachten.
Doch die Surfer schlugen zurück. Sie wollten sich nicht vermarkten lassen und ihre Subkultur bewahren. Also verzichteten sie fortan auf Wettbewerbe und Preisverleihungen und widmeten sich lieber bewusstseinserweiternden Drogen. Während der Vietnamkrieg immer heftiger eskalierte, wurden Wellenreiter zur psychedelischen Avantgarde Kaliforniens.
Noch heute gibt es in der Ikonografie des Surfers diesen Widerspruch: Freiheit, Selbstbestimmung und Sinnlichkeit auf der einen Seite – aber auch Konsum, Wettbewerb und Statussymbole. Ein Beispiel: Surfer fahren gern einen VW-Bus, natürlich nur älteren Baujahrs – der mittlerweile, gut erhalten, so teuer und gefragt ist wie ein luxuriöser Sportwagen.

Until today, the fascination for the lifestyle of the surfer is unbroken.

Bis heute ist die Faszination für den Lifestyle des Surfers ungebrochen.

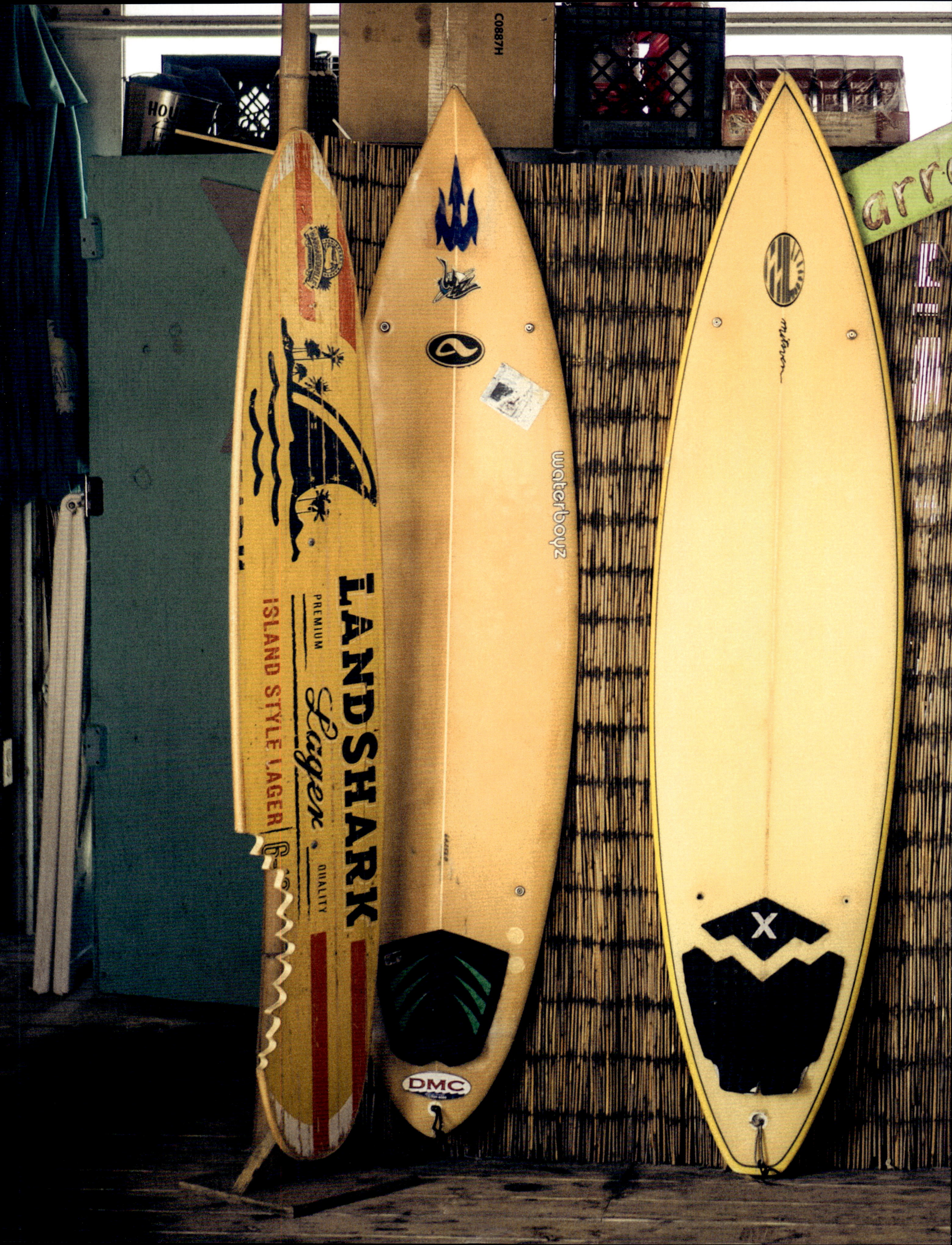
C0887H
LANDSHARK
PREMIUM
Lager
QUALITY
ISLAND STYLE LAGER
waterboyz
DMC

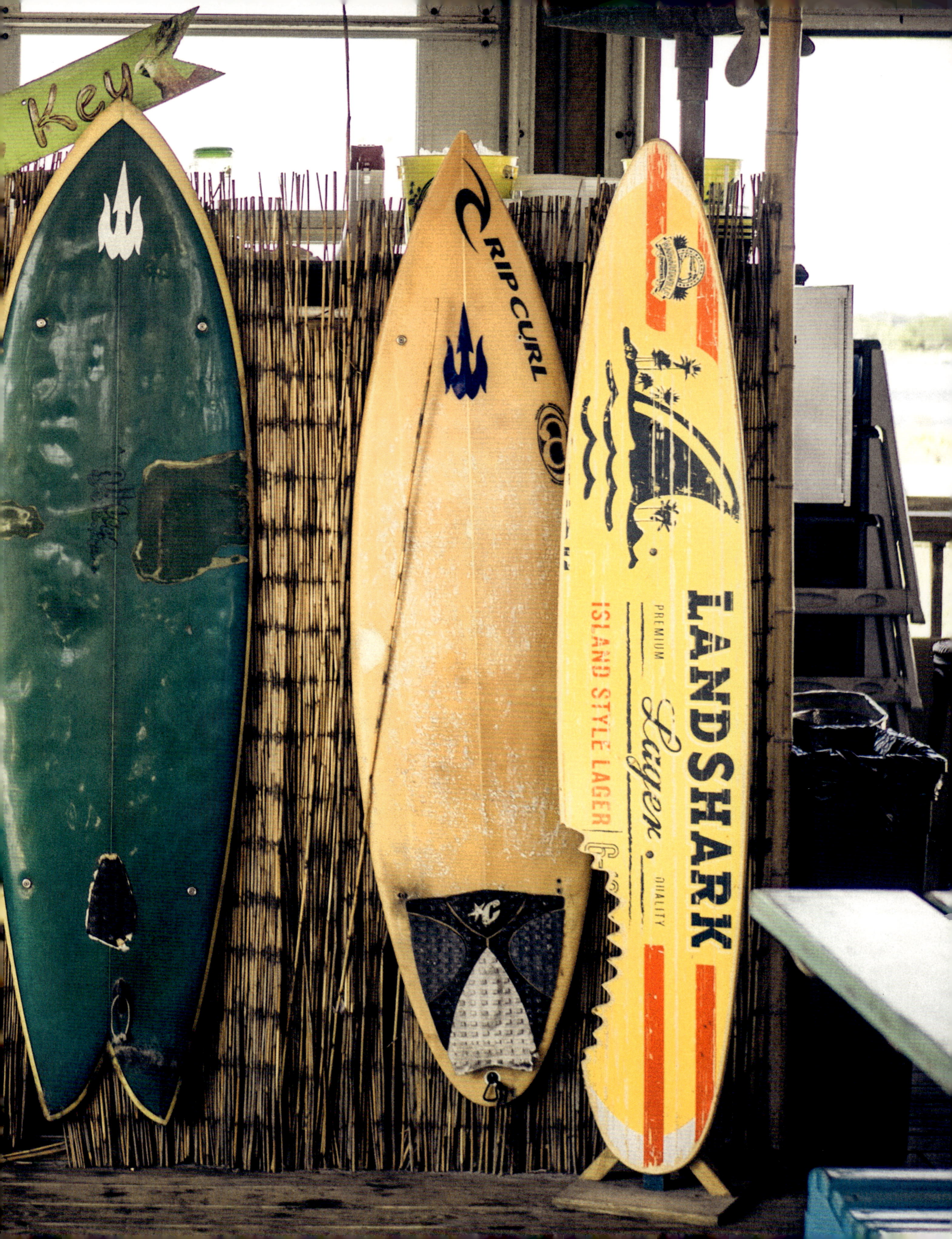
Key
RIP CURL
LANDSHARK
PREMIUM
Lager
QUALITY
ISLAND STYLE LAGER

PREVIOUS PAGE

Surfshack at Pensacola Beach, USA.

Surferhütte am Pensaco a Beach, USA.

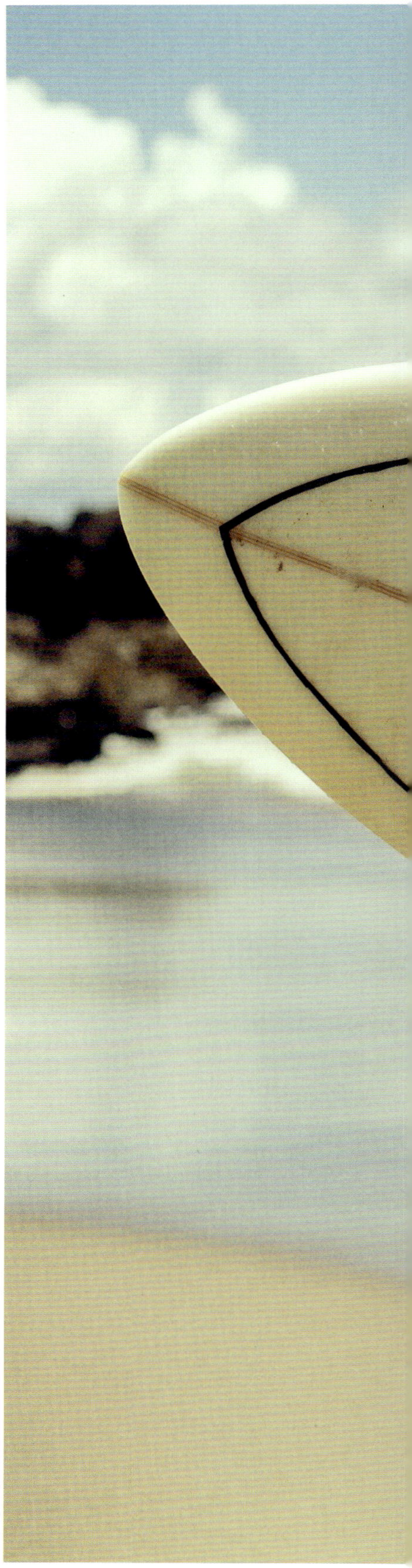

Living along with the wind and waves.
Bali, Indonesia.

Leben mit Wind und Wellen.
Bali, Indonesien.

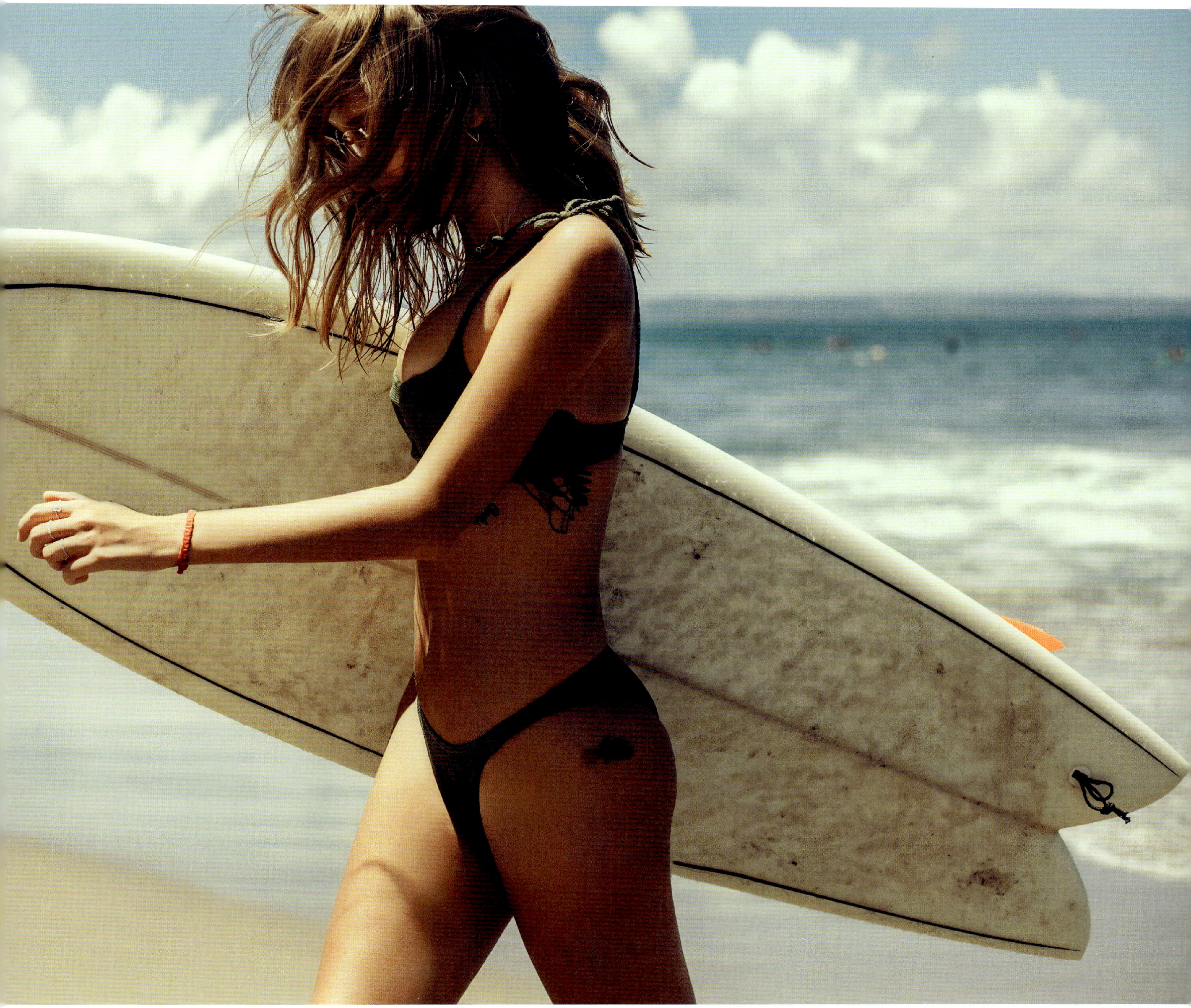

SUBKULTUR UND WETTBEWERB: DIE IKONOGRAFIE DES SURFERS

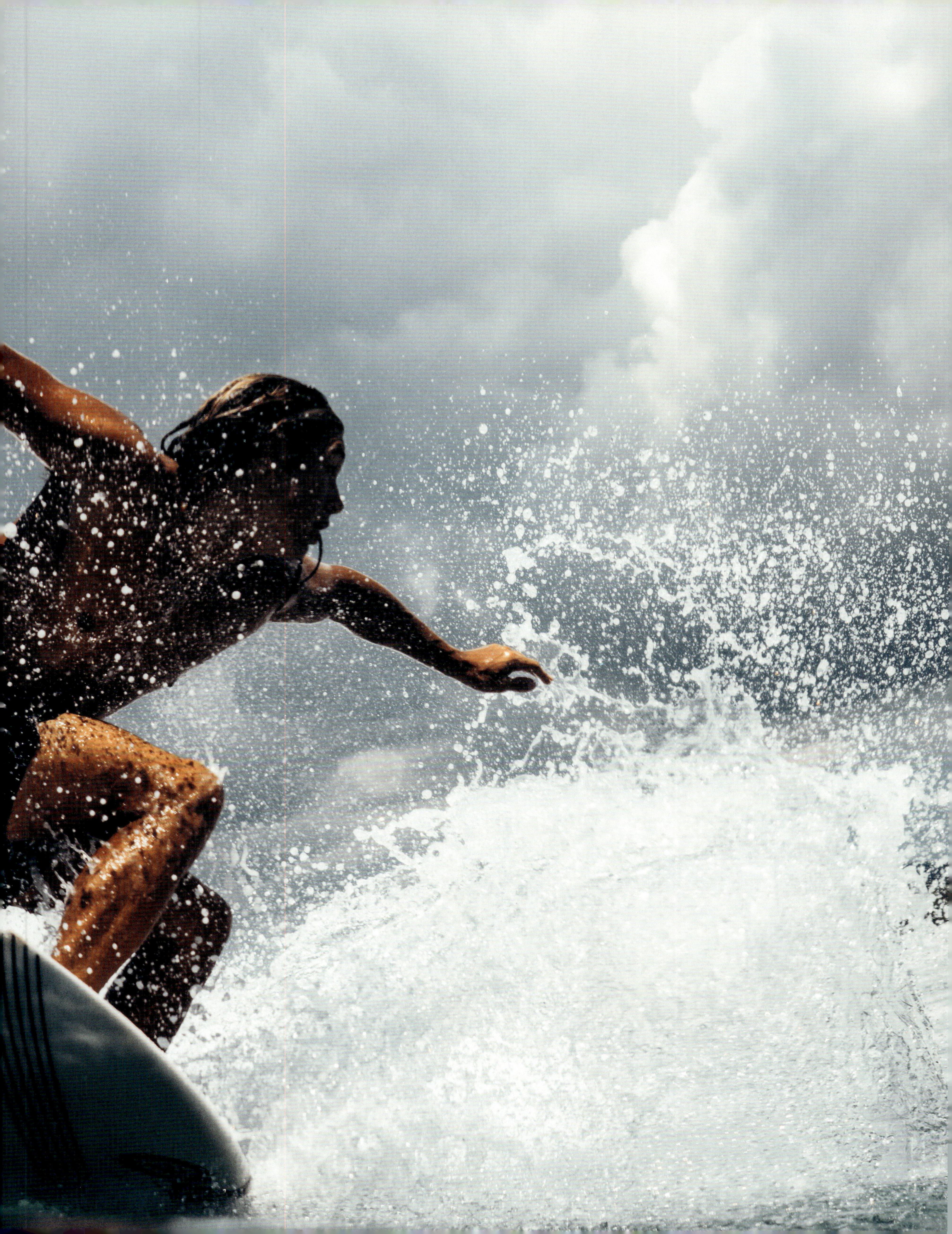

Gerupuk in Indonesia is home to five popular surf spots and draws visitors from all over the world.

Gerupuk in Indonesien hat fünf Surfspots zu bieten und wird gern aus aller Welt besucht.

LIVING

192

LEBEN

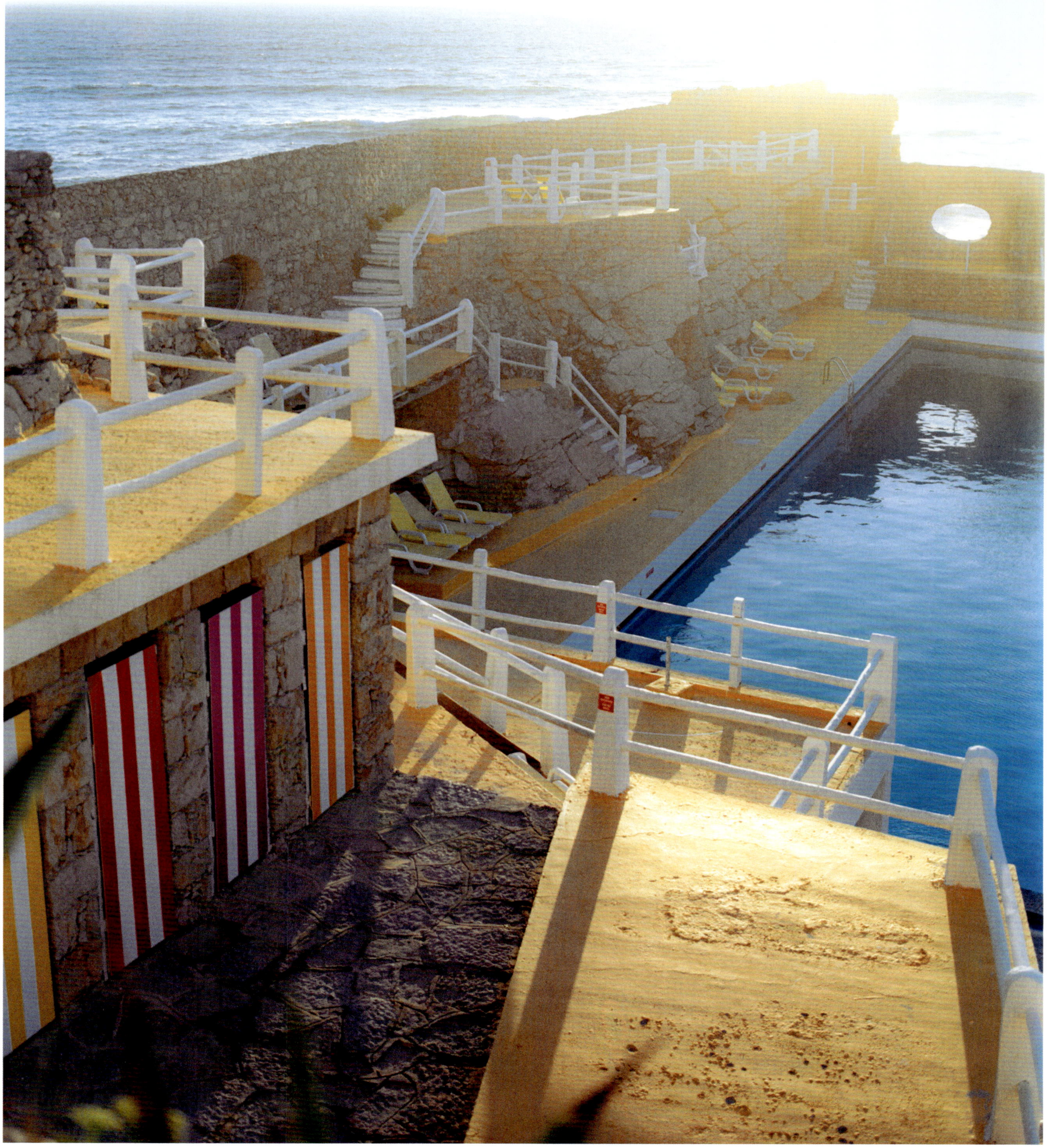

Size doesn't matter. It takes five minutes to cross the town center. 4,000 inhabitants mingle with one another during the off-season. Saint-Tropez is tiny, but it's also larger than life itself. Hardly any other place is so small and yet so gigantic at the same time.
First came the artists. No doubt it was the special light that drew them here a hundred years ago. But maybe artists are just people like you and me, craving lots of sun and plenty of wine and pungent absinthe and beautiful people.
So they came with their easels and their thirst. Soon after came the writers with their typewriters and their even greater thirst. And they started a trend, with the moneyed classes following quickly behind. The VIPs drop anchor right in front of the Sénéquier, that VIP showroom masquerading as a café. A decent berth costs €100,000. Per week. And just painting a 40-meter yacht costs up to €6 million. The old-school playboys were more reserved. Saint-Tropez lover Porfirio Rubirosa was shy, he hated typical men's jokes and he drank champagne in small sips. The only thing he did to excess was drive, until a tree on the Bois de Boulogne in Paris slowed his Ferrari down from 200 km/h to 0. Rubirosa, Soraya Esfandiari, Alfonso zu Hohenlohe, Gianni Agnelli, Samir Sibaei: Those times are gone, and another generation has long since ruled Saint-Tropez. Those were the days when the love story between Gunter Sachs and Brigitte Bardot kept two countries in suspense, and contributed a little bit to the reconciliation of France and Germany. Director Roger Vadim made Mademoiselle Bardot the symbol of Saint-Tropez with his film *And God Created Woman*, and Saint-Tropez the symbol of Brigitte Bardot.

Riviera insider Somerset Maugham described the Côte d'Azur as a "sunny place for shady characters," and this, the old-timers think, has never been as true as it is today. But the playboys of yesteryear were not all that romantic either. "Falling in love," the incomparable Gianni Agnelli once said, "is for waitresses."
But Agnelli always preferred to be in Monte Carlo anyway.

Auf die Größe kommt es nicht an. Zum Durchqueren des Ortskerns braucht man nicht mehr als fünf Minuten. 4.000 Einwohner bleiben in der Nebensaison unter sich. Saint-Tropez ist winzig. Aber auch größer als das Leben selbst. Kaum ein Ort ist zugleich so klein und so gigantisch.
Zuerst kamen die Künstler. Zweifellos war es das besondere Licht, das sie vor hundert Jahren hierher zog. Vielleicht sind Künstler aber auch nur Menschen wie du und ich, hatten Lust auf viel Sonne und reichlich Wein und scharfen Absinth und schöne Menschen. Also kamen sie mit ihren Staffeleien und ihrem Durst. Kurz danach kamen die Schriftsteller mit ihren Schreibmaschinen und ihrem noch größeren Durst. Und sie setzten einen Trend: Der Geldadel folgte rasch. Die VIPs ankern direkt vor dem Sénéquier, jenem VIP-Showroom, der sich als Café tarnt. Ein vernünftiger Liegeplatz kostet 100.000 Euro. Pro Woche. Und allein der *Anstrich* einer 40-Meter-Yacht kostet bis zu sechs Millionen Euro. Die Playboys der alten Schule gaben sich zurückhaltender. Saint-Tropez-Liebhaber Porfirio Rubirosa war schüchtern, hasste Herrenwitze und trank den Champagner in kleinen Schlucken.
Nur das Autofahren betrieb er exzessiv, bis ein Baum im Bois de Boulogne zu Paris seinen Ferrari von 200 km/h auf 0 abbremste. Rubirosa, Soraya Esfandiari, Alfonso zu Hohenlohe, Gianni Agnelli, Samir Sibaei: Tempi passati, längst regiert eine andere Generation Saint-Tropez. Was waren das noch für Zeiten, als die Liebesgeschichte zwischen Gunter Sachs und Brigitte Bardot gleich zwei Nationen in Atem hielt und ein klein wenig zur Aussöhnung Frankreichs und Deutschlands beitrug. Regisseur Roger Vadim hatte Mademoiselle Bardot mit dem Film *Und immer lockt das Weib* zum Symbol für Saint-Tropez gemacht und Saint-Tropez zum Symbol für Brigitte Bardot.
Als „sonnigen Ort für zwielichtige Gestalten" bezeichnete Riviera-Kenner Somerset Maugham die Côte d'Azur, und das, finden die Alteingesessenen, sei nie so wahr gewesen wie heute. Doch allzu romantisch waren die Lebemänner von einst nun auch wieder nicht. „Verlieben", sagte der unvergleichliche Gianni Agnelli einmal, „ist was für Dienstmädchen." Aber Agnelli war immer lieber in Monte Carlo.

PREVIOUS PAGE

View from a beach house to the sea. Praia do Guincho, Portugal.

Blick von einem Strandhaus über das Meer. Praia do Guincho, Portugal

Brigitte Bardot in her villa of La Madrague in Saint-Tropez, France in 1965.

Brigitte Bardot in ihrer Villa La Madrague in Saint-Tropez, Frankreich 1965.

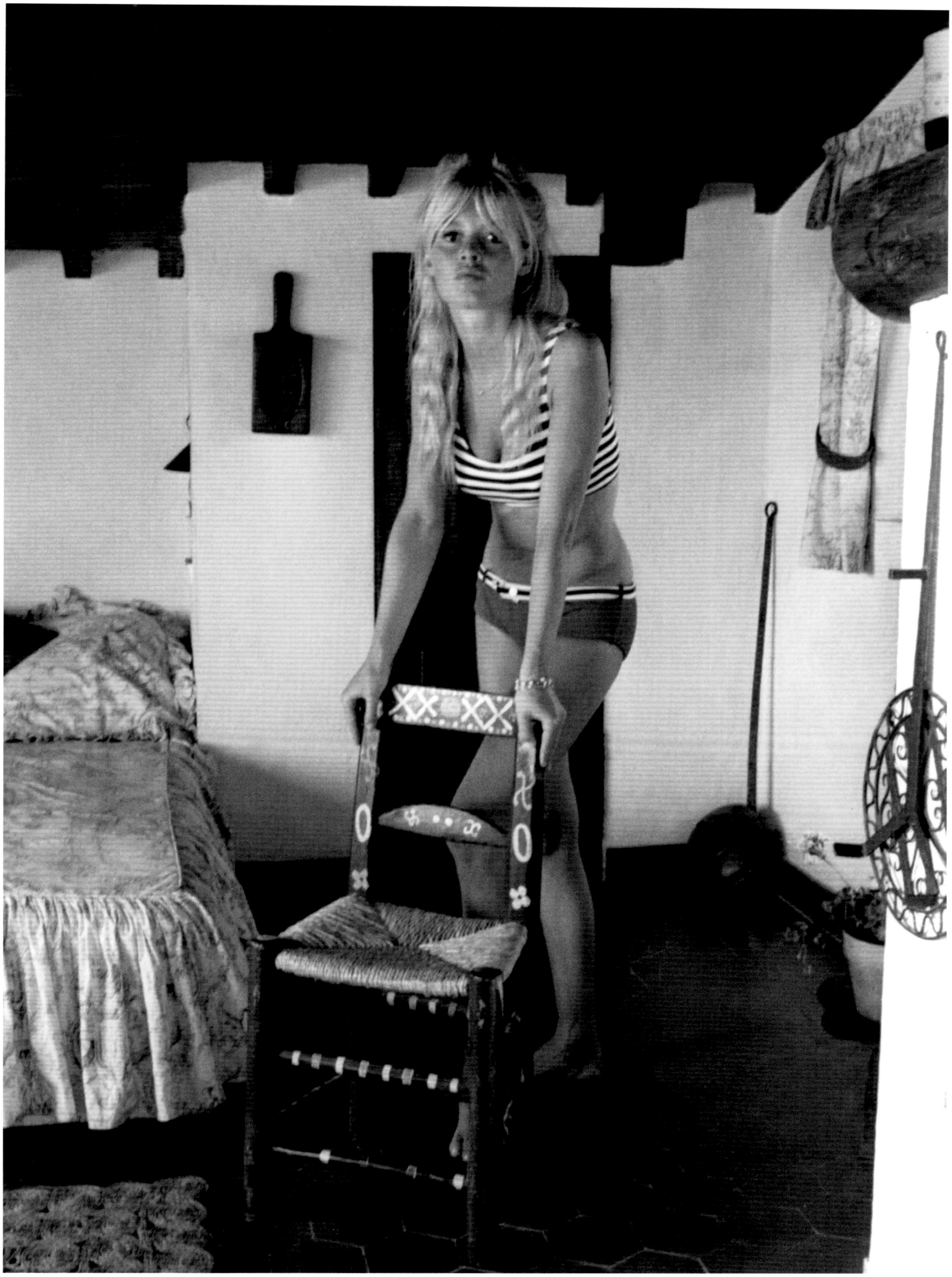

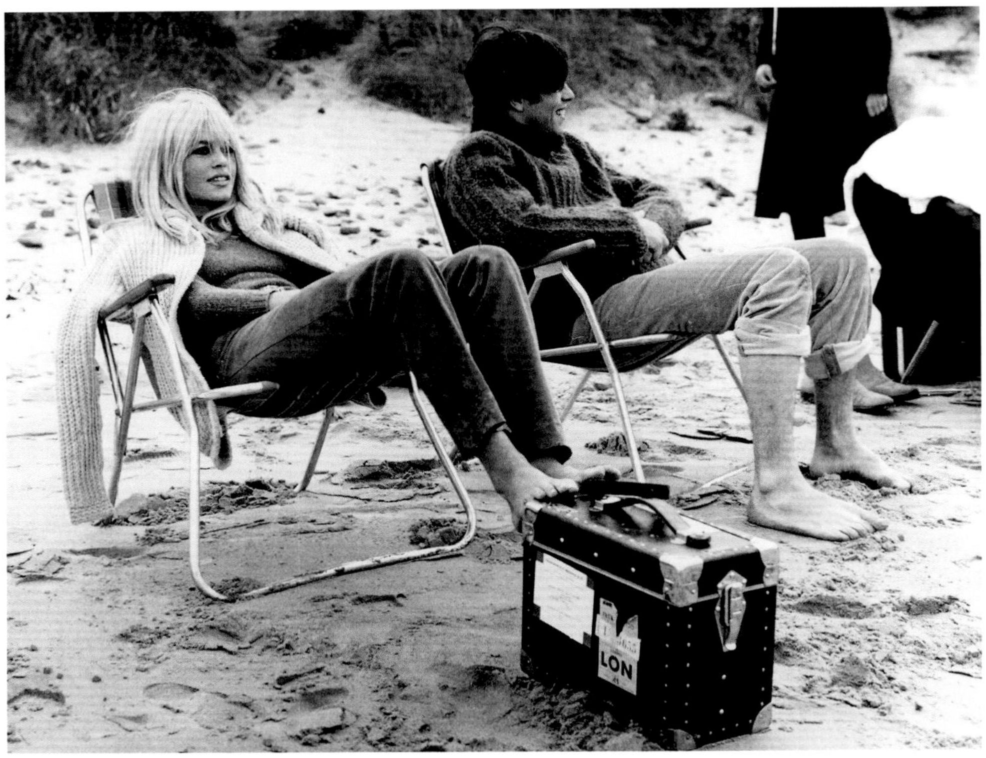

Brigitte Bardot at the beach–here not in Saint-Tropez, but on the set of *Two Weeks in September* in Hollywood, USA in 1966.

Brigitte Bardot am Strand – hier nicht in Saint-Tropez, sondern am Set von *Zwei Wochen im September* in Hollywood, USA im Jahr 1966.

German-French love story: Brigitte Bardot with Gunther Sachs in 1966 in Saint-Tropez, France.

Deutsch-französische Liebesgeschichte: Brigitte Bardot mit Gunther Sachs 1966 in Saint-Tropez, Frankreich.

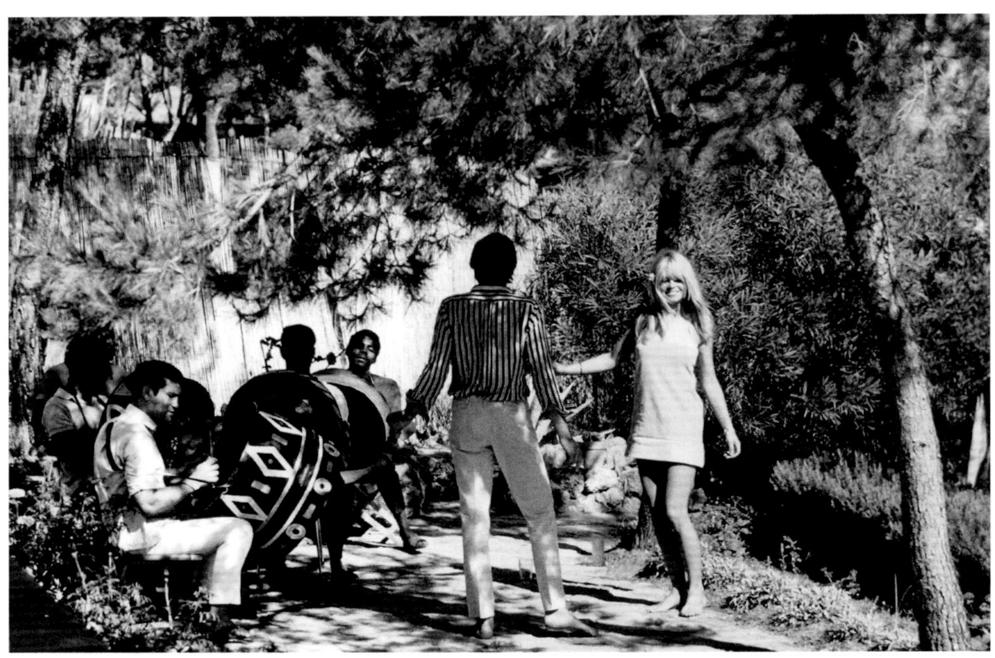

Pier or runway? Both scenarios are possible in Saint-Tropez.

Anlegestelle oder Laufsteg? In Saint-Tropez ist beides möglich.

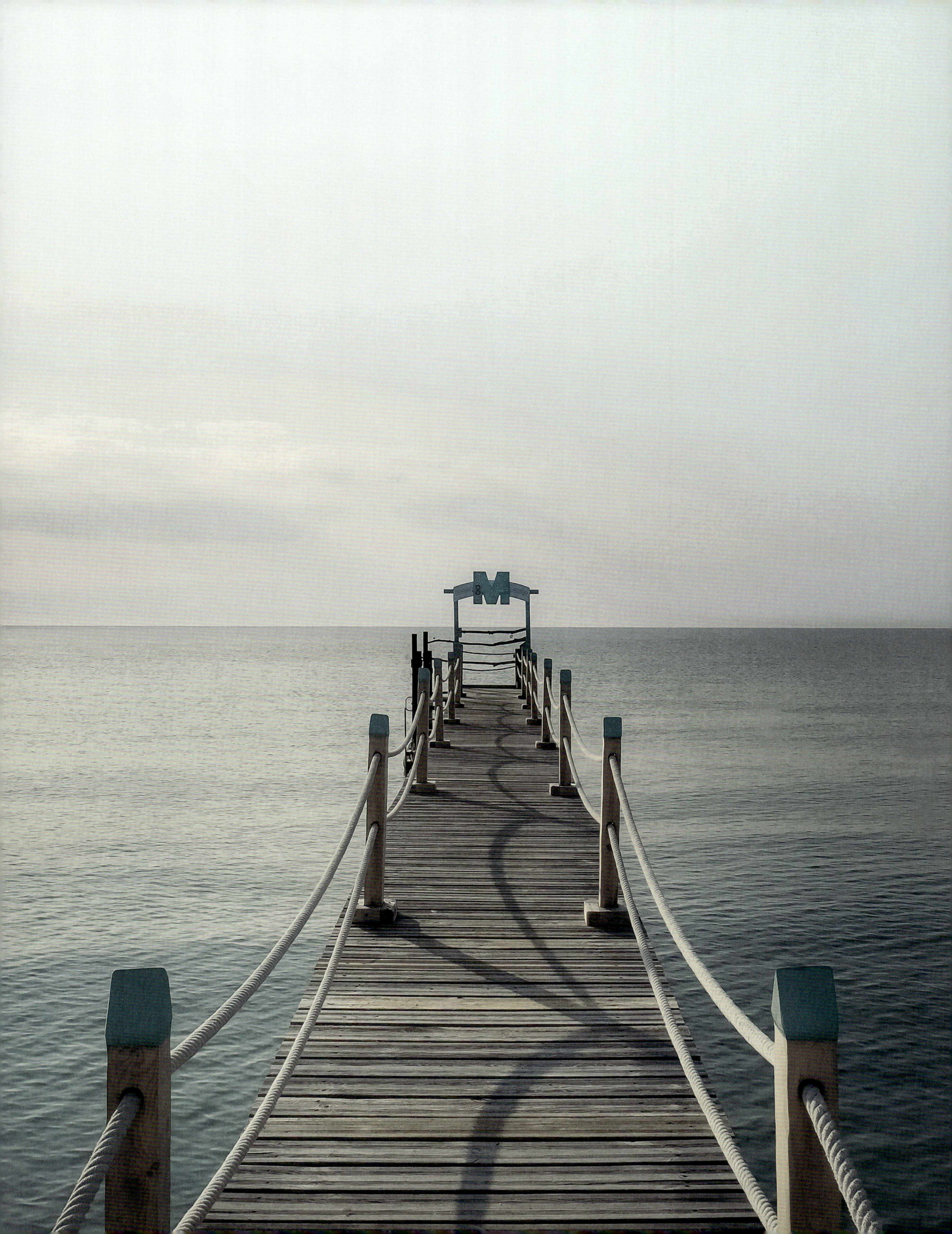
M

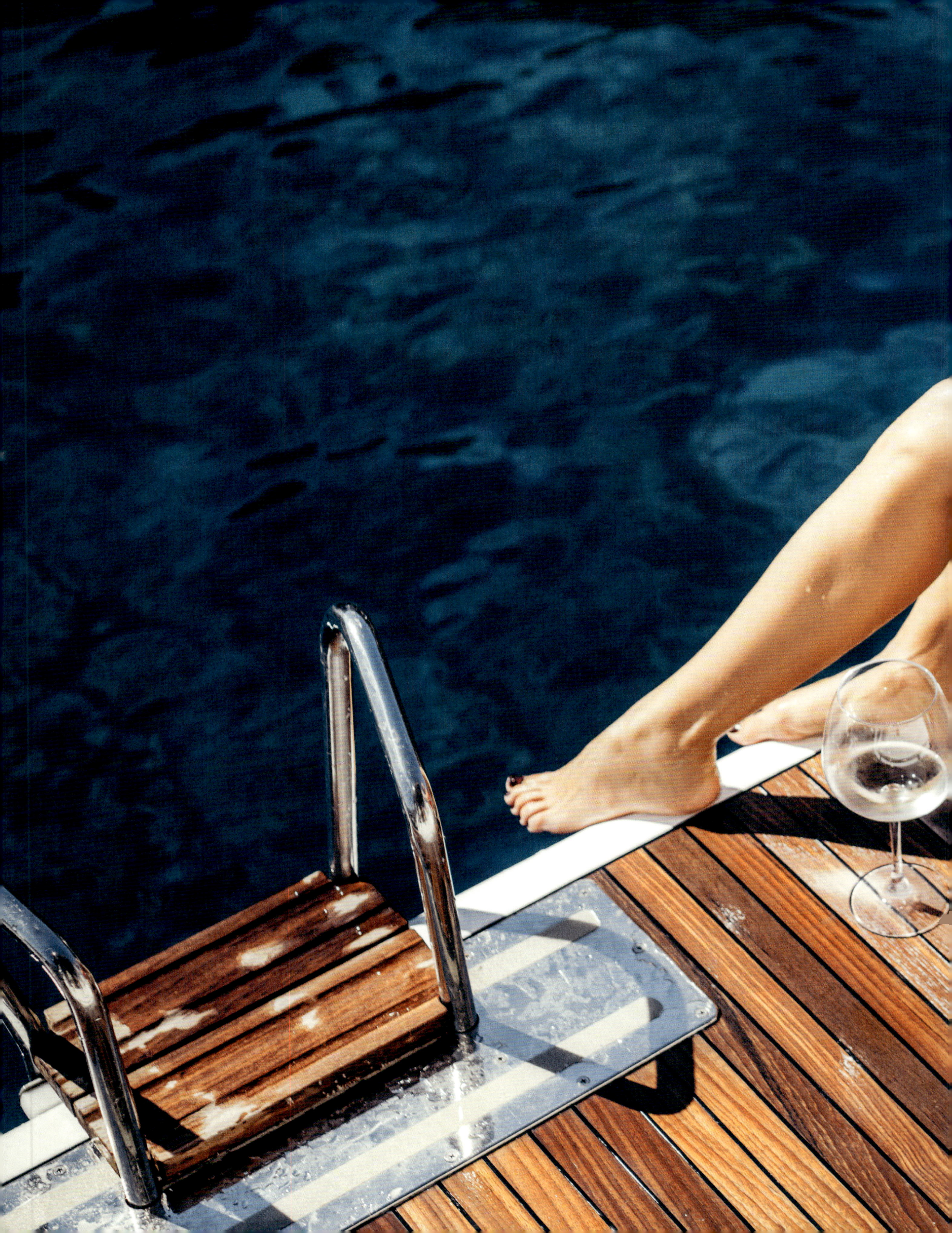

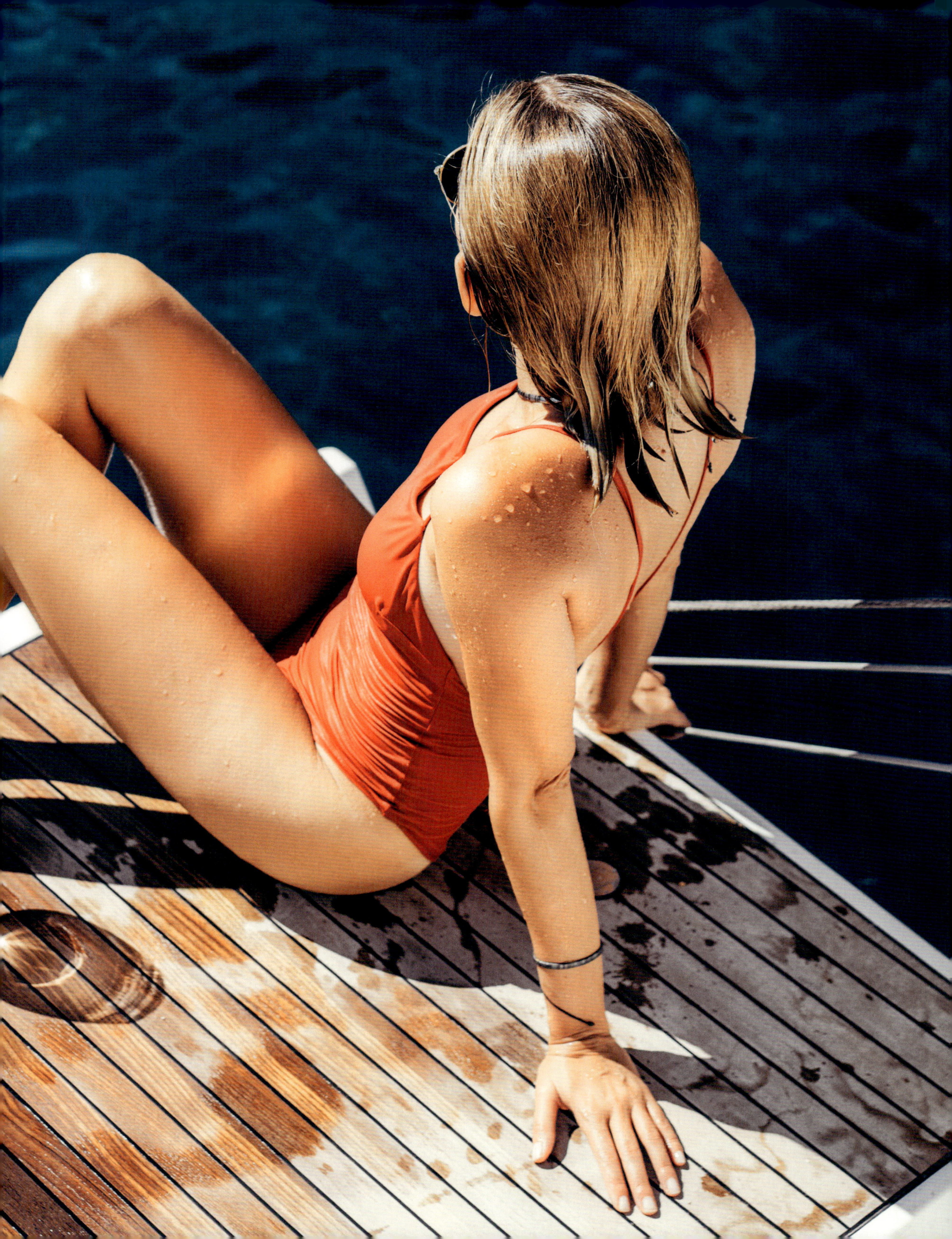

PREVIOUS PAGE

Sunbathing on the yacht.

Sonnenbad auf der Yacht.

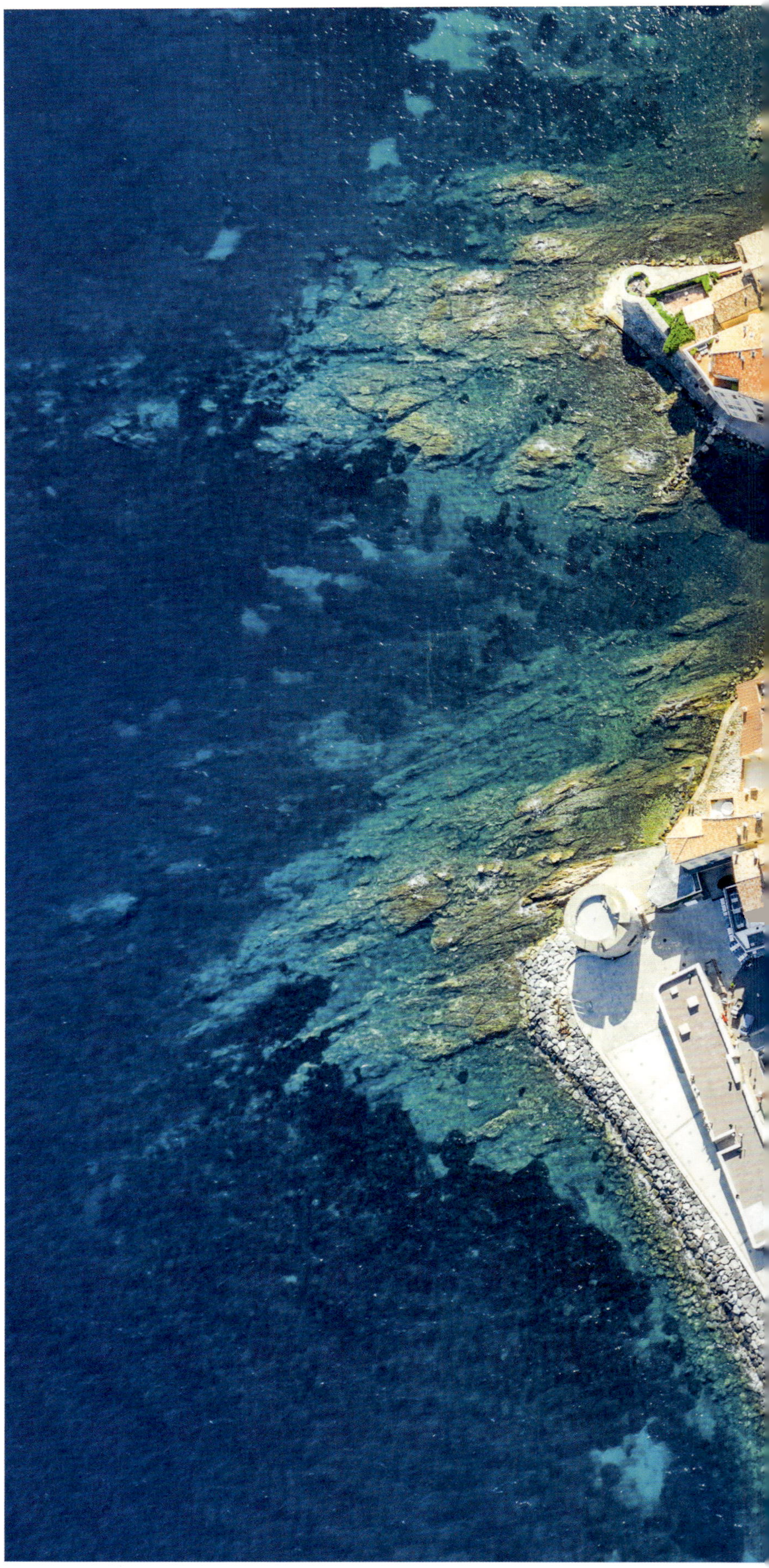

View of the famous marina
of Saint-Tropez, France.

Blick auf den berühmten Yachthafen
von Saint-Tropez, Frankreich.

A TALE OF OLD AND NEW MONEY: SAINT-TROPEZ

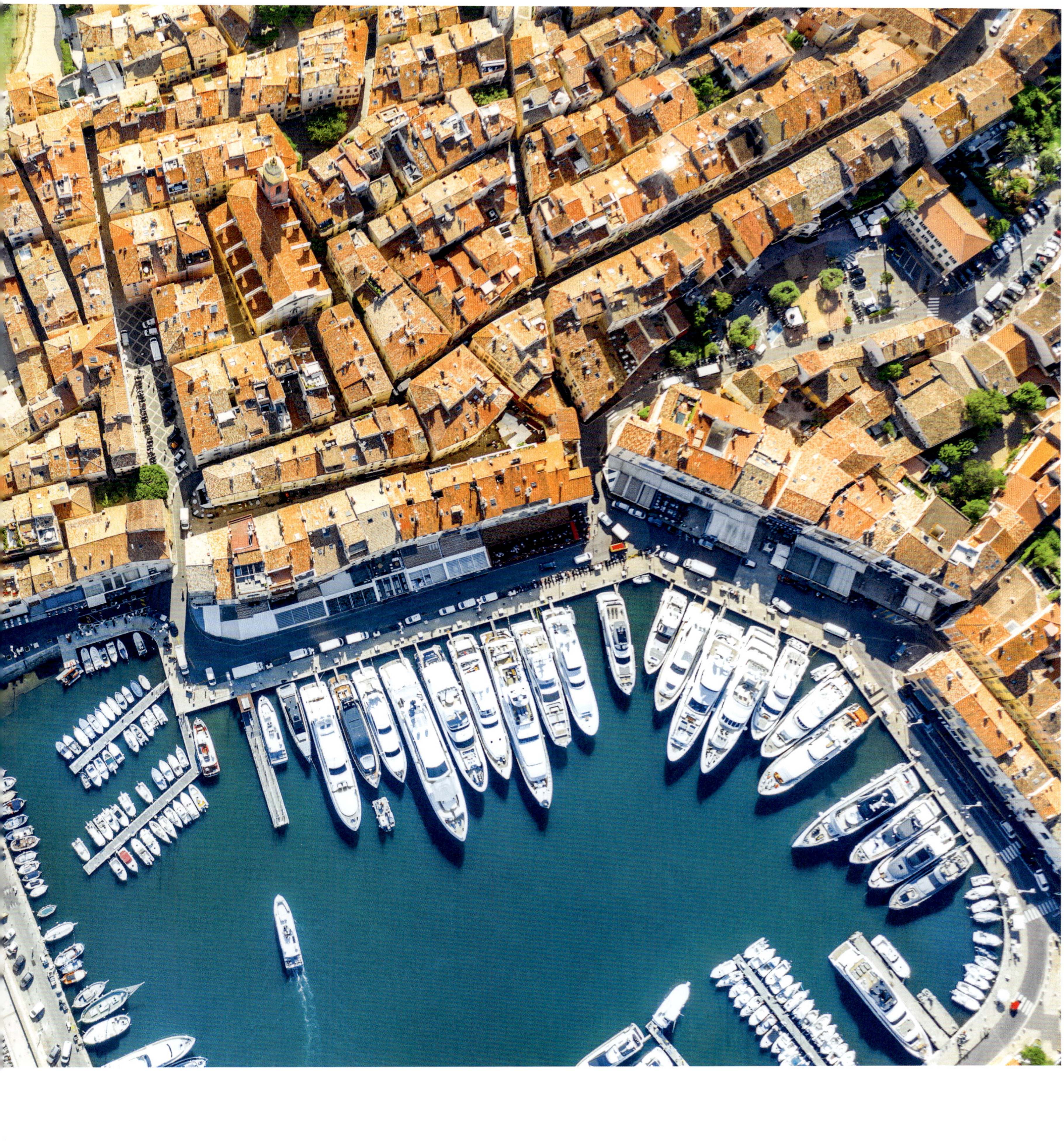

EINE BALLADE VON ALTEM UND NEUEM GELD: SAINT-TROPEZ

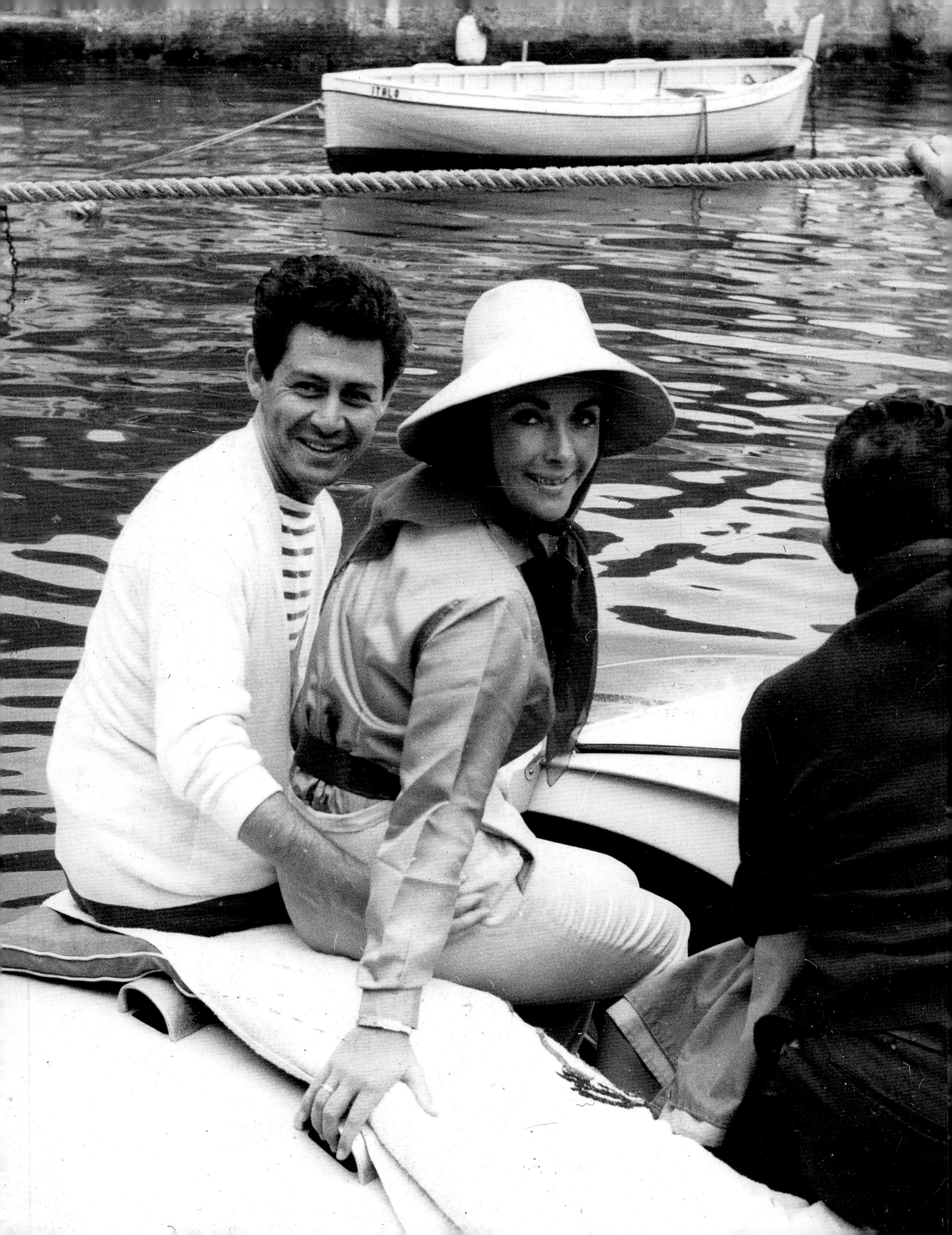
ITALO

It doesn't get any more picturesque than this: Portofino translates as "pretty harbor," but the name presumably comes from the Latin *Portus Delphini* (dolphin harbor) dating back to Roman times. These marine mammals still like to come and visit here today.
Portofino with its enchanting bay was long a sleepy little fishing village; its remoteness kept it immune from exploitation for a long time, and to this day less than 400 people live there. Not until the 1920s did things change: Elizabeth von Arnim wrote her novel *The Enchanted April*, which became a bestseller and drew people's attention to Portofino. In the Italian boom of the 1950s, the hit song *Love in Portofino* and the Austrian movie *Zwei in einem Auto* (Two in a Car) attracted Hollywood stars staying in Rome to the Ligurian coast.
Thanks to the movie *Come September* starring Gina Lollobrigida and Rock Hudson, Portofino finally became a place people around the world longed to visit. The luxury hotel Splendido, located above Portofino and still in existence today, served Robert T. Talbot, alias Rock Hudson, as his private villa in the film. It was also there that Richard Burton asked Elizabeth Taylor for her hand in marriage, and the Splendido's chef created the dish *Spaghetti Elizabeth Taylor*, ordered by the diva on a daily basis and still on the hotel's menu to this day. It is made with fresh tomatoes, basil, garlic, oregano, thyme, lemon zest, and brown sugar. Taylor had also traveled to Portofino with Burton's predecessor, singer and actor Eddie Fisher; the couple had spent their honeymoon here.

To this day, Portofino is a popular film set, seen for example in Wim Wenders' *Beyond the Clouds*, starring John Malkovich and Sophie Marceau.
For the soap opera *The Bold and the Beautiful*, Portofino was recreated at Universal Studios in Orlando, Florida.
Portofino Bay is rocky, its beaches are tiny and difficult to reach, often only by stairs–but the descent is worth the effort for the water quality alone. The bay of Portofino is a protected nature reserve, and the water is sparkling clear at the tiny pebble beaches of Spiaggia dell'Olivetta or Baia Cannone. If you don't want to give up sand, you can travel to the neighboring town of Santa Margerita Ligure; where, along with the beach, you'll also find the world-famous Liberty-style changing cabins.

Malerischer geht's nicht mehr: Portofino heißt „edler Hafen", aber der Name stammt vermutlich vom lateinischen *Portus Delphini* (Delfinhafen) aus der Römerzeit, denn hier schauen die Meeressäuger bis heute gern vorbei.
Portofino mit seiner zauberhaften Bucht war lange Zeit ein verschlafener kleiner Fischerort; seine Abgeschiedenheit machte ihn lange immun gegen jede Vereinnahmung, und bis heute leben keine 400 Menschen dort. Erst in den 1920er-Jahren änderte sich das: Elizabeth von Arnim schrieb den Roman *Verzauberter April*, der ein Bestseller wurde und die Menschen auf Portofino aufmerksam machte. Im Italien-Boom der 1950er-Jahre lockten der Schlager „Love in Portofino" und der Film *Zwei in einem Auto* die Hollywood-Stars, die in Rom weilten, an die ligurische Küste.
Dank des Films *Come September* mit Gina Lollobrigida und Rock Hudson wurde Portofino endgültig ein weltweiter Sehnsuchtsort. Das Luxushotel Splendido oberhalb von Portofino, das es noch heute gibt, diente Robert T. Talbot alias Rock Hudson im Film als Privatvilla. Dort hielt auch Richard Burton um Elizabeth Taylors Hand an, und der Chefkoch des Splendido kreierte die „Spaghetti Elizabeth Taylor", welche die Diva täglich bestellte und die bis heute auf der Speisekarte des Hotels stehen. Sie werden mit frischen Tomaten, Basilikum, Knoblauch, Oregano, Thymian, Zitronenabrieb und braunem Zucker zubereitet. Mit Burtons Vorgänger, dem Sänger und Schauspieler Eddie Fisher, war Taylor ebenfalls nach Portofino gereist, das Paar hatte seine Flitterwochen hier verbracht.
Bis heute ist Portofino eine beliebte Filmkulisse, etwa für Wim Wenders' *Jenseits der Wolken* mit Sophie Marceau und John Malkovich. Für die Seifenoper *Reich und Schön* wurde Portofino in den Universal-Studios in Orlando, Florida, sogar nachgebaut.
Die Bucht von Portofino ist felsig, die Strände sind winzig und schwer zu erreichen, oft nur über Treppen – aber der Abstieg lohnt die Mühe schon wegen der Wasserqualität. Die Buch von Portofino ist Naturschutzgebiet, das Wasser funkelt klar, etwa am winzigen Kieselstrand Spiaggia dell'Olivetta oder der Baia Cannone. Wer nicht auf Sand verzichten will, sollte in den Nachbarort Santa Margerita Ligure reisen; dort gibt es zum Strand auch die weltberühmten Kabinen im Liberty-Stil.

Elizabeth Taylor and Eddie Fisher during their honeymoon in Portofino, Italy. 1961.

Elizabeth Taylor und Eddie Fisher während ihrer Flitterwochen in Portofino, Italien. 1961.

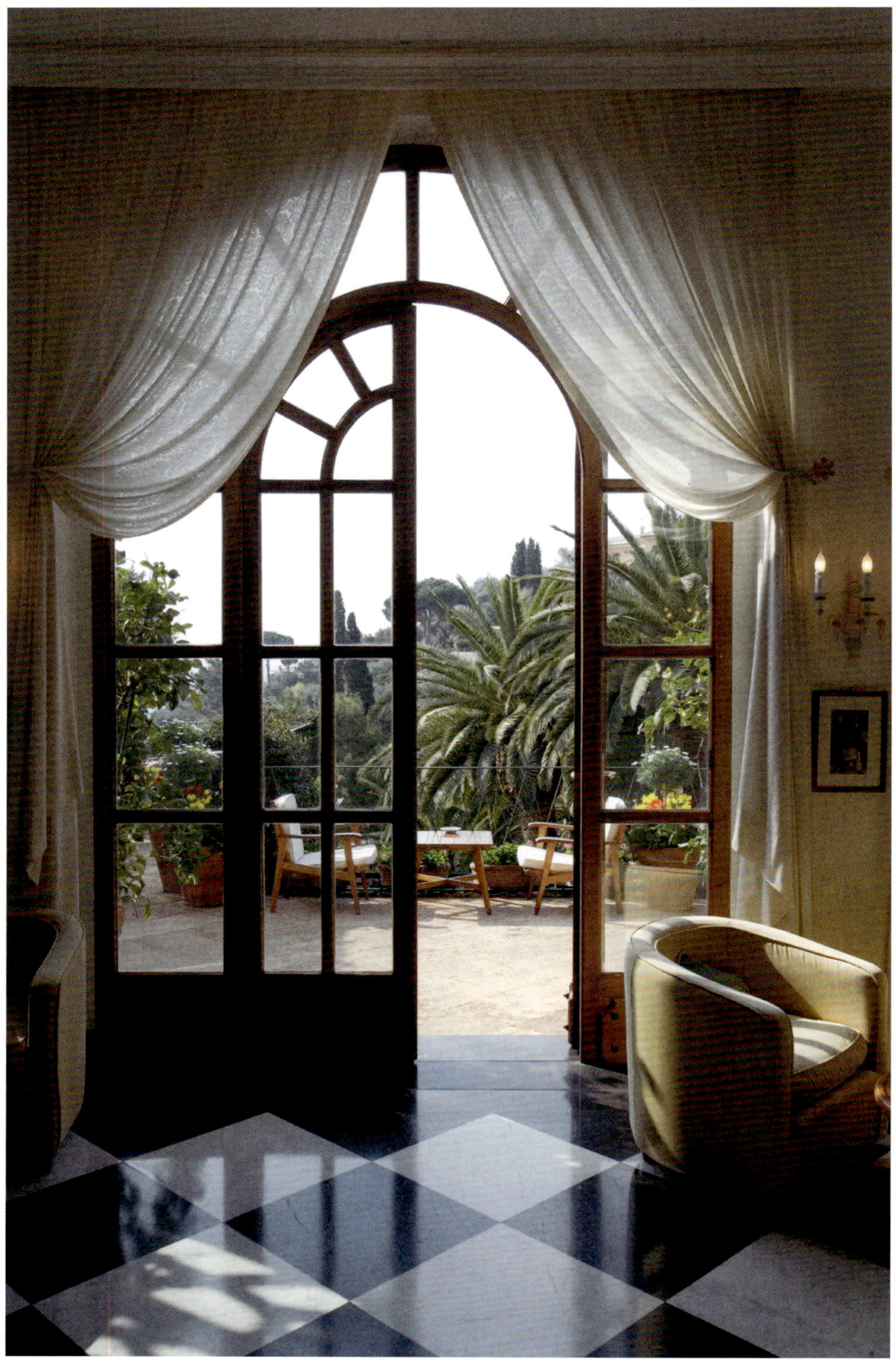

View from legendary Hotel Splendido in Portofino where Burt Reynolds proposed to Elizabeth Taylor.

Blick aus dem legendären Hotel Splendido in Portofino, wo Burt Reynolds um die Hand von Elizabeth Taylor anhielt.

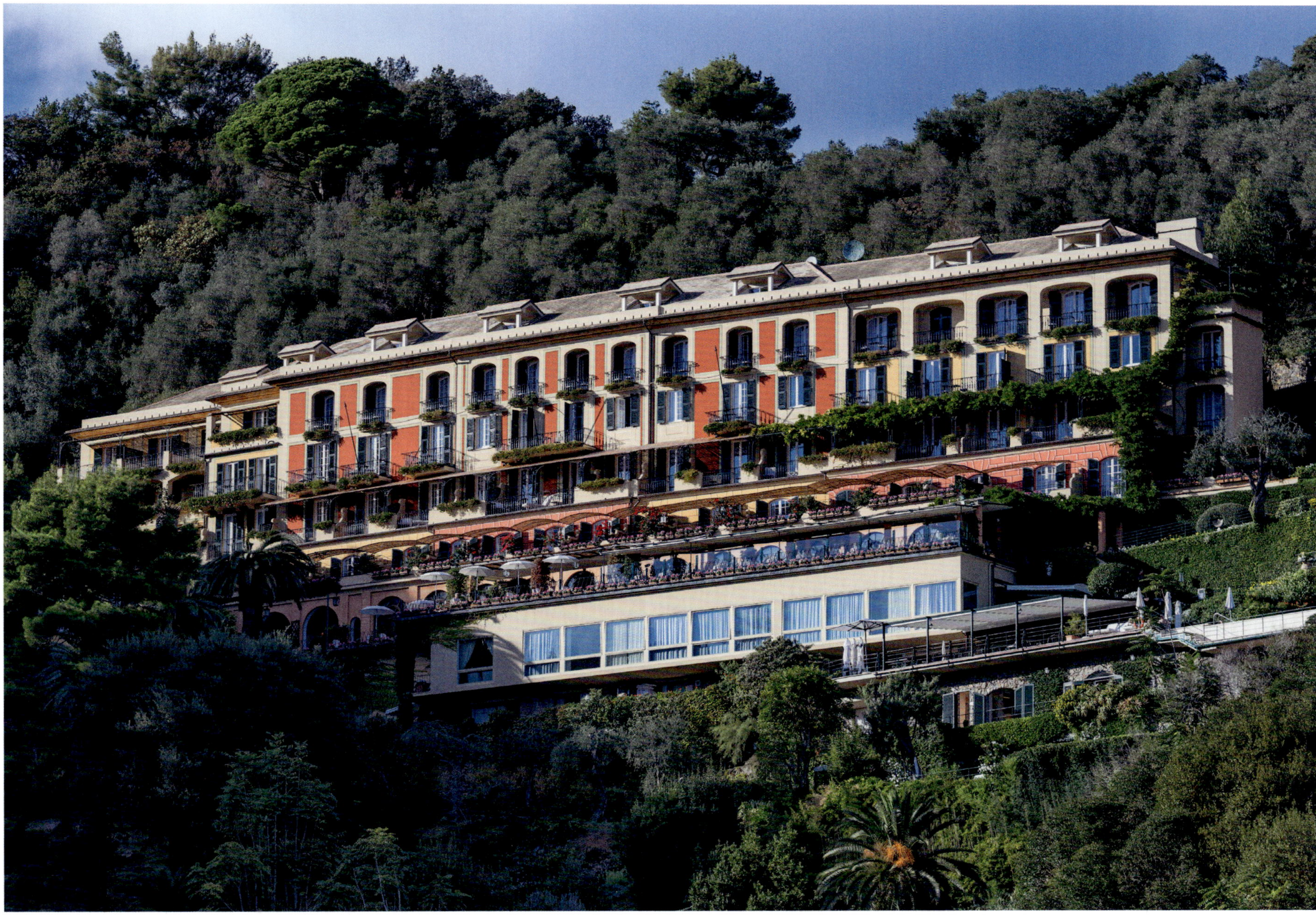

Hotel Splendido in the hills of Portofino, Italy.

Das Hotel Splendido in den Hügeln von Portofino, Italien.

Beach scenery in red and white.

Strandszene in rot-weiß.

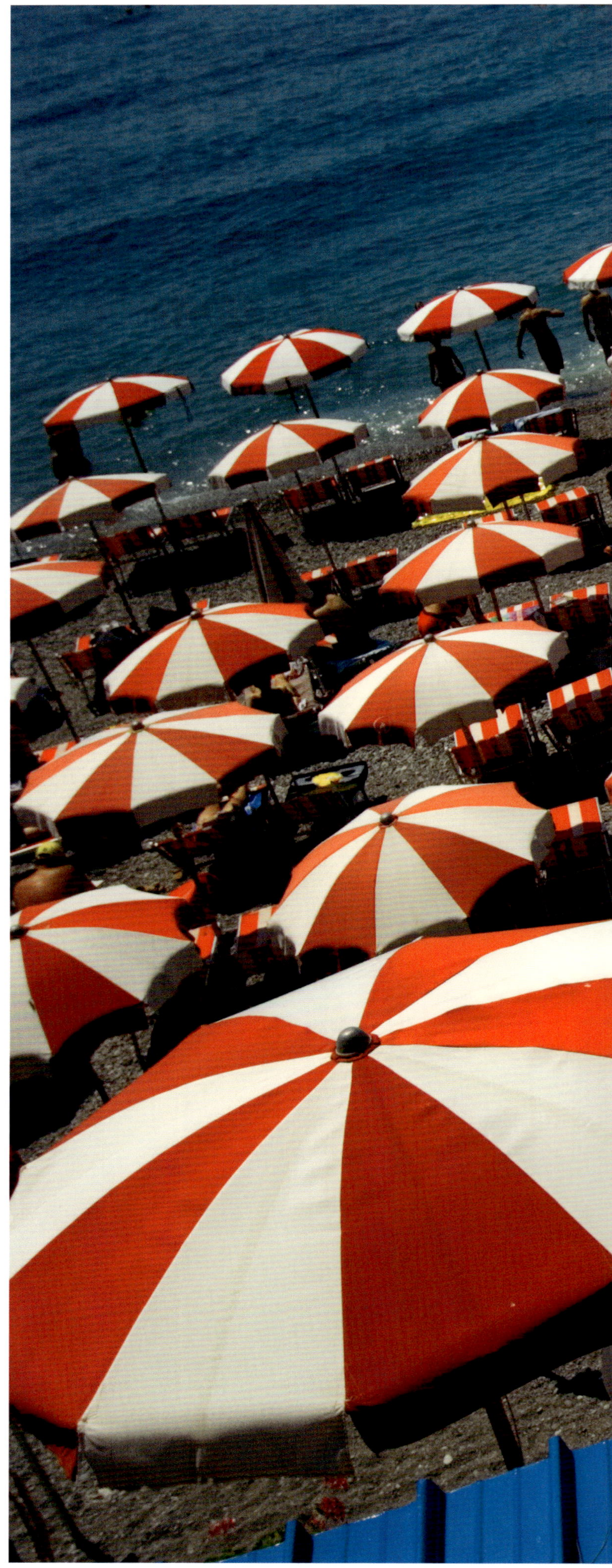

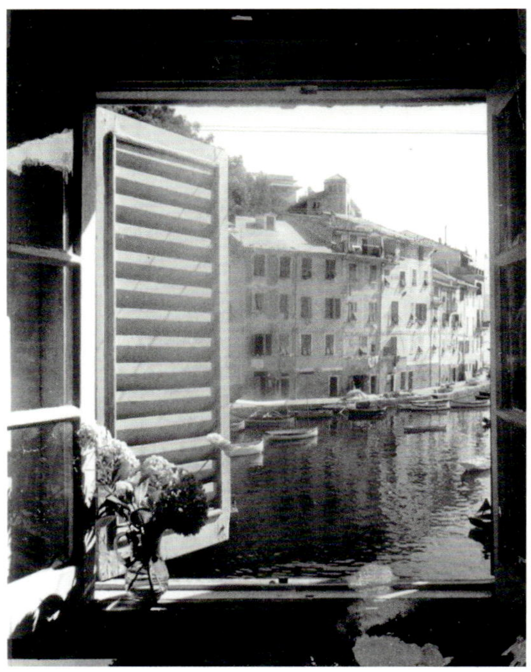

TOP

View of Portofino harbor in 1941.

Blick auf den Hafen von Portofino im Jahr 1941.

BOTTOM

Scene at Portof no harbor in 1969.

Hafenszene in Portofino aus dem Ja hr 1969.

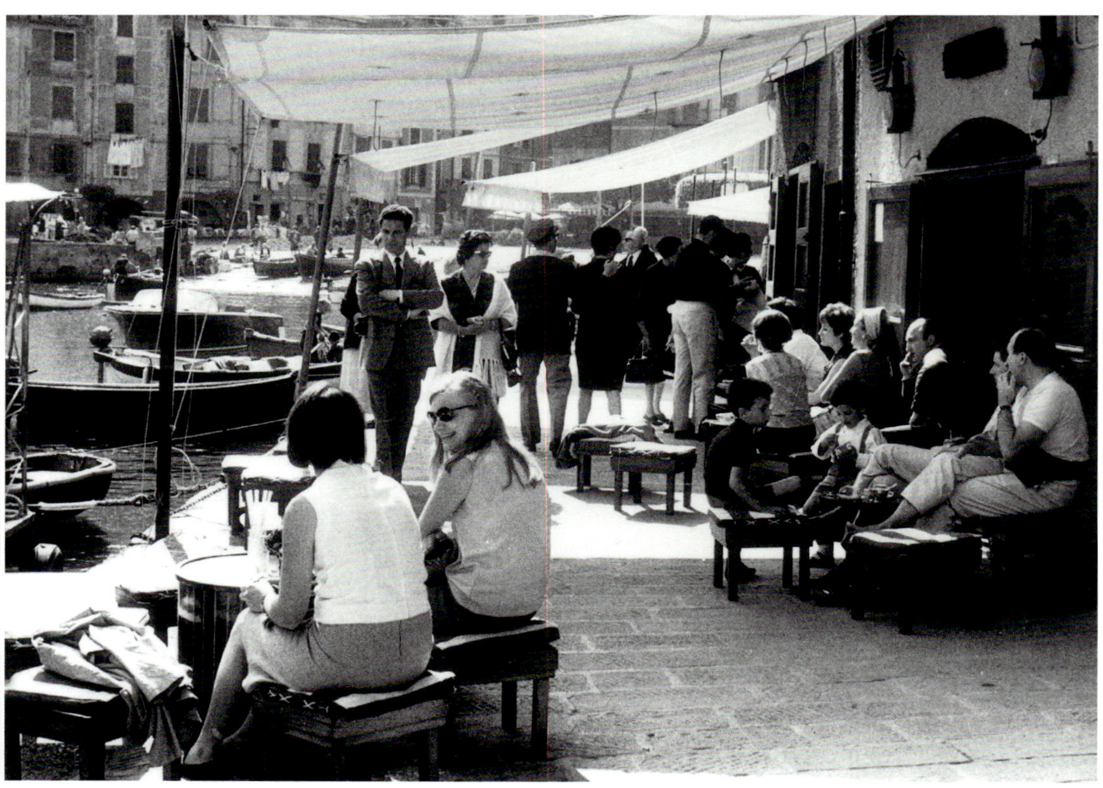

THE HOLLYWOOD OF ITALY: LOVE IN PORTOFINO

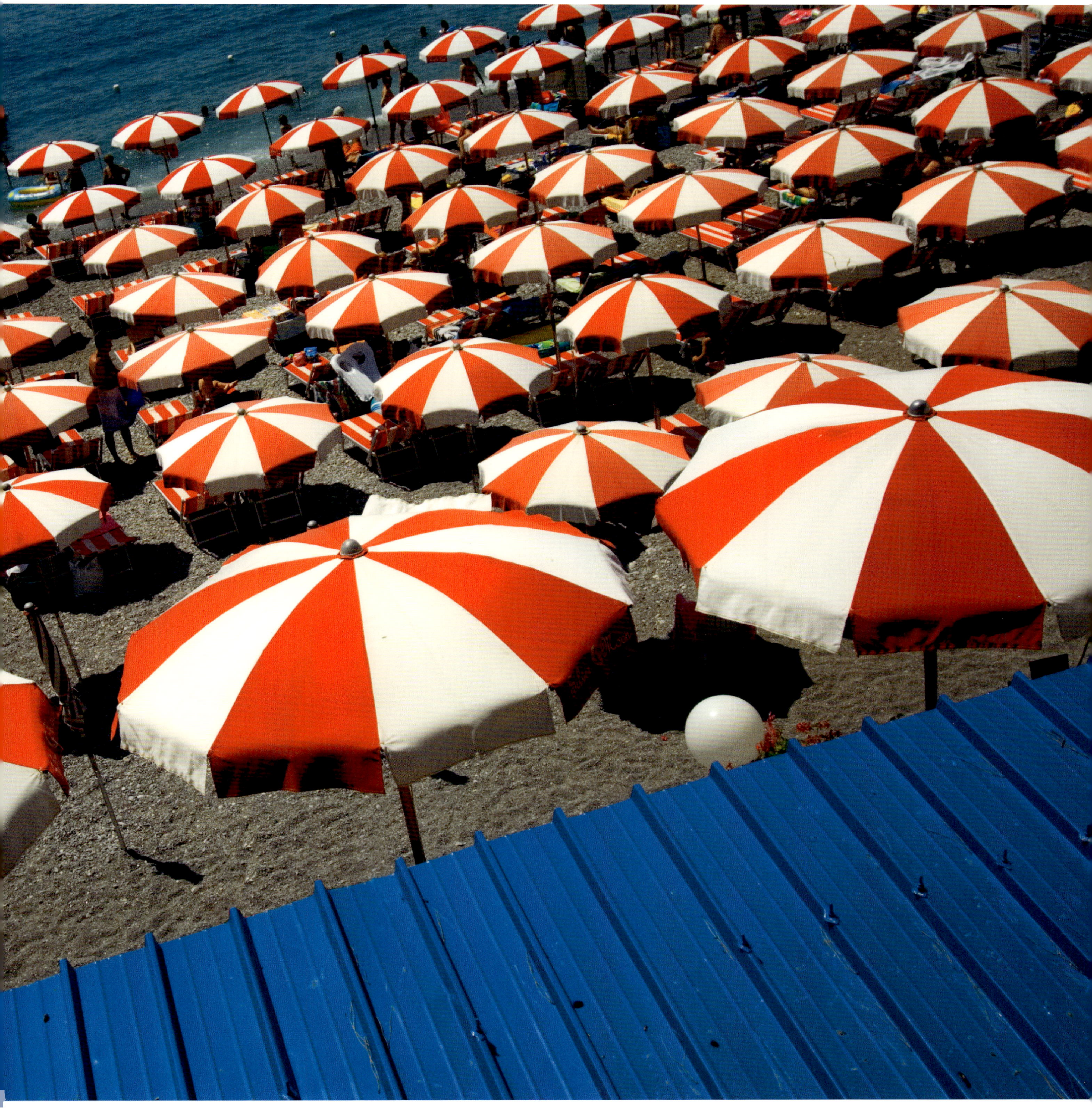

DAS HOLLYWOOD ITALIENS: LOVE IN PORTOFINO

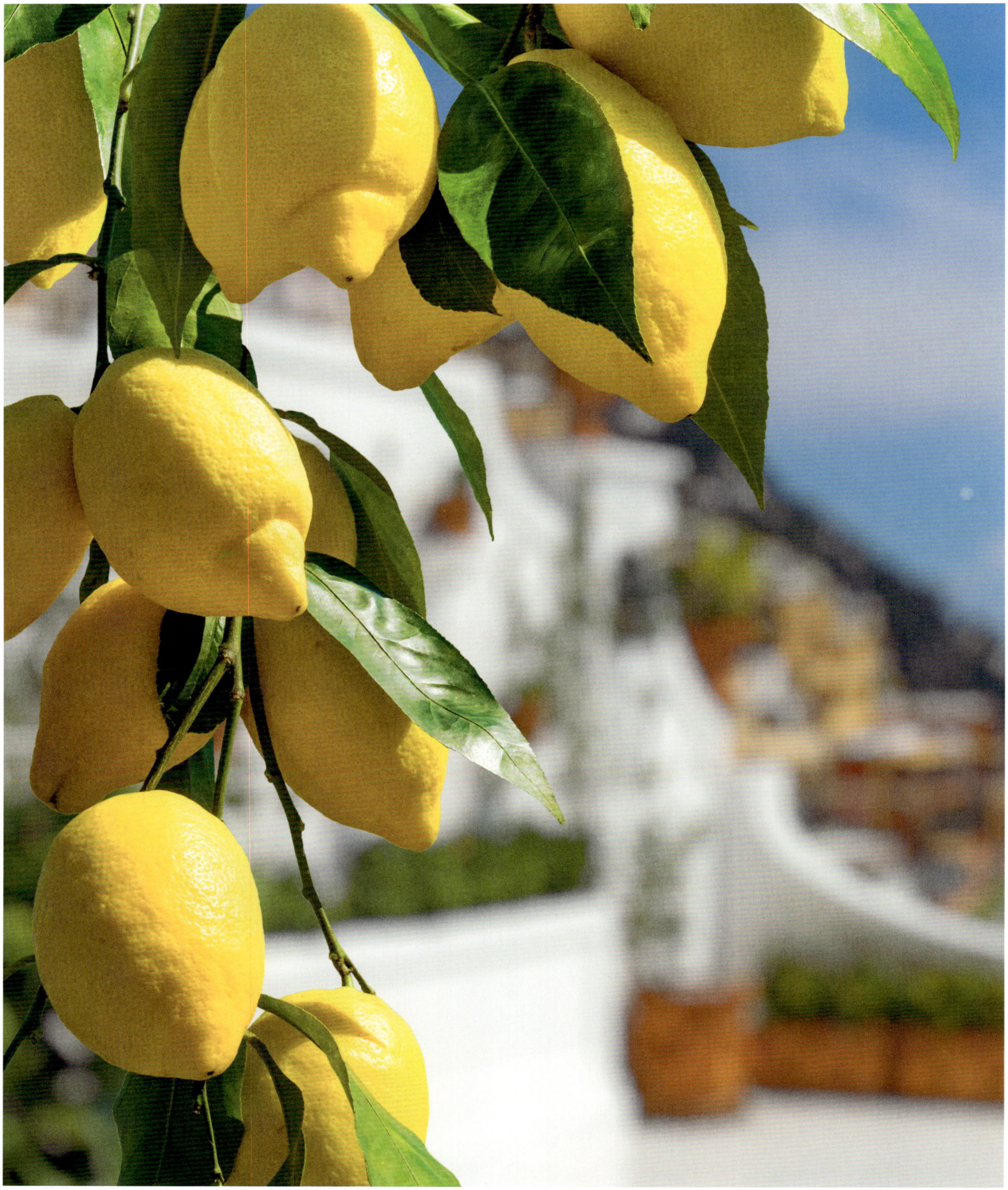

Here, south of Naples and a safe bay's distance away from the high-strung Vesuvius, travelers have to work hard to get to the beach. The Amalfi Coast is undoubtedly Italy's most beautiful coast and is also called *La Divina*, or the Divine. There is also no doubt that the most exquisite delicacies are to be found in the immediate inland, such as the huge, delicious Amalfi lemons, the famous *Anchovy Colatura*, an unusual sauce made from decaying anchovies, already treasured in Roman times. The best pasta comes from here, and the wines that grow with a view of the sea are considered the most interesting varieties beyond the established wine regions.
But you have to conquer the beach. For example, down countless serpentine roads that lead to tiny parking lots where cars are parked as only Italians can–completely chaotically and seemingly without any system, but nevertheless effectively at the same time. Or you can go by foot, following steep staircases carved into the rock leading down from the villages on high.
The sandy beaches in the bays are tiny and, like the parking lots, are used judiciously, every meter of them, mostly by beach clubs. It is 300 steps down to the legendary Club Arienzo to the east of Positano (which, of course, must be climbed again at day's end). Other beach clubs are nestled on cliffs; footbridges lead over the rocks into the sea. But how good life is here, and how clear and clean the water is!
Like so many places, the Amalfi Coast is much smaller than its reputation. The coast is only about 30 kilometers long, and the main town of Amalfi has just 5,000 residents. In the summer there are more–but not many more, because there is not enough space. And that makes the coast very exclusive. For travelers, that is what makes it so charming, because Italian life takes place here unspoiled by groups of foreign tourists. Even during the Roman Empire, wealthy people settled here for the summer. The impressive remains of Villa Romana di Minori, discovered by chance in 1932 during a landslide, show how well the summer residents enjoyed themselves; the villa had elaborate mosaics and a large pool in the courtyard. This coast, by the way, has been a Unesco World Heritage Site since 1997–not many beaches can claim that.

Hier, südlich von Neapel und eine sichere Bucht vom nervösen Vesuv entfernt, muss sich der Reisende den Strand regelrecht erarbeiten. Denn die Amalfiküste ist zweifellos Italiens schönste Küste und heißt auch „La Divina", „Die Göttliche". Und zweifellos gibt es im unmittelbaren Hinterland die erlesensten Köstlichkeiten, etwa die riesigen, wohlschmeckenden Amalfizitronen, die berühmte Sardellen-Colatura, eine gewöhnungsbedürftige Würztunke aus verwesenden Sardellen, die schon von den Römern geschätzt wurde. Die besten Nudeln kommen von hier, und die Weine, die mit Meerblick wachsen, gelten als interessanteste Gewächse jenseits der etablierten Weinregionen.

Doch den Strand: den muss man sich erobern. Etwa über unzählige Serpentinen, die zu winzigen Parkplätzen führen, auf denen die Autos so abgestellt werden, wie es nur Italiener können – völlig chaotisch und scheinbar ohne System, doch effektiv zugleich. Oder es geht zu Fuß über steile, in den Fels gehauene Treppen, die von den höher gelegenen Orten herabführen.
Die Sandstrände in den Buchten sind ebenfalls winzig und werden, wie die Parkplätze, wohlüberlegt und bis in den letzten Winkel genutzt, meistens von Beach Clubs. Zum legendären Club Arienzo im Osten von Positano sind es 300 Stufen abwärts (die am Ende des Tages natürlich auch wieder bezwungen werden müssen). Andere Strandclubs haben sich gleich auf Klippen eingenistet; Stege führen übers Geröll ins Meer. Doch wie gut ist das Leben hier, wie klar und sauber das Wasser!
Wie so viele Orte ist die Amalfiküste viel kleiner als ihr Ruf. Sie erstreckt sich nur etwa 30 Kilometer den Golf von Salerno entlang, und der Hauptort Amalfi hat lediglich 5000 Einwohner. Im Sommer werden es zwar mehr – aber nicht viel mehr, dazu fehlt der Platz. Und das macht die Küste ungeheuer exklusiv. Was für Reisende den Charme ausmacht, denn hier findet das italienische Leben unverfälscht durch ausländische Reisegruppen statt. Schon im Römischen Reich ließen sich Wohlhabende über den Sommer hier nieder. Die beeindruckenden Überreste der Villa Romana di Minori, die 1932 bei einem Erdsturz zufällig entdeckt wurden, zeigen, wie gut es sich die Sommerfrischler gehen ließen; die Villa verfügte neben aufwendigen Mosaiken auch über einen großen Pool im Innenhof.
Die Küste ist übrigens seit 1997 Unesco-Welterbe – das können nicht viele Strände von sich behaupten.

The famous lemons of the Amalfi Coast, ready for harvest.

Die berühmten Amalfizitronen, bereit zur Ernte.

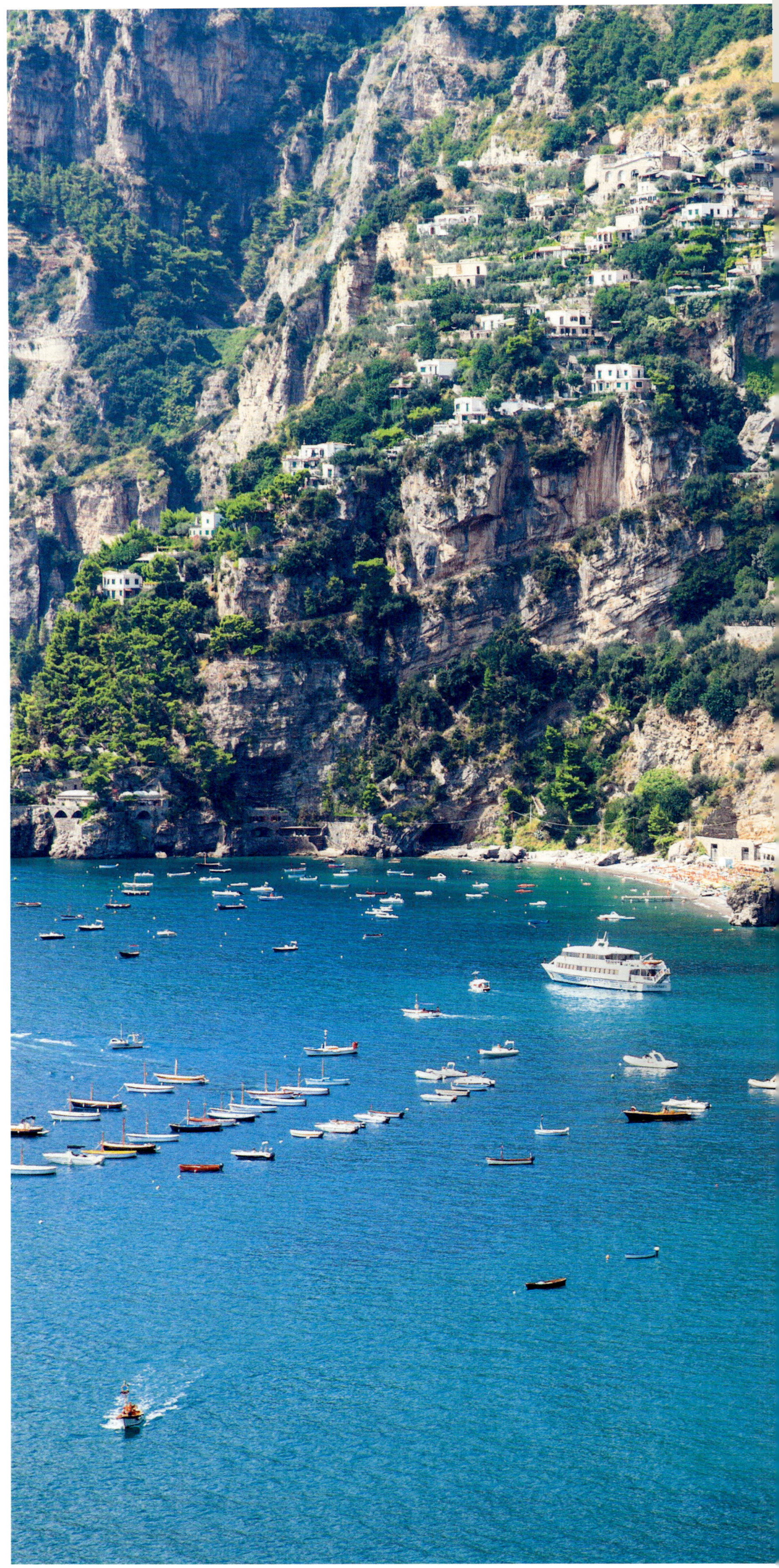

Beautiful view across to Positano, Amalfi Coast, Italy.

Wunderschöne Aussicht auf Positano an der Amalfiküste, Italien.

THERE ARE MANY STAIRWAYS TO HAPPINESS ON THE AMALFI COAST

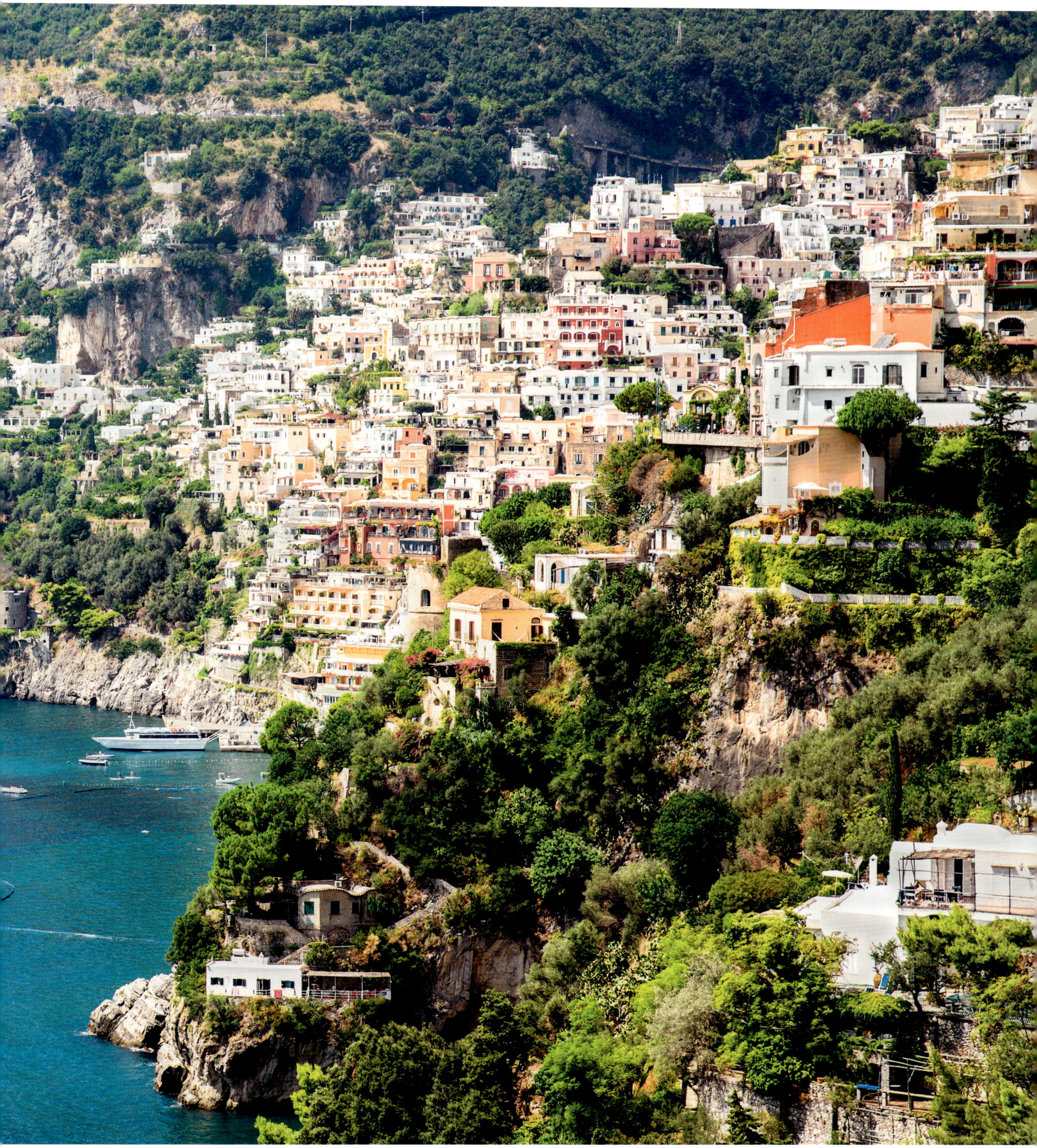

VIELE TREPPEN ZUM GLÜCK: DIE AMALFIKÜSTE

212

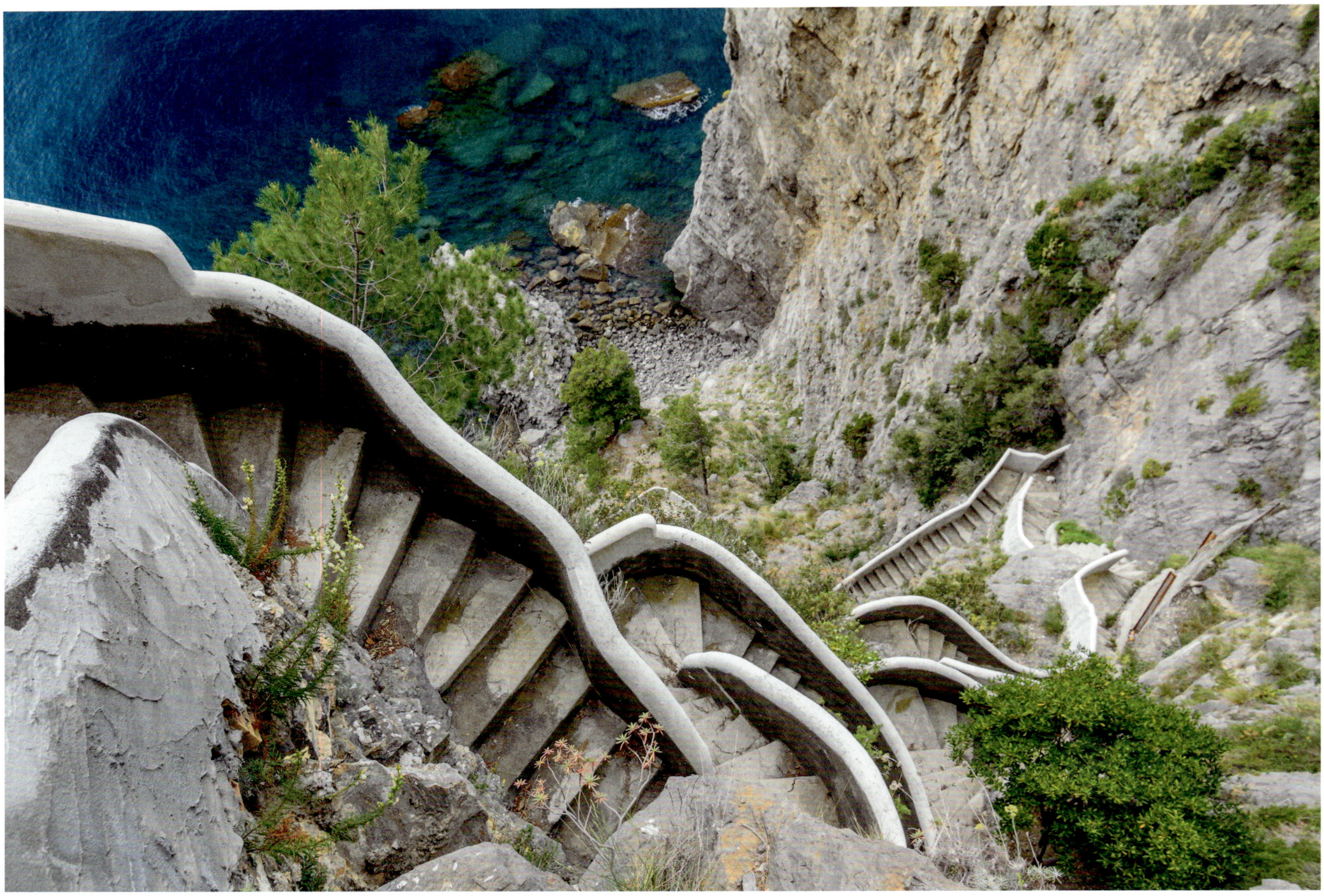

Steep descent to the shore.
Amalfi Coast, Italy.

Steiler Abstieg zum Wasser.
Amalfiküste, Italien.

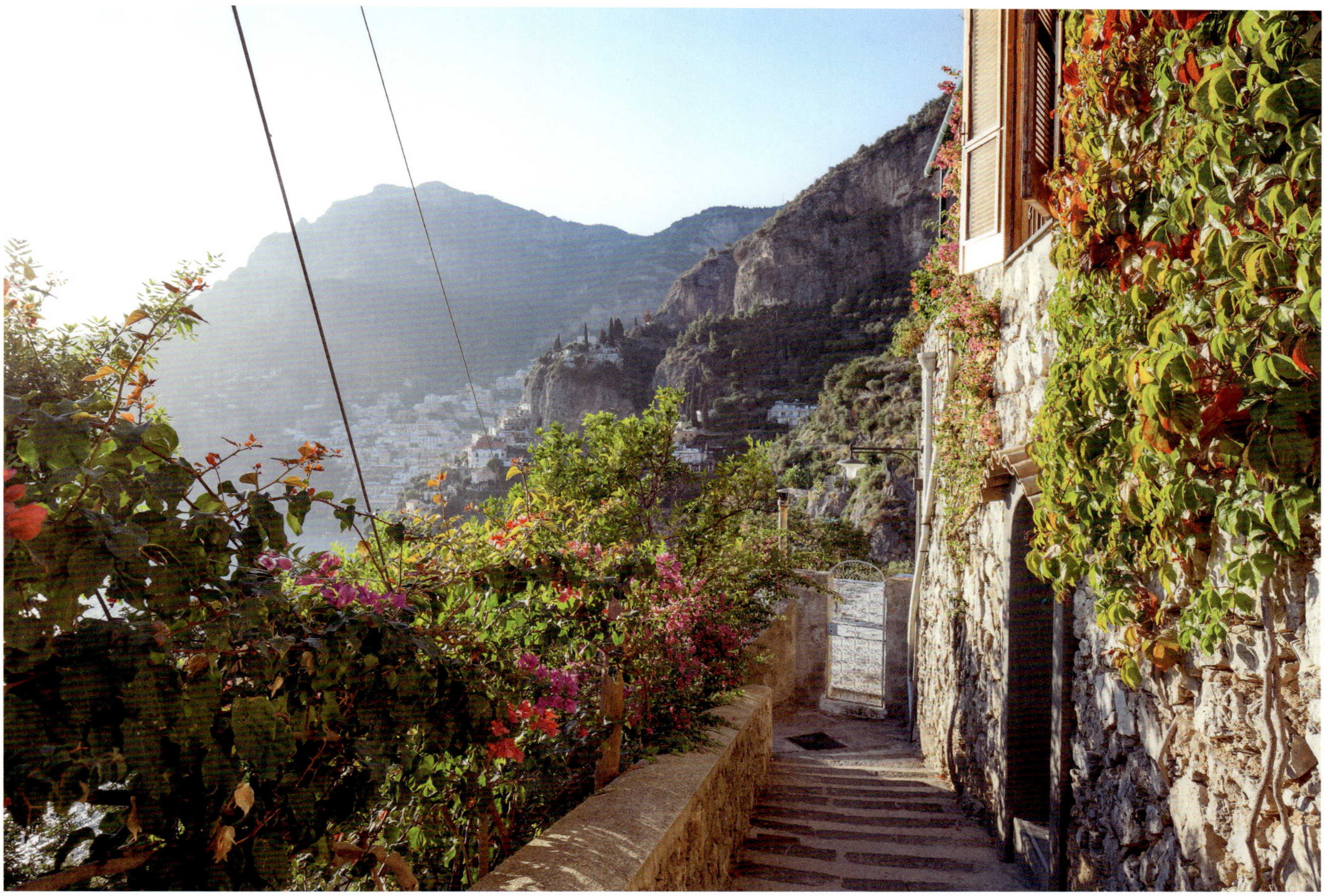

Path of the Gods with end on Positano, Amalfi Coast, Italy.

Der Pfad der Götter an der Amalfiküste nach Positano, Italien.

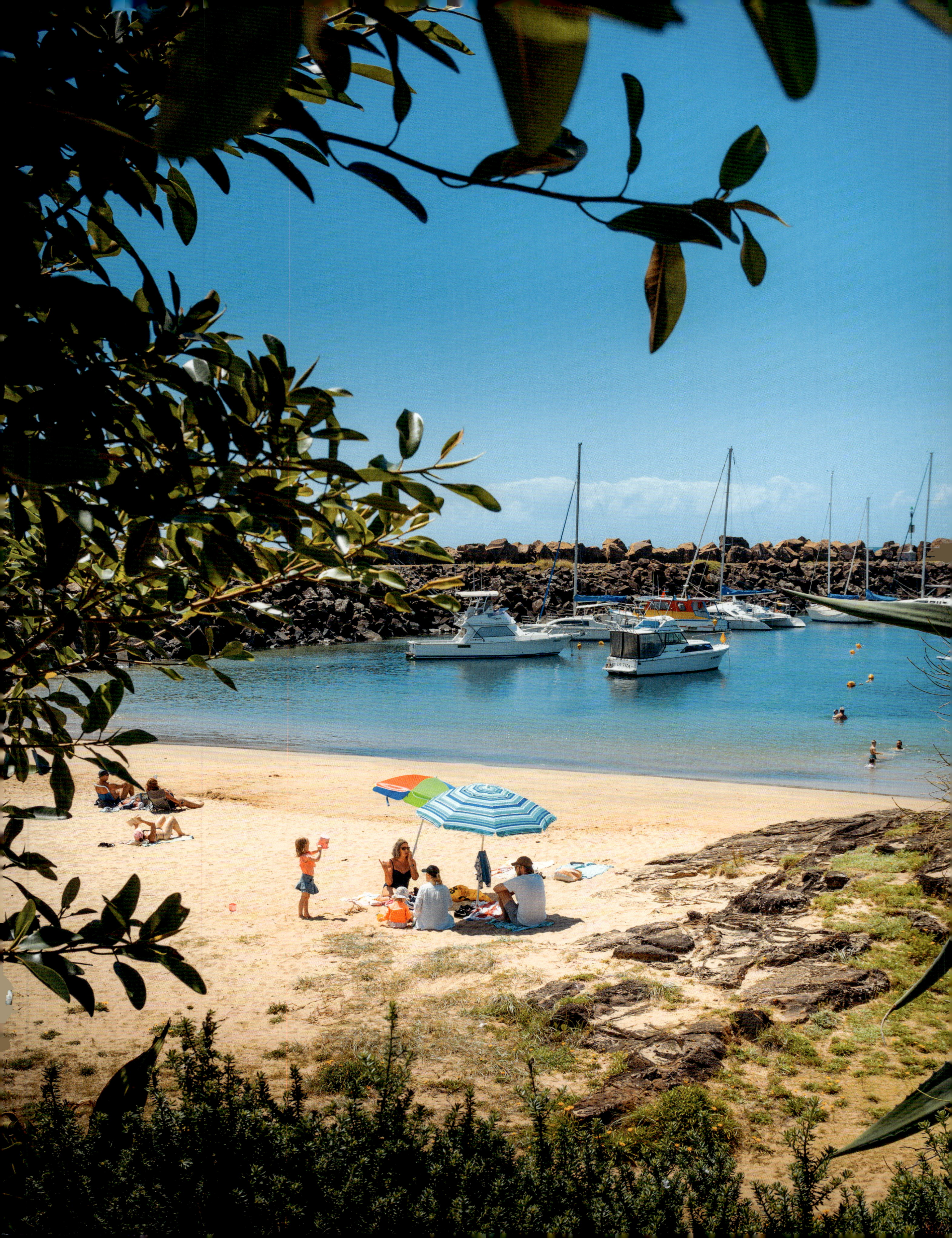

British researchers at the University of Exeter found that living near the beach makes us healthier, both physically and mentally. "Is the coast an underappreciated-public health resource?" the scientists ask. For their study, titled *Coastal proximity and physical activity*, they examined the vital statistics of nearly 200,000 people. Studies in Australia and New Zealand also reached similar conclusions. There are many reasons for this. Just one example of many: People who live near the coast have more opportunities for extensive walks in clean air, preferably on the beach–an opportunity simply unavailable to inland residents or those living in large cities. Sea air can prevent respiratory diseases, and exercise out in nature increases self-esteem and reduces tension, anger, and depression, researchers report in the *Annual Review of Public Health*.
It is also interesting to note that the famous Blue Zones–those areas where a particularly large number of 100-year-olds live and reach old age in good health–are almost all located on islands, such as Okinawa in Japan, Sardinia in Italy, and Ikaria in Greece, along with the Costa Rican peninsula of Nicoya.
Walks on the beach arguably also promote better and longer sleep, even more so than walking inland. In a comparative study, beach walkers slept 47 minutes more than those subjects who hiked through parks and heathlands. The researchers also hypothesize that there are psychological effects behind this: Coastal walks are more likely to evoke childhood memories than wandering inland, and the ocean seems to convey the restful feeling of being particularly far away from everyday life.

Other studies show that the mere sight of the ocean increases our well-being; most people also consider the sound of the waves to be relaxing–virtually every meditation app comes with ocean sounds.
But what are apps compared to real life? And so we dream our way to the beach every day. Counting the days until our vacation. And packing our suitcases much too early. We're almost there.

Englische Forscher der Universität Exeter haben herausgefunden, dass uns das Leben in Strandnähe gesünder macht – körperlich und mental. „Ist die Küste eine unterschätzte Ressource für die Allgemeingesundheit?“ fragen die Wissenschaftler. Für die Studie mit dem Titel „Coastal Proximity“ untersuchten sie Lebensdaten von knapp 200.000 Personen. Auch Untersuchungen in Australien und Neuseeland kamen zu ähnlichen Ergebnissen. Die Ursachen dafür sind vielfältig. Nur ein Beispiel von vielen: Wer nahe der Küste wohnt, hat mehr Möglichkeiten zu ausgedehnten Spaziergängen in klarer Luft, gern am Strand – eine Möglichkeit, die ein Bewohner im Inland oder gar in einer Großstadt eben nicht hat. Seeluft kann Atemwegserkrankungen vorbeugen, dazu erhöht die Bewegung in der Natur das Selbstwertgefühl und baut Spannungen, Zorn und Depressionen ab, wie Forscher im *Annual Review of Public Health* berichten.

Interessant ist überdies, dass die berühmten „Blauen Zonen“ – also jene Gebiete, in denen besonders viele 100-Jährige leben, die auch noch weitgehend gesund das hohe Alter erreichen – nahezu alle auf Inseln liegen, darunter Okinawa in Japan, Sardinien in Italien, Ikaria in Griechenland und die costaricanische Halbinsel Nicoya.
Spaziergänge am Strand sorgen wohl auch für besseren und längeren Schlaf, noch mehr als das Spazierengehen im Binnenland. In einer Vergleichsstudie schliefen Strandspaziergänger 47 Minuten mehr als jene Probanden, die durch Parks und Heidegebiete wanderten. Die Forscher vermuten dahinter auch psychologische Effekte: Küstenspaziergänge wecken häufiger Kindheitserinnerungen als das Durchstreifen des Binnenlands, und das Meer scheint das Gefühl zu vermitteln, besonders weit weg vom Alltag zu sein.
Weitere Untersuchungen zeigen, dass allein der Anblick der See das Wohlbefinden steigert, auch der Klang der Wellen wird von den meisten Menschen als entspannend wahrgenommen – kaum eine Meditations-App kommt ohne das Meeresrauschen aus.
Aber was sind Apps gegen das echte Leben? Und so träumen wir uns täglich an den Strand. Zählen die Tage bis zum Urlaub. Und packen viel zu früh den Koffer. Wir sind schon fast da.

A moment to pause: family picnic at Belmore Basin in Wollongong, Australia.

Ein Moment zum Innehalten: Familienpicknick am Belmore Basin in Wollongong, Australien.

216

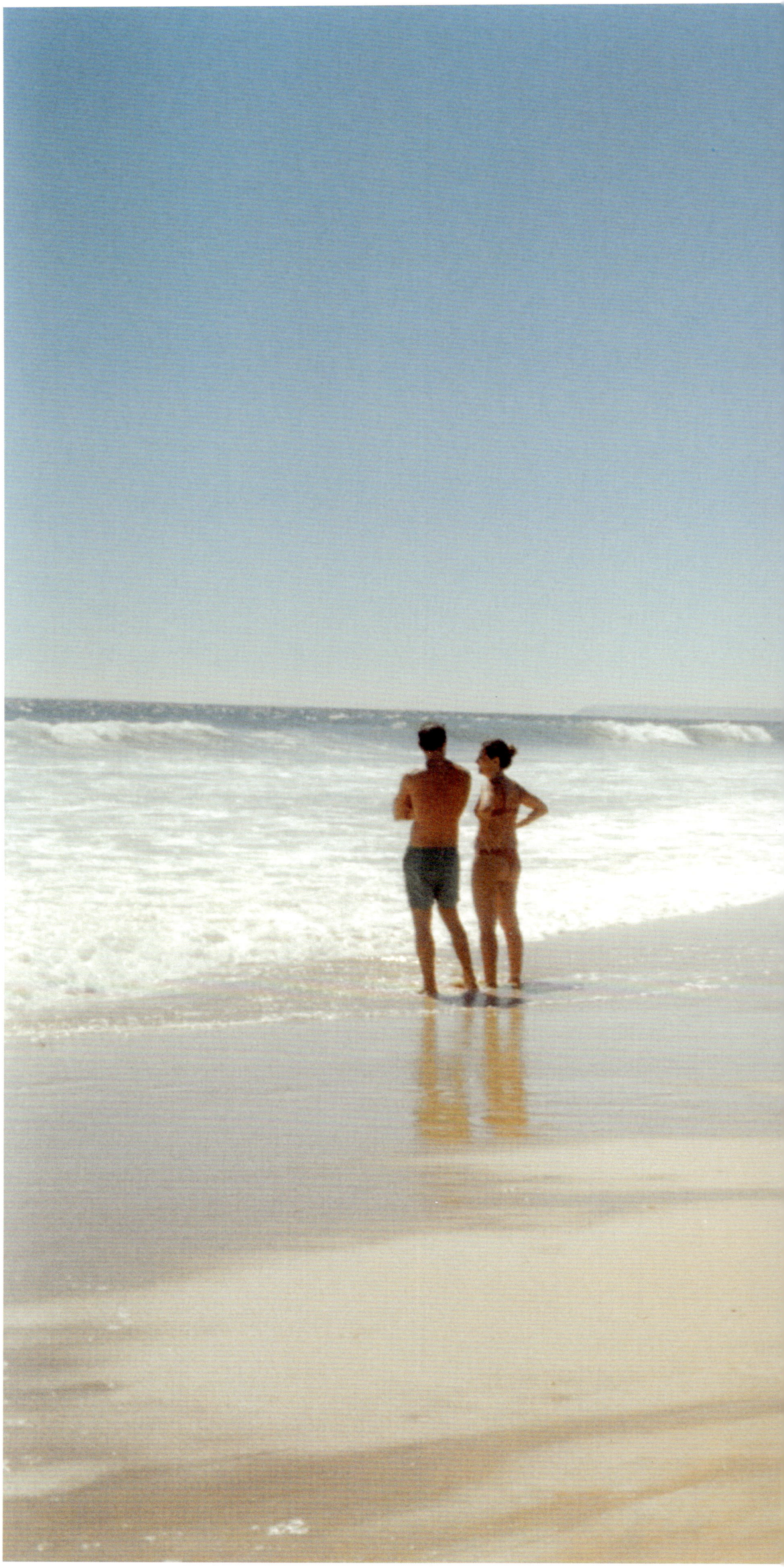

Looking over the endless sea: feeling carefree on a summer day. Praia do Guincho, Portugal.

Blick in die endlosen Meeresweiten: das unbeschwerte Gefühl eines Sommertages. Praia do Guincho, Portugal.

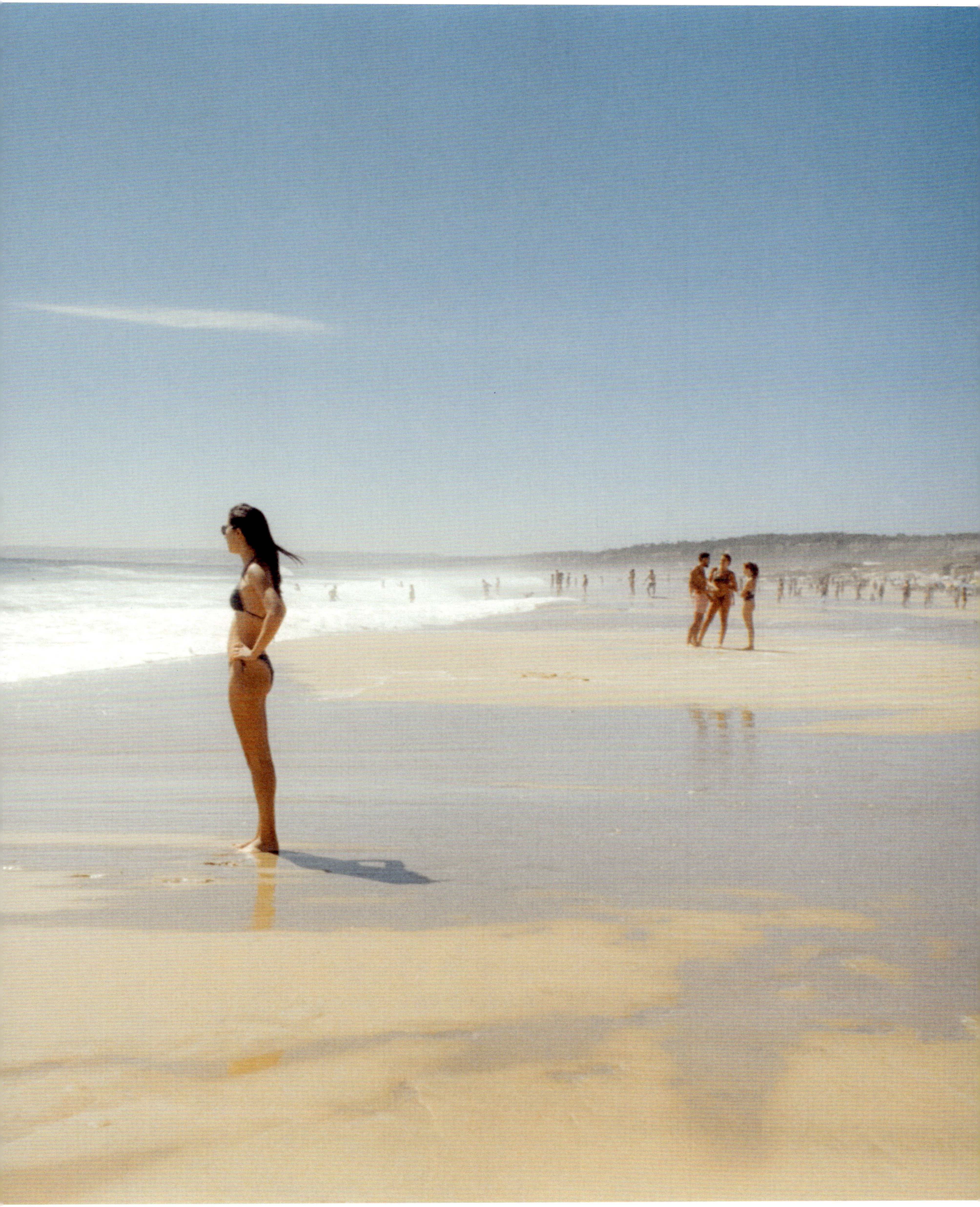

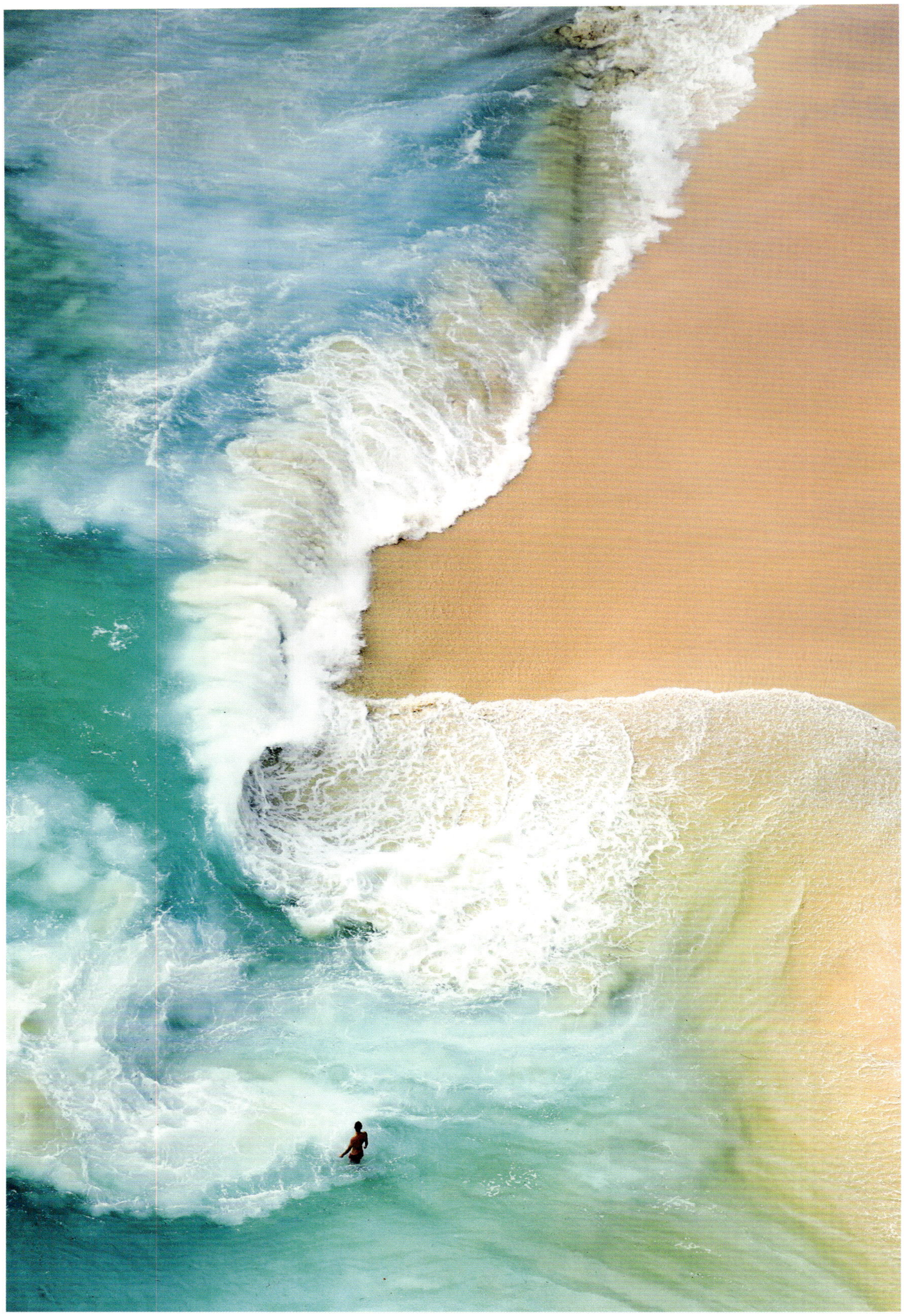

A view of paradise. Kelingking Beach,
Nusa Penida, Indonesia.

Blick aufs Paradies. Kelingking Beach,
Nusa Penida, Indonesien.

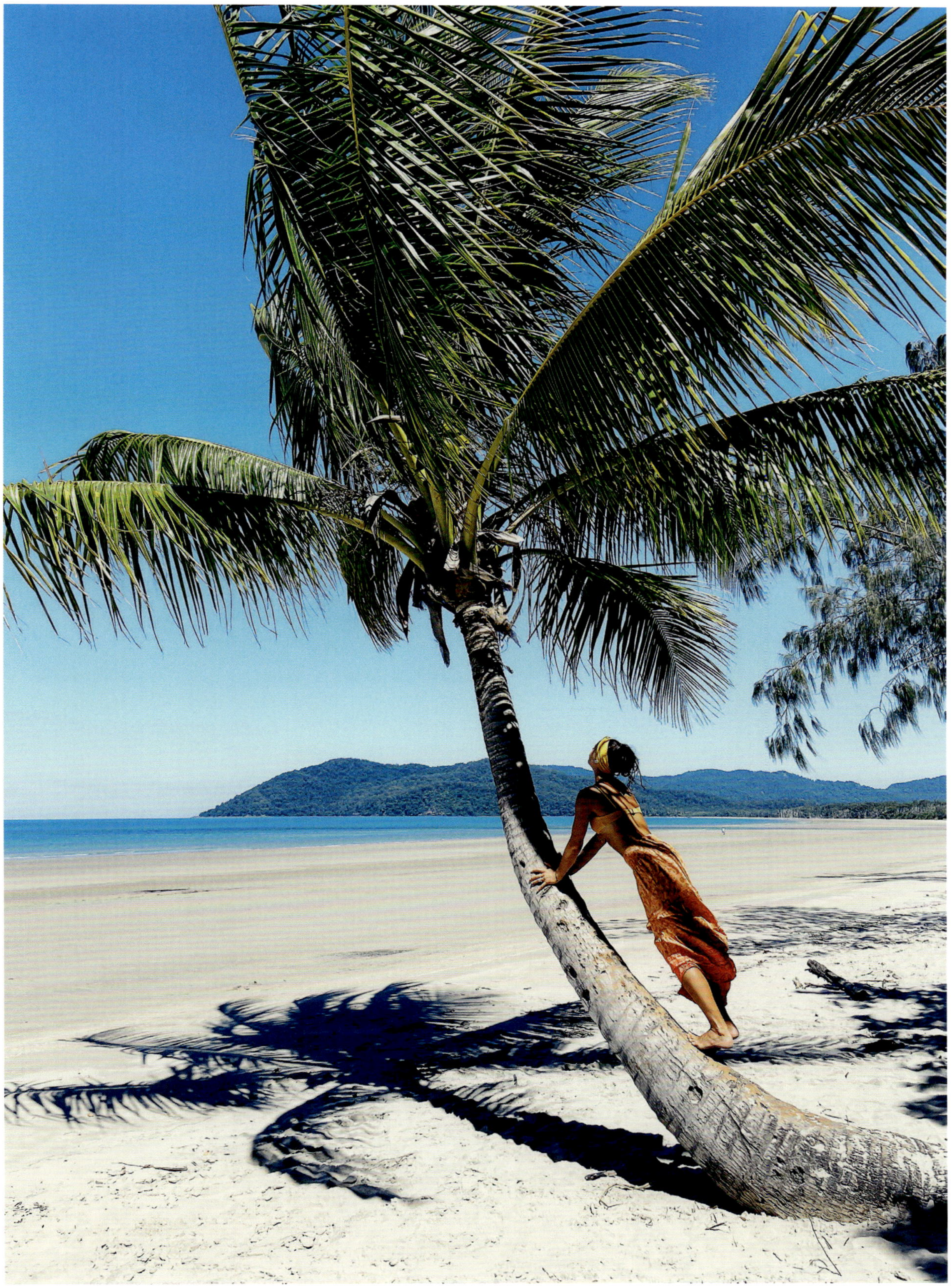

Paradise In The Palm Of Your Hand: Thornton Beach, Far North Queensland, Australia. This is where one UNESCO World Heritage Site meets another–the ancient Daintree Rainforest and the spectacular Great Barrier Reef.

Das Paradies zum Greifen nah: Thornton Beach, Far North Queensland, Australien. Hier trifft ein UNESCO-Weltnaturerbe auf ein anderes – der uralte Daintree-Regenwald auf das spektakuläre Great Barrier Reef.

Moment of enchantment: a couple enjoying the sunset over the sea in Senj, Croatia.

Wie verzaubert: Ein Paar genießt gemeinsam den Sonnenuntergang über dem Meer in Senj, Kroatien.

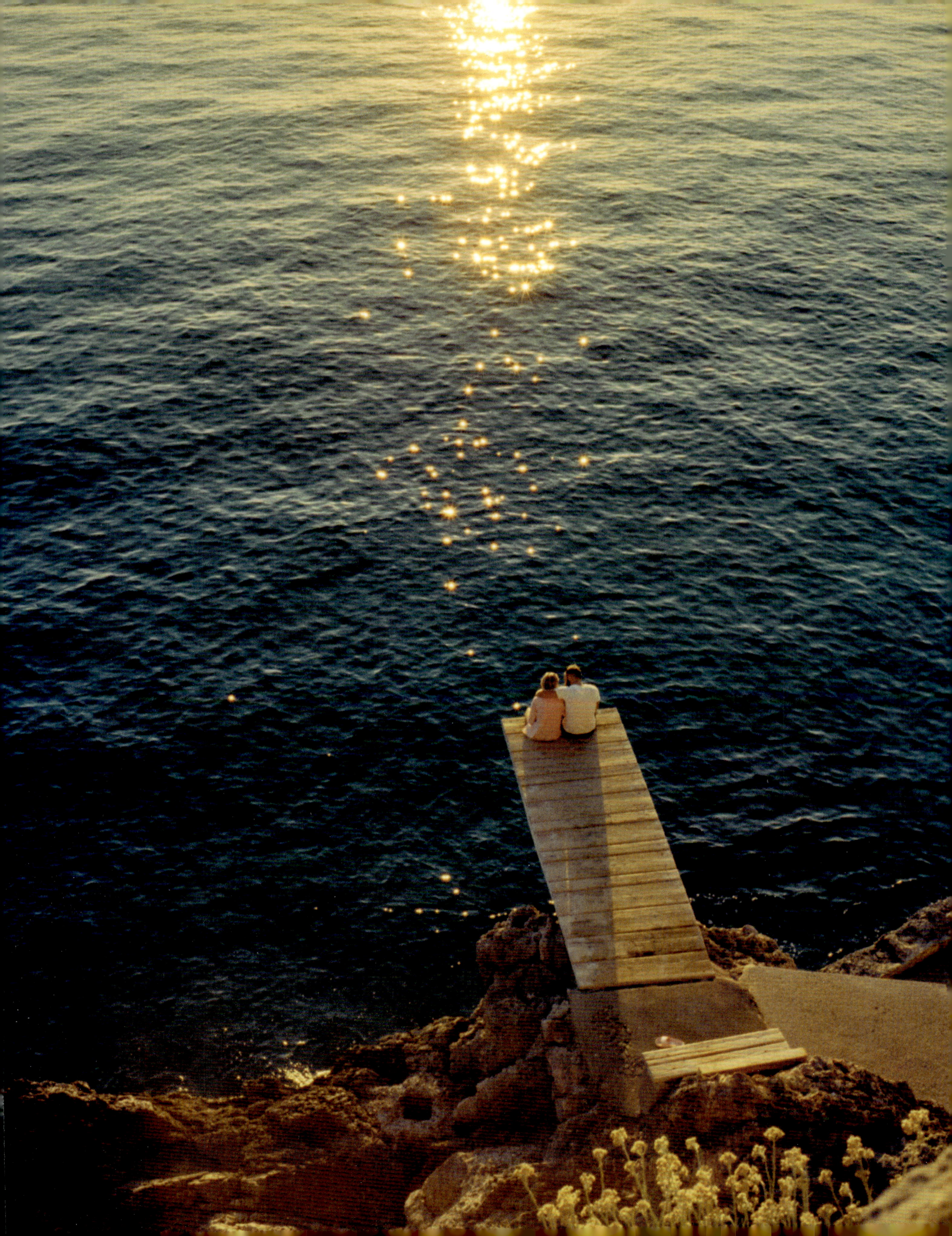

Cover: © David Leto
Back cover: © Shutterstock – travelwild
4, 5: © David Leto; 8, 9: © Irena Menk; 13: © Bogdan Ianosi; 15: © Eva Stadler; 17: © Ullstein Bild – Ullstein Bild; 18: © Ullstein Bild – Martin Munkacsi; 19: © Getty Images – Lucia Mauri; 20: © David Leto; 21 above/oben: © Getty Images – Jena Ardell; 21 below/unten: © Getty Images – Edlira Shuli/EyeEm; 22/23: © Léa Mancuso; 24: © Ullstein Bild – adoc-photos; 26: © Irena Menk; 27: © Ullstein Bild – Ullstein Bild; 29: © Léa Mancuso; 30: © Getty Images – Nicolette Wells; 31: © Shutterstock – Floral Deco; 32: © Unsplash – Afeeq Nadzrin; 35: © Ullstein Bild – Regine Relang; 36 above/oben: © Unsplash – Finn; 36 below/unten: © Shutterstock – Olesya Kuznetsova; 37: © Unsplash – Gabriella Clare Marino; 38: © Léa Mancuso; 40: © Shutterstock – Volles; 42: © Getty Images – Ingo Rösler; 43: © Getty Images – Henglein and Streets.; 44: © Getty Images – Marcelo Nacinovic; 45: © Unsplash – Domenico Daniele; 47: © Getty Images – Chris Tobin; 48: © Getty Images – Barry Thomas/EyEm; 50/51: © Getty Images – Thomas Barwick; 52: © Unsplash – Ilyuza Mingazova; 53: © Unsplash – Lisha Riabinina; 55: © Unsplash – Pip Christie; 56: © Airbare; 57: © Unsplash – Alex Nicolopoulos; 58: © Ullstein Bild – Zander & Labisch; 59: © Unsplash – David Boca; 60: © Getty Images – Sam Edwards.; 62: © Irena Menk; 63: © Léa Mancuso; 64: © Bogdan Ianonosi; 67: © Unsplash – Colton Duke; 68: © Shutterstock – Sharpshutter; 69: © Shutterstock – David P. Baileys; 70/71: © Unsplash – Odd Fellow; 72: © Unsplash – Alexandre Trouvé; 75: © Unsplash – Rafael Romero; 76: © Unsplash – David Cain; 78 above/oben & below/unten: © Mauritius Images – Yuri Arcurs/Alamy/Alamy Stock Phtotos; 79: © Picture Alliance/dpa/MAXPPP/Pierre Berthuel/Le Pictorium; 80/81: © Shutterstock – Ground Picture; 82: © Unsplash – Vincenzo Vesprini; 75, 87: © Bogdan Ianonosi; 88: © Ullstein Bild – Ullstein Bild; 89: © Ullstein Bild – Wilms; 90, 91: © Bogdan Ianonosi; 92/93: © Mauritius Images – Image Source/Guido Cavallini; 95: © Picture Alliance/Ullstein Bild – Ullstein Bild; 96: © Ullstein Bild – Ullstein Bild; 97 above and below/oben und unten: © Eva Stadler; 98: © Getty Images – Lucyna Koch; 99: © Getty Images – Jasper James; 100: © Shutterstock – Phoenixproduction; 102/103: © Getty Images – Fraser Hall; 104: © Getty Images – Dan Kitwood/Staff; 105 above/oben: © Getty Images – Gary John Norman; 105 middle/Mitte: © Getty Images – PhotoAlto/Milena Boniek; 105 below/unten: © Getty Images – Robert Reader; 106/107: © Shutterstock – Alexey Smyshlyaev; 109: © Ullstein Bild – United Archives; 110, 111: © Mauritius Images – Cinema-Legacy-Collection/The Hollywood Archive; 112/113: © Getty Images – Kyolshin; 115: © David Leto; 117, 118: © Irena Menk; 119: © Léa Mancuso; 120: © Unsplash – Letizia Agosta; 121: © Léa Mancuso; 123, 124: © David Leto; 126/127: © Eva Stadler; 128: © Airbare; 129: © Getty Images – Alan Powdrill; 130, 131, 132/133: © Bogdan Ianonosi; 134: © Irena Menk; 136: © Getty Images – Tina Terras & Michael Walter; 137: © Getty Images – Manuel Breva Colmeiro; 138/139: © Getty Images – Miguel Sotomayor; 141: © Shutterstock – Jom Nicha; 142 left/links: © Getty Images – Nastya Poburynna/EyeEm; 142 middle/Mitte: © Unsplash – Roberto Arias; 143: © Unsplash – Philipp Deus; 145: © Unsplash – Mourad Saadi; 146: © Getty Images – Philip Thurston; 148: © Léa Mancuso; 149, 150: © Irena Menk; 153: © Shutterstock – Fashionstock.com; 154 left and right/links und rechts: © Ullstein Bild – United Archives; 156/157: © Getty Images – Peter Cade; 158: © Ullstein Bild – dpa; 159: © Imago – Heritage Images; 160: © Unsplash – Marvin Meyer; 162: © Getty Images – Ghislain & Marie David de Lossy; 163: © Getty Images – Stefanie Hiebl; 165: © Getty Images – Joos Mind; 166: © Getty Images – Johner Images; 167: © Getty Images – Plug Pattarin; 168, 169 left and right/links und rechts: © Shutterstock – Fashionstock.com; 170: © Ullstein Bild – IBERFOTO / ColeccionGasca; 171: © Shutterstock – Paolo Gallo; 173: © Shutterstock – Luisitaphoto; 174: © Léa Mancuso; 177: © Irena Menk; 178: © Getty Images – Rosmarie Wirz; 179: © Getty Images – Jeff Bergen; 181: © Getty Images – Faranoss; 182: © Picture Allicance – Everett Collection/20thCentFox; 183: © Getty Images – Riccardo Testolin; 184: © Shutterstock – Just Life; 186/187: © Unsplash – Nico Bhlr; 189: © Shutterstock – Sergey Causelove; 190: © Unsplash – Oliver Sjöström; 193: © Bogdan Ianosi; 195: © Ullstein Bild – Adoc-Photos; 196 above/oben: © Imago – Cinema Publishers Collection; 196 below/unten: © Ullstein Bild – Photo12; 197: © Unsplash – Fabrice Villard; 198/199: © Getty Images – Maria Korneeva; 201: © Shutterstock – Prosign; 202: © Picture Alliance – Leemage; 204: © Getty Images – Maurice Rougement; 205: © Picture Alliance – Loop Images/John Greim; 206 above/oben: © Ullstein Bild – Regine Relang; 206 below/unten: © Ullstein Bild – Karl Zimmermann; 207: © Getty Images – Reed Kaestner; 208: © Shutterstock – IgorZH; 211: © Mauritius Images – Justine Kibler/Alamy; 212: © Shutterstock – Sogodel Vlad; 213: © Shutterstock – fotolupa; 214: © David Leto; 217: © Bogdan Ianonosi; 218: © Shutterstock – Travelwild; 219: © Airbare; 221: © Bogdan Ianonosi; 223: © Heike Bogenberger

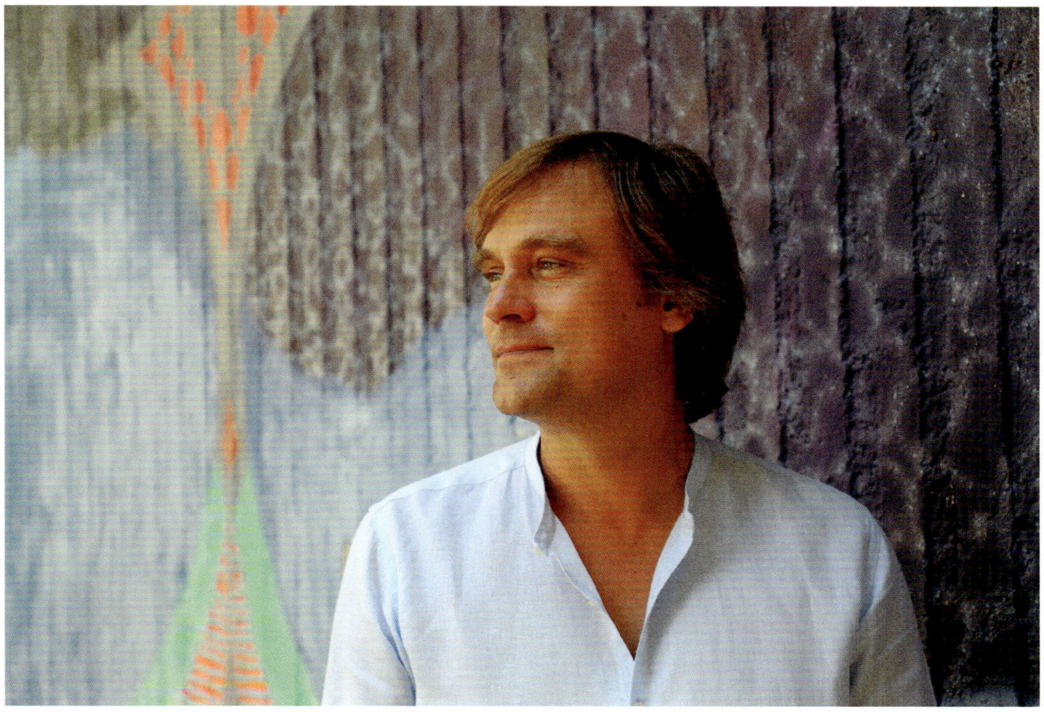

Born in 1971 in Braunschweig, Stefan Maiwald is one of Germany's most versatile authors. For 20 years now, he has lived on the island of Grado between Venice and Trieste, where he writes novels, crime thrillers, and non-fiction, including the bestsellers *Laura, Leo, Luca und ich – wie man in einer italienischen Familie überlebt*, *Wir sind Papa!* and *Das Italien-Prinzip: So geht Glück!*
His historical crime novel set in Venice, *Der Spion des Dogen*, was voted best debut of the year by HOMER, the association of authors and critics, and his blog, www.postausitalien.com, is the most widely read German-language site on Italy and was named the best Travel Blog 2022 by the Italian National Tourist Board.
Stefan Maiwald spends most of his free time at the beach; in Grado there are seven kilometers of fine sand.

Stefan Maiwald, geboren 1971 in Braunschweig, ist einer der vielseitigsten deutschen Autoren. Er lebt seit 20 Jahren auf der Insel Grado zwischen Venedig und Triest und schreibt Romane, Krimis und Sachbücher, darunter die Bestseller *Laura, Leo, Luca und ich – wie man in einer italienischen Familie überlebt*, *Wir sind Papa!* und *Das Italien-Prinzip: So geht Glück!*. Der historische Venedig-Krimi *Der Spion des Dogen* wurde von der Autoren- und Kritiker-Vereinigung HOMER zum besten Debüt des Jahres gewählt.
Sein Blog www.postausitalien.com ist die meistgelesene deutschsprachige Italien-Seite und wurde von der Italienischen Zentrale für Tourismus zum besten Travel Blog 2022 gekürt.
Die Freizeit verbringt Stefan Maiwald weitgehend am Strand; in Grado gibt es davon sieben feinsandige Kilometer.

© 2023 teNeues Verlag GmbH

1st printing

Texts: © Stefan Maiwald. All rights reserved.
Editorial Coordination, Picture Research and Copyediting
by Birthe Vogelmann
Translation by Robin Limmeroth
Design by Eva Stadler
Production by Sandra Jansen-Dorn, teNeues Verlag
Photo Editing, Color Separation by Jens Grundei,
teNeues Verlag

Library of Congress Number: 2023932517
ISBN: 978-3-96171-446-9

Printed in Slovakia by Neografia

MIX
Paper | Supporting responsible forestry
FSC® C020353
www.fsc.org

Picture and text rights reserved for all countries. No part of this publication may be reproduced in any manner whatsoever.

While we strive for utmost precision in every detail, we cannot be held responsible for any inaccuracies, neither for any subsequent loss or damage arising.

Every effort has been made by the publisher to contact holders of copyright to obtain permission to reproduce copyrighted material. However, if any permissions have been inadvertently overlooked, teNeues Publishing Group will be pleased to make the necessary and reasonable arrangements at the first opportunity.

Bibliographic information published by the Deutsche Nationalbibliothek:
The Deutsche Nationalbibliothek lists this publication in the Deutsche Nationalbibliografie; detailed bibliographic data are available on the Internet at dnb.dnb.de.

Published by teNeues Publishing Group

teNeues Verlag GmbH
Ohmstraße 8a
86199 Augsburg, Germany

Düsseldorf Office
Waldenburger Straße 13
41564 Kaarst, Germany
e-mail: books@teneues.com

Augsburg/München Office
Ohmstraße 8a
86199 Augsburg, Germany
e-mail: books@teneues.com

Berlin Office
Lietzenburger Straße 53
10719 Berlin, Germany
e-mail: books@teneues.com

Press Department
presse@teneues.com

teNeues Publishing Company
350 Seventh Avenue, Suite 301
New York, NY 10001, USA
Phone: +1-212-627-9090
Fax: +1-212-627-9511

www.teneues.com

teNeues Publishing Group
Augsburg / München
Berlin
Düsseldorf
London
New York

teNeues